PHILOSOPHICAL SOLUTIONS

IIAS

PHILOSOPHICAL SOLUTIONS

IN PHYSICS, MATHEMATICS AND THE SCIENCE OF SENTIENCE

Ted Silverman

International Institute for Advanced Studies

Layout and design, interior and cover, by the author.
Front cover image is a reproduction and adaptation,
by the author, of a public domain reproduction of
William Blake's *Newton*. The original was created
in 1795 as a color engraving, embellished with pen
and ink as well as watercolor.

All interior illustrations by the author.
Text set in 11 point Minion Pro Medium.

International Institute for Advanced Studies

ISBN:1452886636   EAN-13: 9781452886633

Printed in the United States of America

10 9 8 7 6 5 4 3 2 1

To my father,
Murray Silverman,
*who had the faith and generosity
to initiate a very young boy into
the secret rites and mysteries of
Electromagnetic Theory,
and so set my feet upon
the path of discovery.*

Contents

Philosophical Solutions

V A Framework for Physics

Preface

THOSE physicists, mathematicians and cognitive scientists who occupy themselves with questions of a fundamental nature know that profound logical difficulties surround the theoretical conceptions central to each of their respective disciplines – difficulties which, in mystery and depth, call to mind the perennial paradoxes of philosophy; "*Antinomies of Pure Thought*," seemingly beyond the power of that thought to comprehend.

In physics, the foundational problems of quantum theory, and its formal disparities with relativity, block unification – and thereby consistency, completeness and sapient comprehensibility – of these "twin pillars" of modern physical science. Moreover, problems stemming from the so-called orthodox or *Copenhagen Interpretation* of quantum theory arise in fields well beyond microphysics. As a consequence of Copenhagen and its close variants, the specter of Cartesian dualism is once again palpable; right at the heart of hard science. It haunts researchers in every arena: from cosmologists seeking to model a self-contained universe, absent an "external" cosmic ghost – or a non-denumerable infinity of ghost-like "parallel universes" – to cognitive scientists pursuing naturalistic methodologies; seeking to correlate mental and biological processes without ontological justification – singular targets of harassment by the philosophical police.

1

Perhaps most significantly, the erstwhile quest of Meta-Mathematics to develop a "proof theory" – that is to say, a logical framework for mathematics itself – is demonstrably unattainable. The *Gödel Incompleteness* theorems, taken to invalidate the axiomatic reduction of arithmetic to a unified, logical basis, raise questions not only about mathematics but about the significance of formal systems generally; any deductive theory as complex as arithmetic, and therefore every non-trivial scientific enterprise.

As it turns out, these crises are more than merely reminiscent of the infamously intractable, ancient concerns of philosophy. Rather, they are modern manifestations of them. But they need not stand for centuries. Careful consideration reveals, in each case, that the underlying impediment to clarity is not insoluble enigma but deeply ingrained misconception, and that the foundational problems of the sciences are closely, albeit subtly interrelated. They can be effectively approached as a group and, as such, are susceptible to critical analysis and satisfactory resolution.

Accordingly, the present treatise is concerned, for the most part, with those fundamental aspects of theory in physics, mathematics and psychology, which bear most saliently on this general, philosophical issue; the elaboration of which, it is hoped, will alight pathways to progress and deeper unification, both within each of these sciences and, ultimately, among them as well.

Although the main thrust of the work is epistemological, it is not philosophically "technical"; which is to say that no special knowledge of technical terminology is required in order to understand the arguments. When a special purpose term is invoked or invented, it is defined in the context of its usage. And while the work is aimed largely at researchers in theoretical physics, it is not heavily mathematical. Rather, it should be intelligible to anyone who is familiar with any of the sciences discussed, as well as the non-specialist generally.[1] Toward this end, one of the novel features of the presentation is a fairly extensive use of graphics to illustrate "philosophical" concepts, and narrative descriptions of salient mathematical ideas.

The first three chapters are concerned with an epistemological investigation of various general issues that, as far as the author can discover, have

---

1 Particularly inasmuch as the author is such a non-specialist.

not received adequate treatment elsewhere to date. This is followed, in chapter four, by a detailed analysis of the theoretical conceptions central to physics and the problems particular thereto. The two subsequent chapters establish, respectively, renewed conceptual foundations for each of physics and the science of consciousness.

With respect to physics, the immediate goal is a framework for a quantum theory of gravity and unification generally; with respect to the study of sentience, a similarly general framework by which it may be possible to increase the detail, comprehensiveness and quantifiability of phenomenological description and, eventually, extended contact with physiology and thus physics. The overall aim of the work is to bring into sight the possibility of such an over-arching unification.

It should perhaps be mentioned that the rather grandiose-sounding nature of these claims is not lost on the author; nor the universal fallibility of the human intellect – though the provisioning of a framework and an equation or two stands in a somewhat modest relation to the development of a definitive theory. Rather more to the point, should the research efforts of any sympathetic reader be in any measure hereby facilitated, then the proper object of the exercise shall have been fulfilled.

Introduction

THERE is a sort of mythological creation story about the evolution of modern physics, which runs approximately as follows. Newtonian mechanics represents the quintessential, theoretical embodiment of intuitive ideas about how the world works – a methodological elaboration of "common sense" notions, which originally evolved to deal with ordinary, everyday experience. But while logically and empirically unexceptionable in application to that realm, these so-called *classical concepts* are inapplicable to the more recently discovered phenomena that comprise the subject matter of quantum and relativity theory, with respect to which great velocities, vast or minute distance and/or time scales, or intensive gravitational and/or other fields come into play. To adequately describe the reality revealed by these modern discoveries, Newtonian notions must be superseded by purely mathematical, otherwise non-graspable concepts, which necessarily lack comprehensibility of the sort associated with ordinary physical intuition and visual perception.

In other verses, the transition from mechanical models to more abstract representations is traced to the idea of non-material yet energetic fields, usually credited to James Clerk Maxwell in connection with his extensive contributions to electrodynamics. Indeed, Maxwell's work in this area, together with his kinetic, statistical-mechanical treatment of thermodynamics, set the stage for the historical, contradictory convergence of classical concepts, *viz.*: first, with respect to the failed experiments to measure the velocity of the earth relative to the aether, which led directly to relativity; and second, via the incongruity of kinetic theory with electrodynamics as revealed in the spectra of blackbody radiation, which precipitated Max Planck's reputed discovery of the quantum of action. These are the infamous "two dark clouds" of natural philosophy, obscuring, at the turn of the last century, what was otherwise widely believed to be the imminent "end of physics" – eerily akin to current prognostications along like lines.

These brief narratives, which do capture, cursorily, certain essential aspects of early twentieth century developments in physics, are nevertheless somewhat misleading. In point of fact, Newton's laws are neither intuitively obvious nor free from logical inconsistency. More importantly, it is not necessary to forsake intuitive understanding in order to accommodate mathematical description to contemporary empirical knowledge. Rather, all of the phenomena comprehended under Newtonian, relativistic and quantum theory can be satisfactorily understood as emergent from intuitively graspable first principles.

Intuitive understanding is absent from modern physics primarily because of philosophical prejudice, along with certain misconceptions about quantum and relativity theory that took hold early in the last century. The historical roots of the prejudices and misconceptions prevalent today can be traced, in substantial measure, to the many difficulties attendant to late nineteenth century efforts to establish a mechanical foundation for electrodynamics, and widespread misunderstanding regarding Einstein's approach to these difficulties.

Einstein was always unsure about crucial aspects of this issue, and throughout his career had several epistemologically inspired changes of heart regarding the foundations of physics. His first efforts to come to grips with Maxwell's electrodynamics, while in his mid-teens, involved dynamical models of what he then called "magnetic states of the aether." His later putative abandonment of such constructions, famously connected with his 1905 paper *On the Electrodynamics of Moving Bodies*, and the formal unification of the concepts of space and time that followed thereon, is often interpreted to mean that he did not, at the height of his powers, consider the *aether* concept viable. However, his thoughts on the subject were not so black and white. While it is fairly well known that he came to realize, in connection with his work on General Relativity, that the extension of relativity to account for gravitation conferred properties of an energetic substance on space-time, his subsequent efforts to make sense of this aspect of relativity, inextricably tied with his lifelong quest to subsume quantum phenomena under a continuum umbrella, were not widely appreciated; and not at all by most of the so-called *Founding Fathers* of quantum mechanics.

Thus it came to pass that, in the mid 1920's – after three decades of struggle with the mysterious quantum of action – a new paradigm of physics was ushered into the world, largely by way of Werner Heisenberg working under the auspices and driving vision of Niels Bohr; although the vital, unifying interpretation that combined Heisenberg's contribution with Schrödinger's ground-breaking work, which latter appeared shortly after the former, was introduced by Max Born. This paradigm, sprinkled with a few finishing, philosophical flourishes by Bohr, came to be known as the *Copenhagen Interpretation*, after the cosmopolitan center to which physicists the world over flocked in order to work with the zealous Bohr.

Copenhagen abruptly altered the conceptual foundations of physics via the introduction of sweeping, philosophical innovations. This happened without the consensus of all relevant, leading thinkers of the time and, more to the point, absent adequate epistemological analysis. The Copenhagen vision is narrow; and though Bohr's philosophy is more or less logically self-consistent, it is by no means comprehensive. That such an inadequate vision prevailed can be attributed partly to the crisis of the times and the pressing need to find an expeditious way forward. But in the final analysis, Copenhagen was largely informed by the philosophical prejudices of a prominent subgroup of key figures; namely, those associated with Bohr.

By the time these events were unfolding, Einstein's early work had set the stage for a radical break with classical ideas. The mathematical framework of General Relativity is rather abstract, involving a four-dimensional continuum in which space and time labels can be interchanged and space acquires a degree-of-freedom that makes models of the cosmos "difficult" to visualize. This forsaking of visualizability, with its roots in Minkowski's spacetime framework, foreshadows the further radical departure from intuitive concepts brought about by quantum theory.

While Einstein famously opposed the developments of quantum mechanics with considerable vigor, especially from circa 1927 onwards, he nonetheless had a major part in shaping its growth to that time. He was, to an extent not always appreciated, the conceptual father of quantum theory. He was the first to explicitly introduce the concept of discrete energy in his 1905 "photoelectric" paper, *Concerning an Heuristic Point of View Toward the Emission and Transformation of Light*, with extraordinary prescience regarding the meaning and implications of the innovation – at best implicit if not altogether absent in Planck's earlier treatment of blackbody radiation employing discrete quantities. He also was the earliest, clearest, and generally most forceful proponent of the wave-particle dichotomy, throughout the two subsequent decades leading to Copenhagen. He discovered many of the key implications of these concepts, from the consequences for the heat capacity of solids to the possibility of boson condensates.[1] Moreover, the minds behind the Copenhagen "non-world"-view were inspired by the possibility of extending the radical break with classical physics associated with Einstein's theory of relativity, as well as by the rejection, on principle, of unobservable quantities – a principle which they believed had guided Einstein to his results, as they believed it was guiding their own footsteps.

Unfortunately, by 1927 Einstein was so deeply engrossed in his unified field program that he was relatively uninterested in the quantum theory of Heisenberg and Schrödinger. He was so convinced that no formulation of the new quantum mechanics could provide a satisfactory way forward

---

1 Einstein speculated, very early on, about some of the most radical possibilities implied by quantum phenomena. Consider the following quotation from a conversation that occurred in early 1920: "Hitherto we have regarded physical laws only from the point of view of *Causality*, inasmuch as we always start from a condition known at a definite cross-section of time, that is, by taking a time-section of phenomena in the universe, as, for example, a section corresponding to the present moment. But, I believe that the laws of Nature, the processes of Nature, exhibit a much higher degree of uniformity of connexion than is contained in our time-causality! This possibility suggests itself to me particularly as the result of certain reflections concerning Planck's Quantum Theory. The following may be conceived: What belongs to a definite cross-section of time may in itself be entirely devoid of structure, that is, it might contain everything that is physically conceivable, even such things as, in our ordinary physical thought, we consider impossible of realization, for example, electrons of arbitrary size, and having an arbitrary charge, iron of any specific gravity, etc. By our causality we have adjusted our thought to a lower order of structural limitations than seems realized in Nature. Real Nature is much more limited than our laws imply. To use an allegory, if we regard Nature as a poem, we are like children who discover the rhyme but not the prosody and the rhythm." pages 159-160 of: Moszkowski, Alexander. Conversations with Einstein. New York: Horizon Press, 1971.

that, although he forcefully and consistently voiced his disagreements with Copenhagen and indicated tentative support for de Broglie's first foray toward a causal counter-interpretation – along with Schrödinger's ongoing efforts to furnish a physical interpretation of his wave equation – he was not adequately supportive with respect to these efforts, particularly the former. As a result, it would be another quarter century before the world would learn, from David Bohm, that a different interpretation of quantum mechanics was even possible. By then, the philosophical framework introduced by Born, Heisenberg, Bohr *et al* had taken firm hold.

Bohr's conceptual straightjacket effectively bound – and von Neumann's "impossibility proof" gagged – virtually every dissenting thinker; a circumstance that continued for many years after Bohm's incontrovertible existence proof appeared in 1952 (and his corresponding critique of the previously unchallenged *impossibility proof*), which demolished von Neumann's theorem and the exclusivity of the orthodox philosophy. At least part of the reason for this is that Einstein was again not sufficiently supportive of the alternative that had appeared, even though all acknowledged – and none more explicitly than Bohm himself – that it was best understood as offering a heuristic, tentative position; a foothold on at least one alternate path forward, which might lead to others.

The orthodox viewpoint permits no mental maneuvering beyond blind manipulation of mathematical formulae. Absent physical modelisation of experimental arrangements – except for models confined, literally, to the macroscopic operations of experimental apparatus *per se* – the capacity to find new relationships among such operationally represented phenomena is severely handicapped. Physical intuition – spatial imagination generally – is one of the most powerful capacities of the human mind. It is an extremely important compliment to analytical thought, especially inasmuch as it becomes increasingly difficult, in proportion to the complexity of mathematical construction, to find relevant relationships purely by "play with equations," as Paul Dirac once expressed his own approach to physics. Moreover, an upshot of Gödel's Incompleteness theorems is that inconsistencies, and the probability of intra-theory incompatibilities, grow in proportion to axiomatic and deductive complexity. And, although many are averse to raise the issue, it is yet a salient objection that a science whose subject matter can be described as "that with respect to which one never knows what one is talking about" – once the exclusive province of the *Queen of the Sciences* (i.e., pure mathematics) – cannot be expected to yield much explanatory satisfaction.

In addition to the interpretive issues introduced by quantum theory, physics is plagued by severe mathematical difficulties that have developed in conjunction with them. Problems with infinities arise in both quantum theory and general relativity, and especially in straightforward efforts to combine them. Together with the problems of interpretation, these difficulties comprise a mutually reinforcing, vicious circle. This dilemma has catalyzed the development of increasingly tenuous, diverse mathematical apparatus, which – though not undesirable *per se* – have outpaced ideas that might anchor a unified formalism and span the divided edifice that has become a wavering, bipolar superstructure.

While foundational problems have always existed in the sciences, the extent to which they currently impede research, particularly in physics, is well beyond a tolerable limit. Recent efforts to push past these problems without directly addressing them, as exemplified by *String Theory*, have little chance for long-term success. A tangle of subtly linked conceptual cords must be carefully dissected before fruitful efforts toward unification can proceed; both within each of the major scientific disciplines with which this treatise is concerned and between them – if such a wide ranging merger should turn out to be possible.

Interestingly, resolution of these core difficulties, in each of these distinct fields, cannot be achieved independently. Rather, it is found that they comprise a group of logically related obstacles that must be overcome together. This approach – that is, the analysis of these problems from a group perspective – is essentially epistemological in nature, and thus a substantial portion of what follows comprises epistemological considerations. In addition, new frameworks are proposed, for each of both physics and the science of sentience, with an eye to their possible if not likely unification at some future time. The balance of the treatise is concerned with the elaboration of the conceptual foundations of these frameworks, with the majority of attention devoted to the physical paradigm, which is discussed in some detail.

The upshot of the epistemological investigation is that widespread, deeply rooted misconceptions about the nature and interrelation of theory and experience underlie crucial tacit beliefs held by researchers working in both the physical and cognitive sciences. Ironically, this confusion tends to cause contemporary theoretical physicists to distance themselves from conceptions that smack of a classical physical flavor, and too often to a search for ever more "bizarre and surreal" ideas – while having the opposite effect on cognitive investigators; whose naturalistic tendency, with a few notable exceptions, is to ground psychological phenomena in neurological processes describable in more or less classical physical terms. Neither tendency is informed by an adequate understanding of the salient issues involved. Accordingly, the purpose of the epistemological section is to remedy this defect and thereby clear the foundations, in both fields, for the establishment of new frameworks.

With the removal of several additional misconceptions specific to the subject matter of quantum mechanics and relativity theory, and the consequent clarification of these fundamental matters, a simple and elegant vision emerges for the unification of the foundations of theoretical physics. In particular, the connections between special and general relativity, quantum theory and Newtonian dynamics become apparent in an entirely new light. Gravitation, the quantum of action and Lorentz invariance can be explained on the basis of a simple set of closely related physical principles, along with Newton's laws of motion in the non-relativistic/non-quantum limit.

An elementary framework is also proposed for the description and investigation of the phenomena of sentience. Here, as with physics, the primary obstacle to progress is found to be philosophical prejudice and lack of epistemological clarity. Although the concepts invoked for this development are of a purely phenomenological, and thus psychological or "mental" nature, they nevertheless provide, explicitly, for the description of comprehensive correlations with physiological processes on a level of detail hitherto not achieved; and, accordingly, are intended to augment unification with the physical sciences.

It has been noted, by philosophers and scientists of great renown, that the sociological forces active in every field of human endeavor have, in the contemporary period, become particularly decisive, especially in the sciences with which this treatise is concerned. And it does seem to be the case that such factors had much to do with the early twentieth century developments alluded to above. But lack of epistemological lucidity, now as then, is the primary obstruction to progress. Sociological factors manifest as ideological dogmatism, which does not thrive under the light of earnest, critical analysis. It is hoped that, if nothing else of value derives from what follows, at least one or two dark clouds will, at last, be lifted.

Chapter I

Epistemological Considerations

§ 1.1   World Views Old and New

EVERY theoretical construction is framed within the context of some ontological perspective, or *paradigm*, by virtue of which its conceptual elements are connected with experience, and thus acquire meaning. In the theoretical models of physics, one of the defining yet often tacit aspects of the operative paradigm is the nature of the distinction posited between those attributes of experience that are deemed "objective" in origin and character and those, in contradistinction, which are considered "subjective." It is a premise of the view here to be developed that the perspective on which this distinction is drawn is the most essential aspect of such a paradigm, and that changes in this perspective mark major turning points in the history of philosophy and science. From the early metaphysical speculations of ancient Greece to the empirico-mathematical innovations of Galileo-Newton and the abstract positivistic conceptions of quantum mechanics and relativity, the evolution of natural philosophy is punctuated by such paradigmatic changes of perspective.

By this criterion, *Mediaeval Scholasticism* – a narrow re-interpretation of select ancient Greek ideas based largely on the compilation of works attributed to Aristotle – can hardly be classified as natural philosophy. It is a collection of ideas, rather, not much concerned with the distinction

between various instances of "things in the world" and the multivariate attributes of those things. So, for example, the meaning associated with the
word *home*, considered as "physical structure," takes its place in the Scho-
lastic scheme alongside all other meanings that may be associated with the
concept, including such teleological or intentional attributes as shelter or
dwelling; familial or social gathering place; associations with architectural,
religious or cultural aesthetic, etc. In more modern thought, such meanings are considered to have significance only as semantic properties of the
concept *per se* – that is, as *mental construct* – as opposed to independently
existing referents, or again, *things in the world*.

On the other hand, ultimately all conceptual meaning can be interpreted
as referent-independent in this sense, even – perhaps more to the point,
especially – that associated with the concepts of modern physics, which
are often considered, ironically, the most specifically "referent-directed."
Moreover, there is a subtle spectrum of rather fine lines along which the
distinction between *objective* and *subjective* attributes can be made. Appreciation of this distinction as it relates to the theoretical constructions
of science is crucial to a proper understanding thereof.

§ 1.2 Naïve Realism

The origins of the modern development of the *subject-object dichotomy*,
from what might be called mediaeval literalism to the first stirrings of empiricism, can be traced to several prominent, early-Enlightenment figures,
among them Bacon, Descartes and Galileo. In its nascent state, the notion
of physical causality is based on the view that all things in the world are
mechanically connected as the parts of a machine, and thus all events are
constrained within a deterministic framework mediated by contact action.

On this view, a chain of physical causes must always intervene between
an event in the world and its observation by a knowing subject. In other
words, the direct object of perception can no longer be thought of as a
physical thing, but rather as representative of such a "thing," which is indirectly sensed, and to which physical qualities are attributed. In accordance
with this view, if a tree falls in the wood, sound of the fall is propagated as

pressure waves through the air, which in turn set the ear drum vibrating, which in turn, through some unknown and possibly unknowable process in the brain, causes the conscious experience of the sound of a falling tree (the difficulty associated with this last step in the process is of course one of the problematic roots of Cartesian dualism). A similarly mechanistic interpretation holds for visual perception, via the mediation of light falling on the eye; tactile sensation by the pressure of touch; etc.

The evidently instinctive worldview that is displaced by such considerations is known, in philosophic parlance, as *naïve realism*, while its deconstruction, in the manner described, is sometimes referred to as *Mechanical Philosophy*. But this overcoming of naïve realism spawns a new group of quite profound logical difficulties, as a host of apparent paradoxes arises from the seemingly simple notion that the phenomena of nature reduce to the movement and interaction of inert material objects in otherwise empty space, with cause and effect bridged by physical contact. The underlying paradigm of mechanical philosophy (in the most general sense, and thus of modern physics as well), is hereinafter referred to as the "subject-object dichotomy," while the vexing group of conundrums and persistent misconceptions that arise in conjunction with it is [thus collectively] denoted, for historical reasons, the *Mind-Body Problem*.

§ 1.3 The Mind-Body Problem

The historical difficulties associated with the subject-object dichotomy can be sorted into two broad categories, either of which may seem primary depending upon perspective. From one point-of-view, there is an epistemological dilemma, perhaps first clearly articulated by Descartes, regarding the reconciliation of the human faculties of language and thought with the evidently unconscious, mechanical action of physical cause and effect. The core difficulty from this perspective is the question of how it comes about that insight, volition, and communication concerning such things – things which can be characterized as "meaningful," and which typify the response of an intelligent being to its environment and circumstances, while not being ascribable to any apparent mechanical cause – how, that is to say, such things can arise in a machine.

From another standpoint, there is an often vaguely formulated yet tenacious [apparent] paradox connected with "the notion of *consciousness* itself." Though based on a misconception, this causes a good deal of confusion, even among philosophers, and is thus the subject of endless monographs and discussions. This *problem of consciousness* is poignantly felt whenever the ostensive connection between subject and object, taken to comprise an act of observation, is called into question. For it does not seem possible for a connection to exist between processes considered to be "physical" and their concomitant subjective reflection "in the mind."

Consciousness cannot be clearly conceived as contained within physical space, nor as a container of spatially extended objects, because consciousness is not defined in terms of spatial characteristics. Rather, as the substrate or carrier of subjective experience – as "pure awareness" – it is deemed to be exclusive of, in a sense outside of, everything that is *object*-ive – i.e., physical objects extended in space, which, under the point of view that gives rise to the paradox, is all that is deemed to exist in the absence of an observing subject. Yet the image of existence as a space-time manifold of physical events presents itself as something completely comprehensive, inasmuch as everything that exists, including sentient beings, is conceived to exist within it. An apparent paradox thus arises in the dualistic conception of the conscious being, at once physical and not physical – that is, both one thing and its logical opposite.

Again, while this confusion is tenacious it is only that – *viz.*, confusion, as will readily become apparent in what follows. On the other hand the first-mentioned aspect of the problem, which has been elaborated at some length by Descartes and others, while also based on misconception is yet somewhat subtler. And though it also arises with the notion of the universe as machine, its implications are more general.

In accordance with the Mechanical Philosophy that had taken hold of the imaginations of Descartes, Galileo and virtually all Enlightenment thinkers, the whole of nature seemed explicable on the premise that the world is composed of inert matter responding to contact forces only – all of nature, that is, except the human mind. Although this enigma is closely related to the difficulty associated with the concept of consciousness described above, it addresses a different aspect of the core problem. In particular, the creative aspect of thought, which is reflected very clearly in the use of natural language,[1] seems impossible to explain on the basis of mechanical or, more generally, algorithmic principles. And so Descartes postulated the existence of a second, "thinking substance" in addition to the extended substance that ostensibly constitutes physical space and the things within it.

These two aspects of the mind-body problem – the apparent incompatibility of mechanical action with human thought and the closely related "hard problem of consciousness" (or *problem of qualia* as the latter is sometimes called, because sensate qualities such as color and sound are not attributable to inert objects or empty space) – though largely based on misconception are not for that reason trivial. To the author's knowledge, the conceptual problems at the root of this discord have not been adequately addressed elsewhere to date. Accordingly, satisfactory resolution of this group of problems is one of the principal objects of what follows.

§ 1.4 Mechanical Philosophy and Newtonian Physics

It is sometimes suggested that the mind-body problem, at least in the Cartesian sense, came to a formal end with the establishment of Newtonian physics, because Newton's invocation of gravity as a force reaching instantaneously across empty space is incompatible with the notion of mechanical action. Moreover, since Newton, the concepts of physics have grown ever more remote from "common sense" ideas of inert matter responding to contact forces. On this view, there can be no mind-body problem for the simple reason that there is no clear notion of body with respect to which any meaningful distinction can be formed.

---

1 Noam Chomsky is known for the phrase *the creative use of language*, by which he means the ordinary use of natural language, and by which he characterizes articulations that appear to be "without any finite limits, influenced but not determined by internal state, appropriate to situations but not caused by them, coherent and evoking thoughts that the hearer might have expressed and so on": Chomsky, Noam. New horizons in the study of language and mind. Cambridge [England]; New York: Cambridge University Press, 2000. page 17

Of course, the complex of conflicts associated with Cartesian dualism is somewhat larger than the relatively restricted set of problems that arise strictly from the attribution of the quality of "inertness" or "unconsciousness" to matter. Not only is the primary intuition of spatial extension salient *per se* – i.e., aside from any consideration regarding the nature and constitution of matter – but also, again, many aspects of subjective experience evidently elude algorithmic representation generally.

It should be noted in this connection that, at least in his early years, and as a matter of publicly stated opinion more or less throughout his life, Newton did not altogether abandon mechanical philosophy. Actually, he entertained many action-by-contact speculations regarding gravity. One of these was not radically different from Descartes' image of the cosmos as plenum, and another, kinetic theory – proposed to Newton by his associate Nicolas Fatio de Duillier – was identical to the model later championed by Le Sage, and which ultimately furnished the conceptual foundation for the statistical-mechanical theory of thermodynamics. Rather, Newton's hesitancy to publish such speculations regarding gravity reflected, in the main, his empirical conviction that a natural philosopher should not hypothesize gratuitously. Because he felt that the explanations at hand were neither sufficiently developed nor sufficiently compelling to overcome all possible objections, his cautious but pragmatic position was simply to keep to the equations that worked, and "wait and see" (although it is not entirely irrelevant in this connection that Newton sometimes stretched and fudged facts in order to fit some of his published speculations – speculations, as noted below, of arguably less plausibility than, for example, the kinetic theory of gravitation).

In contrast to this initial, conservative position, based largely on skeptical and sound epistemological considerations, Newton later adopted an altogether different objection to exclusively mechanical explanations, as his religious impulses – and/or, perhaps, Mercury poisoning, incurred via his alchemical experiments – led him to believe that gravity must operate in a manner analogous to that in which the movements of an animal respond to its volitions. He evidently came to believe that all physical phenomena must be understood on such a basis – in the case of gravity, via the action of an all-pervading aethereal spirit, which he identified with God – thus explicitly invoking a mind-body duality.

Such a space-filling spirit, at least for Newton, also fulfilled the conditions demanded by his laws of motion, which had compelled him to attribute an elusive physical property to spatial relations vis-à-vis his definition of "*Absolute Space*." It is possible that his personal philosophical prejudices, as well as his mental health, had some effect on his early unwillingness to publish his ideas on mechanical causes of gravitation. While the physiology of animal musculature and neural systems was not understood at the time, so that a conjecture comparing the motion of a planet to that of an organism might have seemed plausible (that is, reducible to simple dualistic considerations, which in turn cannot be further reduced on a physical basis), his late-life preference for such an interpretation almost certainly originated in philosophical bias – again, in his case theological – a tendency, albeit often more secular, that can be discerned in the writings of many prominent physicists, and which has profoundly influenced the evolution of physical theory.

§ 1.4.1 A Brief Aside on Newton and Personal Bias

In the general introduction to this treatise, scant reference was made to the origins of the so-called orthodox formulation of quantum theory. It is important to understand that the conventionalization of the Copenhagen Interpretation and closely related renditions (i.e., characterized by the visions of Bohr and Born and/or by von Neumann's formal mathematical and epistemological treatment of the subject) was shaped by many factors, the most influential of which had little to do with explanatory efficacy or logical consistency.

Although the circle around Bohr harbored strong philosophical beliefs antithetical to those of Einstein and, *ca* 1925-27, de Broglie and Schrödinger as well, it is perhaps more relevant to the course of history that, if a causal interpretation had taken hold early on, the mathematical theory developed by Heisenberg – which was chronologically prior to the work of Schrödinger – would have likely lost its overall essential significance and recognition of primacy. This is because Heisenberg's *Matrix Mechanics* is formally isomorphic to Schrödinger's *Wave Mechanics*, and only Heisenberg's philosophical interpretation, crystallized in the *Uncertainty Principle*, distinguishes Heisenberg's approach from that of Schrödinger in any meaningful sense. However, if the direct, physical significance that Schrödinger sought to attribute to his mathematical theory had been accepted as viable, Heisenberg's theory would have become largely a footnote. Moreover, although each of the two formalisms offers unique calculational conveniences that vary with circumstance, solutions based on Schrödinger's treatment are usually the most tractable.

Not many generations have passed since the development of quantum mechanics, and the author has no wish to offend anyone who, by familial relation or otherwise, might be needlessly distressed by an inquisition impugning the Founding Fathers' scientific integrity. Suffice it to say that, among the best of the best, all are yet human, and behavior cannot align perfectly with intention. And so in place of such an exposé a few relevant points regarding some of the infrequently discussed aspects of Newton's career are here noted, with the hope that they will have a relevant, edifying effect.

While it is fairly well known that Newton threw himself into his work as *Master of the Mint* with perhaps unexpected devotion, it is not usually mentioned that he was in a position to earn a profit based on the percentage of coinage, and that he was not above juggling the books to ensure or enhance that profit. Moreover, from very modest beginnings he made himself quite wealthy via the public trust, and took immodest pleasure in extravagantly flaunting his *nouveaux* riches, as it were.

Another item not often discussed – or at least not in much depth – is that his commitment to the Coin of the Realm's integrity was also informed by a sadistic satisfaction, which he derived from the fanatical persecution of those within his mercy; defenseless people of often questionable guilt. Indeed, he was not above using paid informants to secure convictions in controversial cases, and personally saw to it that those he convicted experienced the most gruesome sorts of executions. For anyone interested, the barbarity of English law during the period of Newton's tenure at the Mint is well documented. While religious fundamentalism is often accompanied by such fanatical intolerance of "criminals" – whom Newton compared to dogs eating their own vomit – the fact that he was clipping the coin, from inside the system that had entrusted him, belies his vehement denunciations of the small-time, often accidental crooks he persecuted – not to mention innocents – and reveals a stunning hypocrisy.

Perhaps more relevant to the matter at hand, he evidently had no qualms about covering mistakes and holes in his scientific work, which he hid in consciously dishonest ways.[2] He made his cover-ups as mathematically complicated and difficult to read as possible, substituted guesses for facts, faked calculations and invented fudge factors to fit specially selected data, while discarding data that could not be made to fit (beyond, that is, interpretive issues involving honest discretion, which always arise). One of the reasons this is known is because he conspired with others – in writing, thus leaving an evidentiary trail – to publish fraudulent data.

This aspect of Newton's character seems to have emerged largely with or after publication of the first edition of his *Principia*, in anticipation or reaction to criticism – much of which was certainly misguided, and included misplaced accusations of plagiarism or denials of priority. Newton was acutely sensitive to and feared this sort of attack, which he gave as a reason for having waited until his mid-forties to publish the salient work and discoveries of his early twenties, including his invention of Calculus.

One must bear in mind that intellectually acute people are no less likely to be disturbed by intense emotional feelings than others – perhaps more so – and understand that one of the most powerful emotions a scientific discoverer can experience is recognition for original work. The thought of losing priority, or acknowledgment generally, can be powerful motivation. Moreover, megalomania is a good balm for the conscience, inasmuch as it can elevate selfishness to a virtue.

When one examines the details of the development of twentieth century science in this light, it is easy to apprehend the manner in which much of the irrationality surrounding current-day interpretations of relativity and quantum theory evolved. But although understandable, and perhaps, at the time of introduction, even expedient and thus – from a pragmatic

2 For a good bibliography of historical documentation of this aspect of Newton's character, see: Ohanian, Hans C. Einstein's mistakes : the human failings of genius. New York: W.W. Norton & Company, 2008. pages 70-72 and notes thereto

viewpoint – beneficial, bad philosophy should not be allowed to impede ongoing research, as it has now for decades.

§ 1.4.2 Newtonian Physics and Idealism

Modern physical science is premised on the distinction between physical attributes of phenomena on the one hand – in the language of the early British empiricists, the "primary qualities" of spatial extension and motion – and the sensate "secondary qualities" of color, sound, fragrance, etc. on the other. Following on this distinction, and in conjunction with several expedient *ad hoc* suppositions about space, time and force that found their way into the *Principia*, the extraordinary structure that Newton bequeathed unto the world is seeded with logical inconsistencies, and thus destined to sprout entangling nettles.

Beside the expediency of *action-at-a-distance*, Newton found it necessary to introduce the concepts of *Absolute Space* and *Absolute Time*, and the ingenious mathematical device of *Fluxions* – i.e., *infinitessimals*. These innovations, crucial as they are, introduced new conceptual difficulties into the foundations of science. One of the earliest and clearest analyses of these difficulties is due to Bishop George Berkeley, who is known primarily for his philosophical challenge regarding the mind-body paradigm. But he also astutely criticized Newton's invocation of Absolute Space, as well as the notion of infinitely small measures of space.

Despite its vital role in mathematics, the concept of the infinitesimal is not entirely unobjectionable. The source of contention, of course, is the conjecture that distance and duration can be indefinitely divided, which recalls the problems of motion famously attributed to Zeno of Elea, vis-à-vis the paradoxes that bear his name. While conceptual problems already plagued pre-Newtonian space and time, Newton's elaboration of them as "absolutes" – i.e., independently existing entities availed of a definite, uniform structure – only amplified the ambiguities inherent to these ideas. The concept of Absolute Space – of space as something substantial – seems to be a *non sequitur*, inasmuch as it means "that which remains after the removal of every-*thing*." Similarly, the concept of Absolute Time, or time as an independent thing, seems to mean "that which is happening when nothing is happening." To make absolutes out of negatives is to court confusion.

Berkeley recognized these problems with Newtonian mechanics, not to mention the obvious "absurdity," in Newton's own words, of action-at-a-distance. He also realized that the belief in physical causality, linking object with observation, is at the root of the mind-body problem and, if only pursued to its logical conclusion, rules out the possibility of direct awareness of the objects of perception. He thus became convinced that the best solution to the mind-body problem – as well as to the difficulty of establishing physics on a mechanical basis – is to simply let go of the belief that physical substance is something real, existing independently of minds that conceive it. In philosophical terms, he adopted the position of *Idealism*.

§ 1.5   Critique of Idealism

David Hume pushed the analysis begun by Berkeley much further, as he was convinced that simply dismissing the existence of a material world does not resolve all epistemological difficulties. Berkeley avoided *Solipsism* by positing the existence of an all-encompassing, universal Mind, in which all subjective experiences inhere, and via which they are ordered. Somewhat loosely, his premise was that the totality of all experiences – across all minds in all places and ages, taken as a whole – comprises one aspect of the *"Mind of God."* Individual human minds, accordingly, are small subsets of the "divine set," so to speak, of the totality of the experiences of all sentient being. An example of a simplified form of such an arrangement is characterized in the popular *Matrix* movies, in which a supercomputer is the ordering principle that establishes and maintains the inter-relations among individual minds (although both the computer and the individual minds are therein represented to have a physical existence as well). In Berkeley's view, God is likewise more than the mere totality of minds, as *He* is also the ordering principle.

However, Hume argued convincingly that, just as with matter, mind is also never perceived to exist *per se*. On introspection, all that one seems able to become aware of is experience – a flow of sensations, emotions, thoughts, and "reflections" on such things, which memory seems to organize "in time." Moreover, the very notions of memory and mind are evidently conceptual constructions, and as such are on an existential par with the "physical substance" of those "things" conceived to constitute the physical world.

Even the sense that there is a causal connection among experiences can be nothing more than such a construction, a belief *about* experience that corresponds to nothing that can be found *in* experience. Hume thus arrived at the conclusion that anything deemed to be a general property of things or events – such as *causality* – or to represent a *continuity of identity* from one experience to another – such as the concept of a person, place or thing referenced by a proper noun – must be, in this sense, a sort of illusory artifact of thought; an idea without a direct, corresponding referent in experience.

And so, if pushed to the limit, empirical analysis seems to invalidate the possibility of any sort of ontology whatsoever. Moreover, the distinction between ontology and epistemology is ultimately vague, inasmuch as those aspects of things that are deemed to be artifacts of thought must nevertheless be attended to as things that exist, because they are unquestionably experienced, as ideas. In this sense, epistemology might be considered the ontology of thought. But if all ontology is suspect, then so too are the conclusions of epistemological analysis. Therefore, even those philosophical perspectives that seem to survive the empirical analysis of Locke and Berkeley must yet be as questionable as their less sophisticated counterparts.

From the most "abstract" to the most "concrete," all concepts are thus ambiguous. Even the seemingly self-evident notion of mechanical contact force – a specific form of the general concept of causation – is not entirely free of the difficulties associated with the Newtonian conception of gravitation, which is deemed to act – "spookily," as Einstein would say – at a distance. Paraphrasing an early-modern critic: "If a red ball strikes a blue ball, why is it that motion is communicated but color is not?" In other words, what is force? Is the agency associated with contact action, in the final analysis, really any less mysterious than that of gravitational action at a distance? Is not the concept of contact action also "rather...a notion hovering in a mystic obscurity between abstraction and concrete comprehension..."[3] as another writer contended of the Newtonian conception of force? (It will be demonstrated in chapter four that the Newtonian definition of force is neither logically nor empirically tenable.)

§ 1.6   The Limits of Skepticism and *A Priori* Knowledge

Kant sought to restore the certainty associated with the fundaments of Newtonian physics and the applicability of mathematics to experience generally. His intention was to demonstrate, in a remarkably thoroughgoing but somewhat misguided analysis, that the basic forms of human experience associated with apprehension of the "physical world" – i.e., perception of causally connected events extended in space and occurring in time – are mental constructs that reflect the structure of the mind as much as they reflect the structure of the "external" world they ostensibly mirror. The twin primary, innate intuitions of space and time, on which geometry and arithmetic, respectively, are deemed by Kant to be based, are thus prerequisites of all human perception. Therefore, because these basic aspects of experience – again, events ordered in space and time, connected by the chain of causality – are due to the activity of the mind itself, it is possible to have assured, absolute knowledge of them (i.e., insofar as these forms of perception and conception are deemed to be permanent aspects of experience that, in principle, cannot change).

There are, of course, some flaws in this reasoning. First, while it is true that the format of physical experience – the context of space and time – is given directly in intuition, it does not necessarily follow that everything that happens in the "external world" will necessarily be conformable to representation in this fashion; a belief specifically rejected, as is well known, by much of modern physics. Although the Kantian premise is essentially mooted by this first point, there is yet a second obvious objection, namely, that human nature is neither well understood nor, for that reason, categorically unsusceptible to change, whereas the pre-condition

---

3 Lange, Friedrich Albert, and Ernest Chester Thomas. The history of materialism and criticism of its present importance ; Volume I. London: Trubner & Co., 1877, 308.

that the forms of perception need be permanent aspects of reality is a tacit assumption that Kant seems to overlook. That is, he assumes that human nature, as he conceives it, is a self-evident reason for the necessary existence of causality among the objects of experience and, in accordance with his conception, this aspect of human nature and its cognitive interpretation of "reality" cannot be otherwise. These premises are no less arbitrary than the hypothesis that causality necessarily exists as a feature of "things-in-themselves," and would no more survive Hume's critical scrutiny than its less subtle counterpart.

Moreover, with space and time deemed subjective constructs, very little seems to remain that can be stipulated about the external world. It would appear that only very abstract notions of order are attributable to that which some writers have called the "skeletal structure of reality." Accordingly, only mathematical relations can be stipulated to have any sort of "objective" significance, with respect to which "one never knows what one is talking about." It is precisely for this reason that some physicists and philosophers of science have adopted a self-described *Platonic* attitude, i.e., a belief in the objective reality of "*mathematical truth*," which exists independently [and perhaps even to the exclusion] of any physical world or mind.

In any case, while it is clear on Kant's view that empirical phenomena must always have a predictable form, it is yet not certain *a priori* that these forms are adequate for the representation of "things-in-themselves." In other words, while the format of the human internal representation of reality – which presumably has evolved through natural selection, as per any other aspect of life – can be attributed a practical survival value, it carries no guarantee of applicability to realms outside its narrow purview. Indeed, it would seem logical to expect imperfections and limitations – more-or-less the position adopted by Einstein in 1905 and, with a somewhat more radical twist, the Founding Fathers ca 1925-27. So it is not clear what sort of certainty Kant believed he had obtained – although in fairness it should be noted that, in his time (i.e., before the theories of Darwinian evolution and non-Euclidean geometries), such innate forms tended to be conceived in absolute, even theological terms.

The result of this overall epistemological train of thought, up to and in-cluding Kant, is that all ideas about physical reality are just that – ideas – and that it is impossible to establish the existence, let alone nature, of anything that might be the referents of these ideas. (Regardless of Kant's personal position regarding the existence of an extra-personal reality, this is the conclusion of the line of thought that he pursued and elaborated.) Kant's analysis does not deal directly with whatever it is that might exist beyond personal experience. Rather, he posits that the representation of extra-personal existence as a physical universe is an ineluctable consequence of the mind's structure. From this, he draws the conclusion that, because mental representations are, by definition, immediately or directly known, knowledge derived entirely from the forms of experience – literally the *format* thereof – can be attributed an absolute certainty. That is, because these representations are all that can be known, then with respect to their format and implications based upon it – such as deductive knowledge of space and time, hence *geometry* and *arithmetic,* respectively – there is nothing that can be "wrong."

It would therefore seem that the only thing that can be said with certainty about the subjective forms of perception and cognition generally is that: (1) they *are* – that is, they exist and have certain qualities – and most significantly: (2) they *work.* This latter point is sa-lient because, assuming that an extra-personal reality exists, nothing other than the fact that its behavior is somehow approximately conformable to human cognition can be attributed to it. Moreover, any effort to treat the forms of perception as objectively or extra-personally "real" – as per the Newtonian concept of Absolute Space – seems to lead to contradictions; "*Antinomies of Pure Reason,*" as Kant called them, referring to questions about the beginning or end of time and the size or boundaries of space.

But despite these limitations of thought, the very intelligibility of so much experience – the apparent conformability of such vast regions of the world to mental representation – and the direct knowledge of that experi-ence (*qua* experience) provides an open porthole through which much can be discerned.

§ 1.7   Beyond the Kantian Critique of Reality

The most significant epistemological upshots of Kant's work do not differ greatly from those of his predecessors'. Primary among them, as discussed above, is that it remains impossible to know anything for certain about the existence and nature of the *"thing-in-itself"* – that which is ostensibly external to the mind, and which perception is believed to represent. Accordingly, nothing about future experience can be predicted, with certainty, on the basis of past experience. Yet for Kant there is one assured aspect of knowledge, *viz.*: its format – i.e., of events occurring in space and time, connected by the chain of causality.

But as also noted above, Kant's judgment must be qualified in the following manner. While the forms of perception, especially space and time, certainly appear to be intrinsic aspects of experience, it is yet not *a priori* certain that these forms can faithfully represent whatever it is that exists independently of the mind, including any relationships or order that might inhere in that extra-personal existence. In other words, statistically or "on the whole," experience certainly confirms that ordinary perception together with common sense conceptions (i.e., "naïve realism") work quite well in everyday life. This only means that, based on the guidance of these representations, people are able to orient themselves with respect to their experience and make predictions that turn out to be correct almost all of the time (e.g., stepping in front of a moving bus is to be avoided, as is stepping out of a skyscraper window, etc.). Indeed, the extraordinary range of phenomena that have been brought within the scope of science, and harnessed via technology, further confirms the pragmatic value of ideas systematically developed in accordance with this general framework, albeit extended quite abstractly.

And yet, one day, what looks like a six-inch step across a doorway threshold might turn out to be a plunge to the bottom of the Grand Canyon, or a transition to another set of physical dimensions, or an awakening/ escape into some 'extra-*Matrix*-like' world. Perhaps just as unlikely, one might cross the threshold of a black hole's event horizon, or find a cat in a superposition of dead and alive states. Of course, it is precisely in the sense of these latter possibilities that most physicists and philosophers of science understand the invocation of extra-classical concepts in relativity and quantum theory. It is commonly accepted that the fundamental ideas of classical physics – exactly those forms and concepts that Kant believed to be unassailable – are generally inapplicable, only suited to a specific subset of experience.

§ 1.7.1 The First Upshot of Epistemological Analysis

But if nothing can be known with certainty, how is science possible? While predictions about experience cannot be certain – even such as are concerned merely with the basic forms of that experience – it is nevertheless an absolute truth that experience, as such, constitutes an ultimate element or subset of reality. That is to say, turning Kant on his head, experience *is* the *thing-in-itself*, in the strictest sense of the Kantian formulation. It is an undeniable element of reality, irrespective of any representational meaning it might be deemed to convey. So, for example, if one stands looking in a mirror and wonders: "What is the '*thing-in-itself*' that comprises the 'conscious brain' of that person in the mirror?" the answer is that the thing-in-itself *is* that very wonder; that image of a reflection; that feeling of standing upright, of hating that horrible hair/absence thereof and all the other ambient sensations that constitute the "sense of the moment" attendant thereto.

In this view, what might be called the content or stream of consciousness is regarded as an independently existing flux of events – an objective thing, absent the superfluous concept of an observer, which latter is thus simply stripped away… and with it, the contradictions of the subject-object dichotomy. Note that it is not logically necessary to invoke solipsism in order to interpret experience in this fashion. Metaphysics must therefore be deemed a viable science, because experience – "sensate qualia" and all – must be reckoned a directly given subset of reality. Its existence is not subject to question, only its implications under one or another philosophical interpretation or paradigm. Such interpretations confer a symbolic, representational meaning that must stand or fall on its own merits, and subtracts nothing from the phenomenology of experience. In fact, it is only the assignation of symbolic meaning that can infuse and thus confuse *experience* with some purport; it otherwise "just is."

The nature and qualities of these really existing "things-in-themselves" necessarily reflect some attribute[s] of the greater reality of which they are a directly known part - even if only the attribute of *existence*. This is the open window through which, as noted above, much can be discerned. It implies that theoreticians might find ways of linking up [some future extension of what today is called] physical science with the phenomenology of personal experience, in a manner that goes beyond the theoretical conceptions of today's physical and cognitive sciences.

The next question that comes to mind is: "How is this to be done?"

§ 1.8   Beyond the Kantian Critique of Cognition

Consider the assertion, "I think, therefore I am." On the basis of this seemingly simple and self-evident inference, Descartes sought to establish a standard of truth that would bring the certainty associated with mathematics into the realm of speculative philosophy, and remove the cause of all doubtful knowledge.

Unfortunately, this simple assertion cannot connote the supreme assurance its author intended, but rather introduces new obscurities and difficult questions. For in what manner is it possible to conceive the "I" that does the "thinking" in this image? Is it some sort of spiritual substratum of subjective experience? If so, how is it to be distinguished from the experiences themselves? How can it be observed? In what ways is it necessary to the existence of these experiences, and how does it condition them?

If "I" is defined as something that underlies thinking – something, as it were, that "knows that it knows" – then there must be hypothesized a fundamental distinction between the act of knowing and the act of knowing about knowing. But this would seem superfluous, for the reason that reflective knowledge bears no characteristics that distinguish it, in a qualitative way, from any other type of experience. Thus, knowing that two plus two equals four is not substantially different from the reflective awareness of possession of this knowledge, which is simply a different experience. Inasmuch, then, as there appears to be no assured means by which "I" can be identified within the realm of possible experience, it would seem to have no logical justification. Whence, then, comes the concept's persistence?

§ 1.8.1 The Creative Nature of Cognition

The two statements comprising Descartes' maxim, "I think" and "I am," clearly display the basic structure of sentences generally – the association of a noun with a verb; a subject with a predicate. Thus in every more mundane statement, such as "an apple falls," the same structure is equally manifest. This syntax would seem to reveal something fundamental about thinking. In fact, it reflects the most basic characteristics not only of thought, but of what is called perception as well. For the general perceptual forms that embrace the maelstrom of experience, and render its order – apparently inherent in the totality of experience itself – more or less comprehensible, seem to identify persistent patterns against a backdrop of constant change. Perceptions and conceptions effectively reduce the number of distinct things that comprise awareness.

Imagine looking at a painting of a human figure. At first sight – presuming the painter competent – the mind tends to see an imaginary visual scene that it was the artist's intention to depict, rather than an arrangement of colors *per se*. Moreover, the arrangement of colors on the canvas can more easily be manipulated mentally – recognized, remembered, reproduced – as a perception of a visual scene than as an arrangement of colors; as, for example, one finds with a 'Paint-by-Number' picture (figure 1-1), with respect to which an indexical numbering scheme is employed to effect the organization. Perception plays a role similar to such a scheme, inasmuch as it encodes the order inherent in the distribution of colors – it effectively compresses and thereby represents that information, or order. Perception thus reduces what can be thought of as "a complex of sensations" – daubs of color in the case of the painting; multitudinous "raw data of the senses" in ordinary perception – to a simpler, comprehensive experience.

Figure 1-1 Without a unifying perception, one would need a color index to reproduce an image

And the same is true of concepts, although at a higher level of "abstraction." Thus the words of a poem can be more easily reproduced from memory *as poem* than as an arrangement of ink strokes on paper (or canvas, however artful) taken to represent the words – (e.g., *à la* a Japanese work of calligraphic art that expresses a *haiku*, per figure 1-2).

Figure 1-2 Haiku about four dogs playing poker

In such a case, reproducing the words of the poem from memory is considerably easier for someone who is able to read the characters (i.e., *words/
concepts*) than for someone who cannot. In general, the *visual perception*
of a written word is quite different for someone who can read the language
in which it is written than for someone who cannot (see figure 1-3).

Figure 1-3 There is a significant difference between what someone able to read the characters perceives and perception of the characters *per se*

It is in this sense that the experience associated with the observation of a falling apple – a "set of sensations," so to speak, that might be referred to as "*applerama*" – more readily becomes – i.e., more efficiently, in accordance with the nature of *perception* – the experience of "an apple" undergoing "a change in position" in "the course of time." It is this form of perception that underlies the above-referenced noun-verb association, as in "an apple falls," and it is in this sense that the notion of "a physical object" arises.

Now, it is especially important to note the essentially constructive aspect of these processes, because the connection between the so-called "raw sensations" and the putatively emergent "perceptions based on them" is not something that is part of experience – i.e., the process is not "conscious." Therefore, the notion that raw sensations exist absent any particular perceptual form is a conception not grounded in experience; raw sensations are, in this sense, wholly analogous to Kant's unknowable thing-in-itself – something assumed to exist, but not directly known to.

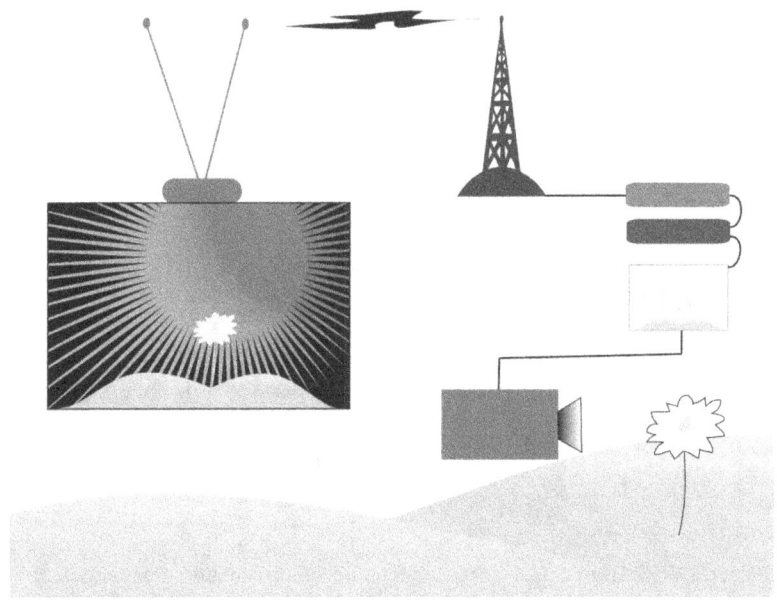

Figure 1-4 Everything other than the displayed image is imperceptible.

Following on this, perception, in its relation to the putative "raw sensations," cannot be conceived to be strictly analogous to the relation in which an image conveyed by a painting stands with respect to the individual daubs of color it comprises, but is rather more like the image an observer sees on a video display in relation to radiation picked up by a television receiver, massaged by video software, digitally woven into the framework of a faux background or setting, and then transmitted to a video screen where corresponding imagery is displayed (per figure 1-4). It is possible to look closely at the screen, see the matrix of dots that make up the image, and believe that the dots constitute "raw elements" of the representation (see figure 1-5 and description of *Gestalt*). But perceiving the dots is just another way of perceiving a previously constructed image – the picture displayed on the screen; the overall picture-forming process is not thereby deconstructed in any meaningful sense.

Figure 1-5 Getting closer to the screen and seeing the dots does not in any fundamental way deconstruct the display of the image. Every perception is given whole to consciousness in what psychologists call a *Gestalt*, and the perception of the dot-matrix pattern that makes up the TV image in the example above is just another, different Gestalt (different, that is, from the overall pattern first perceived on the video display), which says little about the radio receiver or video software behind the construction of the image. Just as with the faces and vase of figure 1-6, these are alternate "ways of looking" – each a distinct creative act, whereby a perception is formed. In this sense, the indirectness that is taken to characterize the awareness of external reality is really a general characteristic of "every-*thing*" that the mind turns its attention to – every thought, memory, etc. involves a constructive process that generates a new Gestalt.

Figure 1-6 Faces or Vase?

§ 1.8.2 Gestalt Wholeness and Figure-Ground Discreteness

Another salient feature of perception and, crucially to all that follows, conception as well, is the so-called *figure-ground* duality that, notwithstanding the just noted unity of experience, manifests as *foreground* and *background*; that is, a distinction between some aspect of experience that is apprehended as the "object of attention" and a background, or context, in relation to which that thing acquires its relative form or meaning.

A familiar example of this attribute of perception is apparent in figure 1-6. Depending upon how the figure is perceived, either the white or the black portion of the image will comprise the foreground as opposed to the background. That this attribute of perception is also a characteristic of concepts is not a novel proposition. Thus, the binary character of logic, and the implementation of such binary logic on digital computers, is a well studied, albeit narrow aspect of this property of concepts. However, the wider import of this proposition seems to be generally unappreciated. For example, the twin concepts of *object* and *space* are as the *yin* and *yang* of the conceptualization of physical percepts. Space is the idealized context in which objects are conceived to be embedded, and the concepts of space and of object are tied together with the same necessity that the alternating foreground of figure 1-6 is tied to its complimentary background. The same is true with respect to the subject-object dichotomy; each is yin to the other, yang.

Now, it is an inherent property of this figure-ground duality, perceptual or conceptual, that there is an apparent but elusive *"boundary"* that separates foreground from background, which is yet a part of both, and neither. This elusiveness is particularly evident with respect to simple mathematical objects, such as an *open disc* or *dimensionless point*. The boundary of an open (e.g., unit) disc cannot be described as a set of discrete points because the number-pairs that define the coordinates of the points are not recursively enumerable. This reflects not only the non-discreteness of the real numbers but also the ambiguous nature of boundaries generally – the difficulty of isolating the parts of something that is inherently whole.[4] Zeno's paradox of the *Heap* is related to this ambiguity, but in the context of concepts rather than percepts. The questions: "When does a group of grains – starting with a single grain and adding one at a time – become a 'heap'?" and "When does a 'heap' – by removal of grains, one at a time – cease to be such?" reflect a general ambiguity of concepts, which, again, do not have direct referents in experience. Rather, they are *ways of thinking* about experience.

---

4 This is not to be confused with the similar-sounding topological principle that "the boundary of a boundary does not exist;" a distinct, quite different mathematical concept.

Every circumstance in which an object of perception or thought can be imagined or conceived to exist involves some background or context with respect to which the object of the perception or thought acquires meaning. But the meaning associated with the object implies that under diverse circumstances – under alternate backgrounds or in different contexts – the object remains the same; which can only be strictly so if it is possible for it to have meaning that is independent of context.

For example, the conception of "a ball," taken to be meaningful for both "a ball in flight" and "a ball in hand," implies that in both contexts *ball* is "the same object," even though it cannot exist independently of specific context. There is no way to unambiguously establish the meaning of such an independent thing – even if only trivially because the boundary between the object and its context cannot be established (that is, it is not clear where "ball" ends and "not ball" begins). In this sense, the very notion that there are discrete elements of experience is an *idealization*; "any-*thing*" – whether deemed "mental" or "physical" – is an ideal, limit property associated with some *perceptual* or *conceptual process*; again, an abstract *aspect* of experience that cannot be directly identified *in* experience.

Perhaps the mathematical concept of a *dimensionless point* is the best illustration of this feature of thought. The *point* is deemed to be without dimension and yet to have meaning in the context of a space of definite dimensionality in which it is embedded and in terms of which its nature and existence is defined – an obvious impossibility. Again, the meaning is associated with a limit that, to be logically consistent, can never be realized. The impossibility of something being absent a certain quality – in the case of the *point*, spatial extension – yet having its meaning defined in terms of that very quality, is an inconsistency that reflects a more general fallibility that all concepts are heir to. Thus, in the case of Zeno's paradoxes, attempting to understand physical motion as composed of an infinite series of time elements called "instants of time" – i.e., each a static incarnation of something both "in time" and "timeless" – is inconsistent, as a result of idealizing – objectifying – the thing called "an instant." An "instant" has nothing of the quality of "time" – the very quality (in this case, temporal dimensionality) in terms of which its existence is defined.

This characteristic of thought – whereby aspects of experience become "*ob-ject*-ified" (hereinafter "*objectification*") – is clearly manifest in the concept of the physical object. And so the quality of *discreteness*, as associated with both physical and mental "*things,*" is a fundamental feature of cognition. It is quite evident in the fundamental forms of [particularly western] natural human languages – even the general notion of "an experience" not escaping its stigma. Yet it is primarily from this quality of *thing*-ness that the concepts of subject and object, and the related notion of "*I*" acquire their persuasiveness.

§ 1.8.3 The Inconsistency and Incompleteness of Conceptual Constructs

In addition to bearing the seeds of logical inconsistency, which is connected with the attribute of objectification attendant to every mental representation, concepts are also necessarily incomplete with respect to the unique qualities of the experiences they are deemed to represent. This is the underlying root of Hume's problematic concerning the notions of causality and of things that can be identified by a proper noun, although he evidently did not recognize this root as such. These universal features of perception and thought are the primary source of the confusing conflicts and ambiguities that inevitably crop up in any serious epistemological analysis. It is a manifestation of this intrinsic *inconsistency* and *incompleteness*, in a particularly striking form, that is experienced when one encounters such apparent paradoxes as arise with the mind-body distinction and Zeno's allegories – although the mind-body problem also involves a context-specific misconception, which is addressed separately below.

Again, incompleteness is related to the representational role that percepts and concepts must fulfill. Visual perception, in ordinary waking life, is deemed to represent specific external objects and the relations among them, whereas concepts are considered general or abstract representations – denotations for entire sets or categories of things and relations, and the meanings that those sets embody.

Perceptions must also, in a related sense, be considered "general," inasmuch as they comprise sets of patterns that, while not strictly identical with one another, are yet perceived as "the same thing." For example, the mental image that a dog (evidently) perceives of a familiar person, and thus experiences from one day to the next, varies in detail but not overall form, which latter is what the animal (evidently) experiences. Or consider a painting created by a well-recognized artist. To the untrained human eye, the difference between an authentic work of the artist and a good forgery goes unnoticed, while only the perception of a general pattern – the object and perhaps style of the painting – will be perceived.

Indeed, it is only possible for a percept or concept to function as a representation because of this attribute. If a perception conveyed all of the detail of the "stimuli" assumed to catalyze it, instead of a connecting function of those details, it would not be a perception – it would not fulfill the simplifying role noted above. Similarly, if a concept were identical to a single, unique member of the set of things it represents, rather than a connection among the members of the set, it would not be a concept.

Conceptions and perceptions thus fulfill a role analogous to that of a map. Representation necessarily involves simplification – i.e., information compression. For a map to be a perfectly complete reflection it would necessarily be identical to the thing that it represents, and thus no longer a map, or simplification. It is because of this generality – literally, nonspecificity – of representational meaning that Hume could not reconcile causality, or any general meaning, with particular empirical things and events. This aspect of cognition also seems to underlie Plato's theory of universal or archetypal *Ideas* or *Forms*. But it is important to realize that every experience carries unique meaning that cannot be fully captured in representation. Reciprocally, and crucially, every representation likewise carries unique meaning or information – meaning which is not inherent but rather supplementary to that of the object of representation. This may seem contradictory in light of the simplifying/information-compressing function of such cognitive constructs – and the common misconception that "abstraction" indicates, exclusively, some sort of stripping away of content/meaning – but it is the case.

Imagine, on the one hand, the act of looking at a painting and, on the other, reviewing the nature of that act. In one case, the experience is of the image represented by the painting, in the other, the concept of the cognitive representation of an agglomeration of colors as a single perception. Or imagine being lost in a house of mirrors (figure 1-7), and then looking at a map showing the location of the mirrors (figure 1-8). In each of these cases, one set of experiences – e.g., perception of a chaotic scene of mirror-reflected images – is representationally-coupled to another set of experiences – e.g., a map, or overview, schematically indicating the relative positions of the mirrors in a pattern. A unique relation binds each pair of these sets of experiences, yet each set is qualitatively quite different from its correlative.

Figure 1-7 Lost in a House of Mirrors

The experience of looking at a painting contains qualities, and thus information, that the concept of perception, raised by reviewing that experience, does not; and *vice versa*. Likewise, the experience of being lost in a house of mirrors has properties that looking at the map of those mirrors does not. And again, reading the map involves awareness of information that was not experienced by looking at the mirrors. Both aspects of the situation are unique experiences; both bear unique sets of information.

Figure 1-8 An "overview" or *map* furnishes a different perception

Therefore, a percept or a concept deemed to represent something else can never carry the full meaning of that something else, but rather a quite different, albeit related meaning that is intuitively understood to be inherent to a relationship of representation. And such relationships of representation can evidently take an infinite range of forms, reflecting the unlimited possible meanings, or perceptions and related concepts, which can be experienced.

These attributes of cognition are at the root of Kant's Antinomies, because all concepts harbor conflicting meanings. (The particular Kantian difficulties that arise with the concept of space – that it can be conceived as neither finite nor infinite because of the paradoxes associated with each [e.g., if space is finite, the question arises as to what is beyond, etc.] – are based on a confusion of the figure-ground context. The appropriate context is: "Object/Space," whereas the paradox arises with the abstract juxtaposition: "Space/No Space," as though space, again, were an object – an independent thing.) In the case of philosophical paradoxes these conflictions are particularly striking, because they are explicitly exposed as logical contradictions. But universally, percepts and concepts carry meanings that are different from the things they are deemed to represent, and are in this sense always incomplete. Moreover, by virtue of the different (and additional) meaning that characterizes the representation, and because it necessarily involves a figure-ground duality, every representation is also inconsistent. The meaning associated with a conception or perception is not logically inherent in any of the infinite experiences to which the construct may apply and, furthermore, the logical implications of that meaning cannot be entirely free from contradiction. All concepts are in this sense logically *incomplete* and *inconsistent*.

§ 1.9   General Properties of Cognitive Processes that are Crucially Relevant to Epistemological Considerations

The upshot of these considerations is that certain characteristics of cognition are of critical relevance to any epistemological critique, and they can be briefly characterized as follows:

   a. *Objectification*: The unification of disparate perceptual elements via a single perception; of disparate perceptions via a single conception; or of disparate conceptions, again, via a single conception – and so, in all such cases, *generalization* of meaning and *compression of information* via the process of representation;

b. *Incompleteness*: The subtraction and addition of meaning/information, which makes the perception or conception different from the elements that it unifies via the process of representation – i.e., *divergence of information content* between experiences connected by a relation of representation;

c. *Inconsistency*: The logical consistency of any representation, perceptual or conceptual, is limited as a consequence of objectification. There is a distinction, between Gestalt wholeness and its concomitant figure-ground discreteness, which imbues these cognitive processes with an inherent dichotomy – i.e., duality of the representation, which manifests as *ambiguity of information content* and *inconsistency of logical implications.*

As a result of these cognitive properties, there is a sort of epistemological analog of the mathematical uncertainty principle associated with the properties of waves and realized so dramatically in quantum theory. Human thought is necessarily limited by these properties of incompleteness and inconsistency, and this limitation must be recognized in the prosecution of any epistemological analysis. The artifacts attendant to these limitations must be teased out of any conclusions, or confusions, that such analyses would seem to reveal.

On the premise that a better understanding of these matters can be derived from a comprehensive, formal illustration of the logical relations involved, the manifestation of these relations in certain well-known theorems of mathematics is examined in the next chapter. In chapter three, the general epistemological discussion continues where it here breaks off.

Chapter II

Meta-Mathematics

§ 2.1 Gödel Incompleteness as Artifact of Representation

IN the early twentieth century, there was a serious meta-mathematical program afoot to establish mathematics on a purely formal, axiomatic basis, and to reduce arithmetic to logic. As with Descartes before them, the mathematicians involved in this pursuit wished to establish the standard of truth, in particular mathematical truth, on as clear and distinct a foundation as possible. But after several seemingly successful efforts along such lines, by some very high profile logicians and mathematicians, Gödel showed that the goal must in fact be unattainable.

Specifically, he proved that any system rich enough to yield the results of arithmetic would necessarily be incomplete or inconsistent, in the following sense. Given a set of propositions (axioms) and an algorithm for generating new propositions from them (deductions), if the system is complex enough to produce the rules of arithmetic it will further produce statements that, while true within the system – that is, confirmable by external human observation as compatible with the axiomatic propositions, and evidently a consequence of them – nevertheless cannot be proven within the system; i.e., by following strictly the procedures dictated by the algorithm. In this sense, the system is incomplete. Put another way, the system can be considered inconsistent, because it can be made to generate a proposition, call it *P*, asserting that "*P* is unprovable."

The method of Gödel's proof is similar to that of Alan Turing's demonstration that there cannot be a general algorithm for deciding the truth of arbitrary mathematical propositions, which in turn is based on the so-called *diagonal slash* device of Georg Cantor's theorem establishing that the real numbers comprise a non-denumerably infinite set. The essence of all three constructions can be briefly characterized via a simple analogy.

The logical inconsistencies that follow from self-referential propositions are reflected in the fact that it is impossible for a model to represent itself. For example, the perfect model of everything that happens in the universe is, necessarily, the universe in its entirety. That is, if the purpose of the model is to predict everything that will happen in the future – not merely to have the capacity, in principle, to make consistent predictions, but to actually represent every event down to the minutest detail – then nothing less than the entire universe will suffice. Put a little differently, the universe must run through all of its processes, in "*real-time,*" in order to produce its outcome, and it is therefore not possible to process such a result "ahead of time" – to get it in advance – by means of any representation. It is impossible to instantiate such a representation because, by definition, the model must also be part of the universe, and therefore its doings must also be described in order to capture "everything that is happening." Therefore it is not possible to represent the universe perfectly, because to do so involves the contradiction of a model modeling itself.

This result, which captures the essence of the self-referential contradiction at the heart of the Gödel Theorems, is equivalent to the statement that a conceptual representation is different from the thing that it is understood to represent. Each of both object and its purported representation conveys unique information that is not conveyed by the other. Therefore, insofar as an axiomatic construction can be considered a model for representing a set of truths, every such construction must be incomplete/inconsistent. Gödel's theorems reflect this intrinsic aspect of cognition.

§ 2.2 Discrete Infinity and the Concept Forming Process

A proof or algorithm is a conceptual construct; a procedure for generating a set of propositions, which thus comprise the "truths" that are deemed to be represented by the axioms and deductive rules. As with all representations, it can be interpreted as a device for information compression. Illustratively, if the label T designates a set of true propositions, then the theoretical conceptions embodied in the postulates and procedures that generate T can be taken to represent T's meaning. In the mathematical example of the open unit disc employed in chapter one, the set of points comprising the disc (i.e., up to but not including its boundary) – here the set T – cannot be generated by an algorithm that terminates, because T is not recursively enumerable. Similarly, the meaning of any ordinary concept, represented by the set of all instances of that meaning, is non-denumerably infinite.

Narrowing the semantic range a bit, consider the numeric concept *Twelve* – and the corresponding set, T, which contains all instances of the meaning of twelve. T includes, among a non-denumerably infinite collection of other things, 12 angels, 12 apples, 12 membrane universes, a pair of dice each showing sixes, 12 pins, 12 angels sitting on the heads of 12 pins and so on. With Kant, we can represent twelve as "5 + 7." But nothing short of the full set, T – or, more to the point, an understanding of the nature of T's contents; the set's "membership qualifications" so to speak, plus all the contextual possibilities – has the full range of meaning that the concept embraces, which a human mind can realize.

This might seem like a trivial example, or perhaps unfairly non-trivial because it involves a numeric conception, but if one applies the same reasoning to any mundane illustration the relevance becomes apparent. For example, consider the concept *Dog*. The cardinality of the set of possible instances of Dog is beyond recursive enumeration – and this applies equally to the case of any given subset of the set. So for example, in addition to every possible *breed* of dog the range of subsets further includes every possible instance of a unique, *individual* dog. Thus a particular Shitzu, a thing referred to by a proper noun, e.g., *Carmine*, can be "the same thing" under an infinitely diverse set of circumstances, such as cutting its hair or letting it grow, changing the composition of its body via diet, changing out its organs or joints with artificial replacements, etc.

And it is clear from other such mundane instances of representation – such as the relation *Canine* to Dog, or Dog to Carmine – that (1) the representation cannot have all the meaning deemed inherent to its object, and (2) conversely, the representation contributes new meaning/information, which is not inherent in its object. Thus, the concept associated with the name Carmine carries meaning in addition to Dog – e.g., "the bundle of beige hair that frequently appears in the kitchen," or "the thing that makes a particular sound under particular circumstances," etc. – as it relates any of these particular meanings and images with other aspects and instances of Carmine. Likewise, the concept Canine carries infinite possible meanings not inherent in Dog or Carmine, inasmuch as Dog relates Carmine to other, similar animals, and Canine relates Dog to others still; for instance, wolves, foxes and coyotes. But while each of these representations adds an infinite set of meanings, each is yet absent, again, that particular infinity associated exclusively with the concept Carmine.

This distinction, between something deemed to be an object of representation and the representation *per se*, is clearly evident in Turing's use of Cantor's diagonal slash procedure. Indeed, some writers have argued, on the basis of this procedure, that it is not possible for human thinking to be a computational process.[1] Less controversially, what such arguments demonstrate is that the conceptual process is in fact creative, as hereinabove described; i.e., that concepts are not identical in meaning to the things they are taken to represent.

Human thought is necessarily its own measure – humans have no extra-human mental capacity by which they can understand themselves. To say that it is not possible for an algorithm, which is a conceptual creation, to represent thinking is not very informative, because it is not possible for *thinking*, in general, to adequately represent "thinking." Not only is it impossible to completely capture human thought algorithmically – which is to say, represent it perfectly by an axiomatic conceptual construction – it is impossible to so capture thought by any conceptual means. Moreover, it is impossible to perfectly represent anything at all, even a rock. The ap-

---

1 Penrose, Roger. Shadows of the mind : a search for the missing science of consciousness. Oxford; New York: Oxford University Press, 1994.

plication of Church's procedure to Turing's argument, regarding the undecidability of mathematical questions, is represented in figures 2-1 and 2-2, and further discussed below – for the sake of completeness and, it is hoped, additional clarity.

	1	2	3	4	5	6	7	8	9	10	...
1	0	12	3	1	0	14	289	0	36	22	0
2	7	34	15	3	23	0	1	44	970	41	2
3	3	0	5	66	2	1	453	21	90	64	5
4	51	0	9	337	215	61	2	101	4786	332	1
5	538	22	134	30	908	376	77	16	279	13	0
6	206	0	2084	1212	243	2	409	732	1	88	50
7	9	4	4545	582	333	182	0	908	2	71	243
8	13	0	1	94	1	33	21	49	11	435	461
9	0	4	1012	29	2	172	4510	56	27	245	609
10	1	0	5	309	0	57	454	1	11	132	71
...	9942	1	0	43	1112	1	0	879	1	324	0

Figure 2-1 Cantor's Diagonal Slash

The numbers in the top row each represent an input to an algorithm. Each of the numbers in the left-most column is a label for an algorithm, which – acting on the inputs across the top – produces in each corresponding cell of its row to the right a unique output. The numbers in both the top row and the left column are the natural numbers, in ascending order from left to right and top to bottom.

The output of every algorithm is also a natural number. However, if an algorithm acting upon some input does not stop – and, for the purposes of argument, it is assumed that there exists some "super algorithm" for deciding all such cases – then a zero is entered in the corresponding cell. And so the matrix has infinite rows and columns, and every cell contains a natural number, given by the output of the algorithm assigned to its row acting on the input corresponding to its column, and – if that algorithm, acting on that input, does not stop – that number is zero. In this manner, the matrix is deemed to represent every possible algorithm acting on every possible input.

Again, it is assumed that there exists some super computational procedure capable of determining whether or not each of the algorithms represented numerically in the first column of the matrix will stop (i.e., return a definite result) when acting on each of the inputs in the top row. But, because it is assumed that the infinite matrix represents every possible computational procedure, the hypothetical super algorithm invoked for this validation must perforce be represented in the matrix. And this establishes the self-referential basis for the contradiction that defeats the scheme and proves that no such super-algorithm exists, inasmuch as it is possible to find an algorithm that is not in the matrix in accordance with Cantor's diagonal slash device.

Consider the set of algorithms D (with members D_i for $i = 0$ to infinity) that will decide whether or not the members of any other set A (with members A_i for $i = 0$ to infinity) will stop given a particular input "j" (with $j = 0$ to infinity). If $D(i, j)$ is the algorithm that checks whether $A(i, j)$ stops – where $A(i, j)$ is the ith A acting on the jth input – then if $D(i, j)$ stops $A(i, j)$ does not. Now consider what this means if the A is a diagonal entry in the matrix – i.e., if $i = j$ [$= n$]. In this case, if $D(n, n)$ stops then $A(n, n)$ does not. But in the case where $D(n) = A(n)$ – which is possible, because every D is represented in the matrix by some A – the contradictory result is obtained that if $A(n)$ stops then $A(n)$ does not stop. This is of course the self-referential paradox, which can be visualized in the well-known manner depicted in figure 2-2 and described in the next paragraph, briefly, for completeness.

Forgetting the algorithmic purport of the infinite matrix, imagine that it is populated randomly with natural numbers. Now, beginning with the cell in the first row and first column, add or subtract the number one, to or from, the number in the cell. Then do the same in the second row and second column, and continue in this fashion diagonally across the entire matrix. The result of this procedure will be a string of numbers – i.e., comprising the diagonal – that cannot exist in any row of the matrix, because it differs, by one, from the first number in the first row, from the second number in the second row, and so on.

	1	2	3	4	5	6	7	8	9	10	...
1	0 +1	12	3	1	0	14	289	0	36	22	0
2	7	34 -1	15	3	23	0	1	44	970	41	2
3	3	0	5 +1	66	2	1	453	21	90	64	5
4	51	0	9	337 -1	215	61	2	101	4786	332	1
5	538	22	134	30	908+1	376	77	16	279	13	0
6	206	0	2084	1212	243	2 -1	409	732	1	88	50
7	9	4	4545	582	333	182	0 +1	908	2	71	243
8	13	0	1	94	1	33	21	49 -1	11	435	461
9	0	4	1012	29	2	172	4510	56	27 +1	245	609
10	1	0	5	309	0	57	454	1	11	132 -1	71
...	9942	1	0	43	1112	1	0	879	1	324	0 +1

Figure 2-2 Diagonal Slash Plus or Minus One

But the matrix is assumed to contain among its rows every computable sequence – in the case of the above example, the output of every possible algorithm acting on every possible input, where a zero might represent the output of a super (validation) algorithm. Accordingly, the procedure by which the diagonal set of numbers is created embodies an algorithm that is not contained in the matrix. Via this contradiction, it is demonstrated by *reductio ad absurdum* that it is not possible for there to be a universal computational procedure, which can decide the validity (stopping) of algorithms generally. It also shows that a model cannot represent itself and that no representation – in the form of an algorithm/axiomatic construction or otherwise – can be complete in Gödel's sense.

But note that it is easy, as a human being, to "see directly" the possibility of constructing the diagonal slash procedure – which is of course also an algorithm, even though the matrix is deemed not to contain it. In this context, the procedure that produces the diagonal entries represents a new *aspect* of the situation, something not captured in the original, albeit infinite algorithm. This is the creative aspect of the conceptual (and perceptual) process as hereinabove argued; the diagonal procedure is no different from any other perception or conception created by the mind – of a mathematician or otherwise.

Viewing the matrix of algorithms as a conceptual model, one can consider the diagonal line of entries, which is taken as the basis for a new algorithm, to be that aspect of an object that only comes into existence via a change of perspective (in the sense of the vase vs. faces diagram), or new representation – here "external," i.e., as when one "steps outside" the model to form a new representation of it (in the sense of the map of the arrangement of mirrors in relation to immersion in the mirror maze). A model cannot model itself; this is the negative aspect or limit of any representation, and the source of certain logical contradictions and much confusion. But forming new models to capture different aspects of a situation is the positive, creative power of the conceptual process, in which new meaning comes into play; thereby distinguishing the representation from its object.

With respect to human thinking, this simply means that the mental capacity to shift perspectives is the ability to create new algorithms and thereby find new ways of viewing things. The human mind is, in a manner of speaking, a computer that creates new algorithms as it goes along; a capacity, moreover, that is recognized by the computer that possesses it, and which comprises, in a substantial sense, the power of understanding associated with "conscious awareness."

§ 2.3   Upshot for Meta-Mathematics

In chapter one, it was established that there exist limits of thought inherent to the conceptual process generally. These limits manifest in many ways; not-surprisingly, with great force and distinctiveness in the fields of logic and meta-mathematics. Bertrand Russell recognized a particular manifestation of this general epistemological truth around the turn of the last century, with respect to a problem that became known as *Russell's Paradox*. It is related to Cantor's diagonal slash procedure and hence to Turing's non-computability and Gödel's incompleteness results.

Russell's Paradox concerns "the set of all sets that are not members of themselves," which involves a logical contradiction because, if this set is not a member of itself then, by the given definition, it is a member of itself. At first glance this seems trivial, and in some sense it is. But the intransigence of the problem, the aspect of the paradox that has not, to date, been successfully addressed (i.e., to the satisfaction of all philosophers), is directly related to the limits of representation that, as hereinabove discussed, are inherent to cognitive processes generally.

The idea that a set of logical entities is "a member of itself" implies that the *meaning* that *links the members of the set* – which is to say, the attribute that is common to and *defines* the members – also defines and so *represents its own meaning*; i.e., models itself, which is the core of the problem. This again is logically congruent to the notion of a physical model of "everything physical that exists." Such a model must represent itself representing itself, *ad infinitum*. The physical impossibility of such a model is fully analogous to the logical impossibility of a self-representing representation.

And so the old debate between mathematical "Intuitionists" and "Formalists" is moot. The concept of *infinity* is subject to the same logical limitations that afflict every, more mundane concept. This particular mathematical problem merely reflects that, with respect to constructions involving the concept of infinity, it is especially easy to overlook the inherent pitfalls and thus become confused. This simply means that one must be very attentive to detail. All axiomatic systems are prone to incompleteness and inconsistency, because these are characteristics of all concepts and thinking. The upshot is that there are limits to the applicability of any formal theoretical construction, but for reasons more general than usually considered. It is thus incumbent on all thinkers to recognize and work with, and within, these limits.

With respect to the Mathematicians' debate, it is not pragmatic to proclaim: "We will do without the concept of *actual infinity*." It makes more sense to say, "we will employ concepts of infinity, as all concepts generally, with care, and thus investigate where they lead us," recognizing that no single model or axiomatic construction can embrace all mathematical "truths." There will always be contingent, overlapping boundaries between what might be called "*Regions of Truth,*" each containing constructs that are "pretty good models" for that region (figure 2-3).

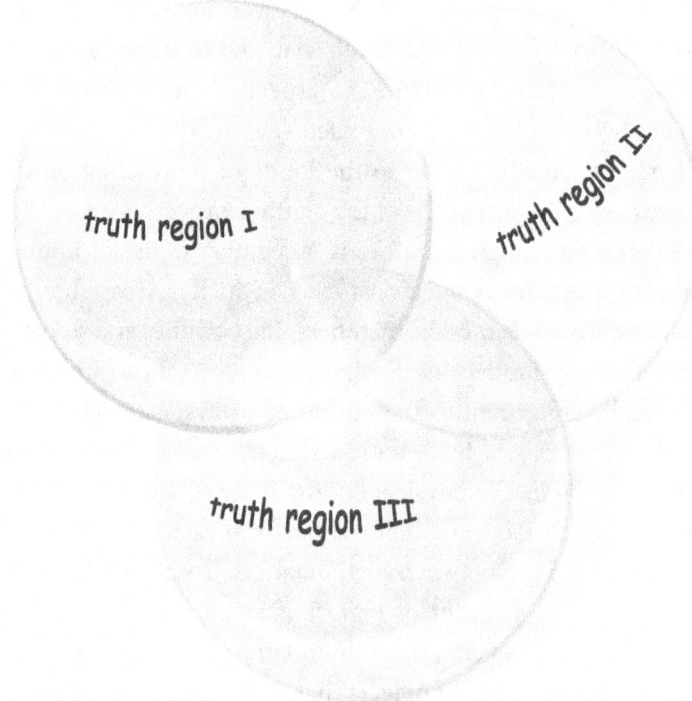

truth region I

truth region II

truth region III

Figure 2-3   Regions of Truth: Spheres, like circles in *Venn diagrams*, here intersect and thus overlap in some regions, indicating sets of propositions that can be reached by deduction from each of the 3 different axiomatic constructions represented by each of the 3 overlapping spheres (each sphere, that is, representing a particular formalism/set of propositions). Where all three intersect each proposition can be reached, independently, by all three axiomatic constructions: i.e., ***three different chains of deduction can reach any proposition in that region, and are thus "dual" to each other. But no single axiomatic construction can reach all truths.***

Chapter III

Intelligibility of Theory

§ 3.1 Asymmetries in the Relation of Materialism to Idealism

THE general epistemological considerations elucidated in chapters one and two, which reveal inherent limitations of the concept-forming process, are key to unraveling the misconceptions that surround the subject-object dichotomy. These misconceptions are analyzed in detail below, as the first step in a larger effort to attenuate the shadows they have cast on the theoretical foundations of the sciences.

One of the unsatisfactory attributes of the Cartesian mind-body distinction, and arguments that attempt to justify and preserve its basic provisions, is the inherent asymmetry in what might be called "the distribution of qualia." Matter is reduced to a very small complement of qualities, essentially *extension* and *motion* (and thus requiring the constructs of *space* and *time*), whereas the infinitude of others – the whole gamut of sensate display, from the colorful to the emotional – is attributed to mind. Moreover, mind is deemed able to conjure a representation of matter, while the suggestion of a reciprocal relationship seems meaningless. While it is possible to stipulate that some physical model is intended to represent the mind – e.g., a theory of the brain comprising neural correlates of consciousness – such a model seems to have meaning only in the context of a pre-existing, model-forming/interpreting "mind in the background," able to create and understand such a representation: just as the conscious observer is a pre-requisite of certain interpretations of quantum mechanics. Thus, Spinoza's assertion that mind and body are simply two different "ways of viewing" the same thing seems to neglect the salient asymmetry that "ways of viewing" connotes a modality of existence pertaining to mind, not inert matter, and so any necessity for the latter is, again, obviated.

Accordingly, *solipsism* seems to be a logically unexceptionable, albeit implausible viewpoint – inasmuch as there is no (unusually damning) contradiction connected with the conjecture that multiple or "other" minds do not exist; merely, rather, a sense of improbability. On the other hand, the thought of an exclusively material world, devoid of the qualities attributed to mind, is a self-evident contradiction. This is why Idealism has often been judged preferable to Materialism.

Meanwhile, there is an insufficiently appreciated upshot of this asymmetry; namely, that contrary to Kant's conclusion regarding the possibilities of metaphysics, human beings do indeed have direct awareness of at least one category of "thing-in-itself" – *viz.*, "first-person" or "sensate" experience, which is an undeniable element of reality; something that undoubtedly exists because it is known directly, even if the purport of what is known is "wrong." Therefore, any consideration regarding the fundamental nature and purpose of science must address this fact.

Again, this holds regardless of any particular interpretation of the nature of experience. That is to say, it is irrelevant whether or not experience accurately reflects an external reality, whether "red is really red," or even whether the notion that the phenomenology of experience "belongs to" or "happens to a mind" has any meaning.

It was in connection with this latter notion that Descartes' analysis went off the rails. He maintained that experience can only occur as a function of a "self" – a thing composed of mental substance, with specific attributes. In any truly adequate, syllogistic formulation of his *Cogito*, this belief must be stated explicitly. It should appear in the *major premise*, as tacitly implied in Descartes' assertion, *viz.*: "*If 'I' (a really existing self) am a prerequisite for the existence of thoughts*," which then must be followed by the *minor premise*: "*and if Thoughts Exist*" (both of which assertions, again, must be considered tacitly implied in Descartes' thus compressed, single statement: "I Think"), then it is legitimate to draw the *conclusion*: "*Therefore, 'I' AM* (i.e., a self exists)."

On the other hand, the simple statement that *experience*, whatever it might be, *exists*, only expresses a logical identity – the mere tautology that "if something exists," such as that which is called *experience*, "then it exists." Therefore, experience can be simply defined as "that which necessarily has the attribute of existence" – like Spinoza's definition of God, except not devoid of content. Again, in Kant's mode of expression, thought (and personal experience generally) can be called a "thing-in-itself" – no mind or self or substrate of any sort being logically required. This is also the ultimate upshot of solipsism, once "mind" is recognized as an artifact of that aspect of experience called "*conceptualization*" or "*the concept forming process*." If experience need not be tied to the idea that it is a representation of something else – if life is a kind of dream, in Berkeley's sense – then the identity of the dreamer becomes moot; it can be one or many minds, or for that matter *no minds*, just a stream or ever-changing manifold of what are called "experiences" for lack of a better word (i.e., one which does not connote the notion of an observer in the background – a subject to whom the experiences belong).

This conclusion would seem, perhaps, to endorse the most radical prescriptions of positivism, or again an extreme, empiricist interpretation of solipsism. However, from a pragmatic/heuristic point of view, the subject-object dichotomy is a necessary orienting perspective for those who do not wish to simply drop off the world; or are at least unwilling, perhaps, to stop "going through the motions" of life, even if a bit skeptical about its meaning.[1] Therefore, in the view that one seems compelled to accept, while experience *per se* is the proper object of metaphysical theorizing, science must yet be especially concerned with those constructions of thought that are taken to represent "the external world" – because they embody rules and relations that make it possible to successfully execute the business of life.

---

1 While according to certain venerable ancient Asian theologico-philosophical traditions human life is not only more sensible absent such a dichotomous perspective but also more fulfilling, it is not unexpectedly an ineluctable principle of all such traditions that experience thus purified is also *indescribable*.

Thus, while it turns out that direct knowledge of the subject of a meta-physical science is in fact possible – *the qualia and totality of experience considered as such* – any interpretation of the ontological status of experience (e.g., that it belongs to some "mind"), or of its representational intent (i.e., that it reflects or models "something else"), must perforce be corrupt. And yet it is precisely the representational purport of a particular, relatively small subset or aspect of experience that is of especial interest, and science can only have meaning if that purport is given some satisfactory interpretation.

§ 3.2 The Subject-Object Dichotomy and the Mind-Body Problem

Consider a schematic illustration of the format-of-experience that seems to follow from the most rudimentary aspects of the subject-object dichotomy. One imagines a body, which represents the self, embedded in a world of other bodies (see figure 3-1). Experience of this external world of physical things, although in turn embedded in a (perceptual) framework of spatial extension, is considered to be somehow "contained within the mind." In the illustration, this idea is depicted by representing the perception of a physical object as an image residing in the head.

The current, usual interpretation of the subject-object paradigm maintains that light is reflected or emitted from external bodies and travels to the retinae of the observer's eyes, which generate electrochemical signals that are transmitted via the optic nerve to the brain and are there interpreted as perception – i.e., a representation of the object that reflected or emitted the light. The problem with figure 3-1, of course, is that there is no such imagery to be found within a real brain. If one looks at a living, ostensibly conscious brain, one sees, superficially, mostly plain "soggy gray" matter (as depicted in figure 3-2). And while increased resolution reveals more intricate structure, qualitatively the picture of the brain never resolves to the image that is "seen" (i.e., from *brain* to *flower*). There is no such imagery to be found in the physical brain, regardless of how delicately one inspects it. This is essentially the "problem of consciousness" or "hard problem of *qualia*" as frequently discussed by philosophers of mind.

Figure 3-1

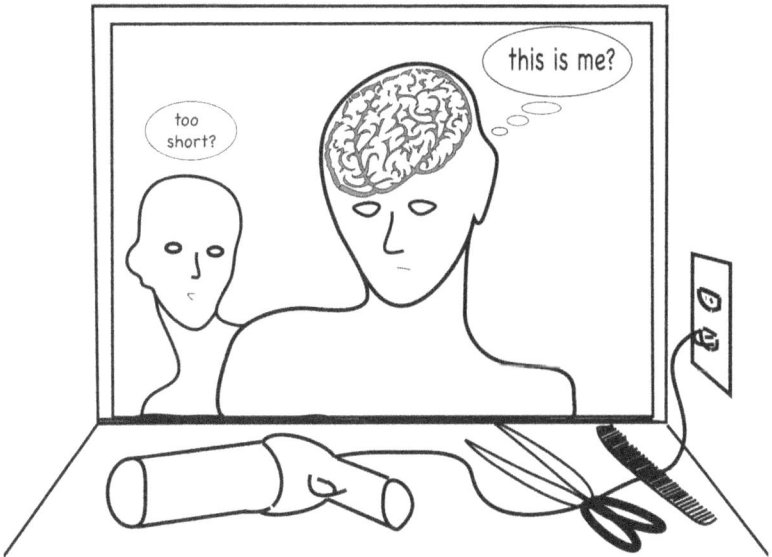

Figure 3-2 In this figure the brain is pictured as a perceptual construct, as seen by the man in the stylist's chair.

Upon more mature consideration, however, it is evident that neither the "soggy gray matter" nor the detailed brain structure thus perceived/conceived can in any sense be identified with the experiences thought to be occurring in the perceived brain-object. More to the point, the observed brain cannot even be "a brain" in the first place. What is *seen* is just another *perception* – "an image of a brain" – which must be understood as occurring in the mind of the observer who is "looking at" it; something very far removed from the hypothetical object of observation (*i.e.*, "thing-in-itself"). In addition, theoretical models of the brain construed along such naïve physical lines (whether as "soggy gray matter" or more sophisticated versions of the same basic idea, such as "neurophysiological processes") can only conjure similarly far-removed illusions; they do not and cannot reference, as demonstrated in chapters one and two, a "really-existing brain," whatever such a "thing-in-itself" might be, if "it" exists at all.

Figure 3-3 *Everything* one sees *is* one's"self" – every flower, every cloud – because everything is, necessarily, *perception*: "First Person Experience."
And so the self cannot be identified with any part or aspect of experience.

Consider figure *3-3*. The terminology employed in the following descrip-tion of the figure reflects the perspective of the man in the barber's chair, which is clouded by two major confusions. First, the perceptual image of the brain, as seen in the mirror, is believed to be a physical object in a world somehow external to the self, whereas the perceived image is really only that: an *image* – a perceptual construct internal to the self – an ele-ment "of the self": *an experience*. Second, this internal image of a "physi-cal brain-object" is further identified with the *entire* self, which [self] is also, in turn, yet another conceptual construct. In this manner, the self, considered as the totality of all personal experience (here pictured as the perceptual image in the mirror plus some fantasy imagery to represent the man's overall mental contents), is paradoxically construed to be some-how contained within what amounts to a fragment of the self – literally, a small "piece" of a single perception, only one among many of the self's experiences – in a thoroughgoing confusion involving (a) a representation representing itself, and the two fictions of (b) the perceived "physical brain object" and (c) the imagined "mental self object" (the latter pictured as an all-inclusive brain in the background).

Thus, if one does not keep alert to the basic implications of the subject-object dichotomy – disregarding, for the sake of argument, the even deep-er representational fault-lines that, per the discussions of chapters one and two, all conceptual constructs must entail – there is a tendency to imagine physical things as really existing in the form directly perceived; i.e., a pow-erful tendency to lapse into naïve realism. Although Kant's overall take on these matters is flawed, his notion of the subjectivity of space has con-siderable validity, inasmuch as perceptions and conceptions of spatially extended things, and the concept of space, cannot be viewed as having any assured meaning beyond the personal/heuristic. It should not be sur-prising, then, that it is not possible to "see" experiences "occurring in a brain" because it is not possible to "see a brain" in the first place – only a *perceptual construction* that is *referred to* as "a brain," and which in turn is conceived to *represent* "such a 'physical thing': *brain*."

So what is to be made of the "thing-in-itself" that, by virtue of the sub-ject-object paradigm, is stipulated to exist independently of the perception that represents it? That is to say, what of the "actual brain" that is imagined, in the sense of Kant, to be an objective, really existing thing in the world, and which is believed to be represented by the perception: "soggy-gray brain-object"? Because it can be stated with assurance that experience is a *thing-in-itself*, which necessarily exists, does it make sense then to say that a brain "really is" such a set of experiences – that what is represented as a physical object does exist, only not as "an object" but rather as a series of experiences that comprise the life of an individual? Moreover, if such is the case, can what is thought of as physical existence at large, the universe, be properly understood as a vast "sensorium," a collection of experiences *per se*?

Imagine that in some future time an elaborate mapping of "subjective" and "objective" processes has been accomplished. That is to say, that every imaginable human experience – every thought, image, feeling, etc. – has been correlated with a concomitant physical process in the human physiology. To know that such a thing had actually been achieved, one must be able, on the basis of a physical model of a given person (not necessarily a physical replica, a computer simulation would perhaps suffice), to emulate everything that the real person does in every conceivable circumstance. So, for example, given a question that requires the individual to draw on memory, think and even feel something about that memory, the model or simulation must process the given input in exactly the same way that the human being does, and produce an identical output. Moreover, it should be possible, based upon the theory behind the model, to understand the entire flow of information, every detail from input to output – i.e., to grasp the process in some sort of deterministic sense. Thus, the model would furnish everything that is necessary to reproduce the information processing that human behavior embodies.

Again, accepting that this has been achieved, could one thus know what aspect of the theoretical model "actually represents" the sensations associated with consciousness in a human being? Well, consider the circumstances this way. It is theoretically conceivable that a certain pattern of inter-neural communication is all that is necessary to model the flow of information associated with a person's response to a given input, and that the processing of that input through to response is in fact correlated with a certain subjective experience – say the perception of the color red (e.g., the question that the individual responds to, the input, might be "Is this patch of color red?"). But what if the conceptual framework under which the theory of neural communication is constructed does not include ideas of anything beyond the electrochemical actions of the neurons, while "in the real world" some as yet unknown processes, perhaps representable as patterns in an "energy field," are affected by the processes called *neural*, and thus are also correlates of the experience *red*. How then could one know what the neural events have to do with the experience? Moreover, even if the existence and response of the field were known, how could one decide which processes are salient with respect to consciousness? Or what if the patterns in the surrounding energy field were in turn affecting another system – far off or even distributed throughout space – and the remote system in question "is" consciousness? That is, what if the brain is more like a *transceiver* than merely an *information processor*?

These questions should have a familiar ring, because they are essentially no different from the arguments of solipsism. Is there any sense in such questions? After all, every theoretical model is nothing but a conceptual representation, which must be different from its ostensive object; both incomplete and inconsistent. What sense is there in asking, "what is a map *really*?" Note that this is a different question from, "what is this map *supposed* to be a map of?" And note how easy it is to slip back into the tacit belief that somehow physical things must be real, which is what lay behind the question "so what is the brain really?" It obviously makes no sense to say "the brain is really a set of experiences," or "the universe is really the sum total of all experience." One may certainly view the universe as a *Sensorium* if that view provides satisfying meaning. Such a view may even prove scientifically fruitful.

It is clear from these considerations that any physical model, any theory, can only have the status of a map, in the sense of chapter one. And maps are not real – except *as* maps. If a model of neural processes is able to furnish predictive information regarding the functioning of a mind, then everything that a representation can do is achieved. This is the only viable, permanent upshot of positivistic philosophy. Therefore, if the subject-object dichotomy is to be maintained at all – as a heuristic/pragmatic guiding vision – then the most natural and logical prescription is to categorize the set of immediately known "things" that comprise the self or mind in the Cartesian sense – experience *as such* – as the "thing-in-itself" of Kant. These are true "existents" that, via theoretical construction, can be cast into relations with "perceived brain processes," as per the general framework of the neuroscientist. Thus, the physical conception of the world provides a way to describe experience in a way that cannot be approached without such concepts. The problem with the physical framework, as currently usually understood, is that naïve notions of brain and mind, as distinct things with qualitatively irreconcilable differences, makes the connection between the two sides of the subject-object dichotomy difficult to comprehend. But this can be remedied.

The world of subjective, waking experience can be viewed as something like an extraordinarily complex dream, which is considered to be closely modulated by whatever exists outside the dream (the "dreamer" and the "dreamer's environment"). This much is commonly conceded. Upon close examination, the dream displays a very high degree of order and consistency, and one mode of interpretation – one that has proven extraordinarily successful in terms of predictive capability – is to conceive the order and connection of experiences to be analogous to an order and connection of extra personal existents. Moreover, if one adopts the position that these personal and extra-personal existents comprise a single existence that can be schematically represented as "an order of physical processes in space and time," roughly analogous to the perceptual world of naïve realism – for example, per the constructs of theoretical physics – one finds that it is possible to map certain experiences to others, for example by neural brain models of cognitive function.

However, it is always to be remembered that the "substance" of the world, if such a concept is even suitable, cannot be "matter" as commonly conceived – such as the "soggy-gray mass of the brain" – but rather is something that has among its attributes the variegated qualities of human experience, which through the optics, so-to-speak, of the channels and constructions of perception and cognition generally, comes to be represented and thought of as "physical stuff." In other words, whatever "the world really is" it necessarily includes sensate qualia, and so the notion of a physical universe can at best be a picturesque but very limited metaphor for an existence that is infinitely complex, both qualitatively and quantitatively.

If this is neglected then confusion ensues, due to the implication of an infinite regress, as the dreamer picks out the "self" as an element "in the dream" – by identifying the *self* with a particular aspect or *part* of the dream, *viz.*: the 'brain' of the *dream-self*, imagined to reside in a tiny *portion of the dream;* the *image* of the '*self*' that the dreamer thus imagines – as, for example, seeing a reflection in a [dream] mirror, as in figure 3-4. This is exactly analogous to the more general problem of self-representation discussed in connection with Gödel Incompleteness – again, it is impossible for a model to model itself.

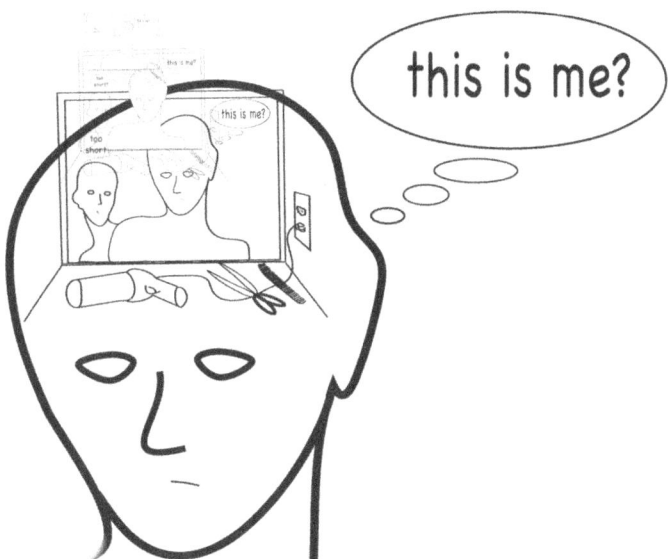

Figure 3-4 The self-representation paradox is illustrated here by a an infinite nesting of the image that the man sees in the mirror, imagined by the man to exist somewhere in his head, which imaginary construct must in turn include the nested perception *ad infinitum.*

With these relations in mind, it is easy to understand how confusion creeps into theory. And it becomes possible to build useful, even beautiful models without being hampered by philosophical prejudices and sidetracked by meaningless questions.

§ 3.3   Paradox and the Limits of the Concept Forming Process

The upshot of these deliberations is that neither naïve realism nor any variant of the subject-object dichotomy that displaces it can provide a complete, consistent framework for the interpretation of experience. Every paradigm must conform to the above-noted features of cognition, and is subject to its limitations. Moreover, in light of the foregoing discussion, it is evident that the mind-body problem, just as many similarly intractable problems of philosophy, is not a logically insoluble paradox but rather *confusion*, based on insufficient appreciation of the logical relations and the constraints of thought. It must be borne vigilantly in mind that complimentary[2] attributes of cognitive constructions, such as subject and object, arise together in a unitary construct, as do the figure and ground of perception. What appear to be individuated objects of thought do not have independent meaning, but rather are logically inseparable from the context of the overall construct.

Thus, the concept of inert matter bears no qualities that can account for consciousness for the simple reason that, by definition, consciousness means "the bundle of all qualities *other than* merely 'taking up space.'" The latter quality – *spatial extension* – is attributed exclusively to "matter" (and is *all* that is attributed to matter), which thus constitutes the side of the figure-ground split opposite to the "mental" side [of the *mind-matter* Gestalt; which, again, is a particular instance of the general subject-object paradigm]. Hence, in a metaphysical sense, the notion of an independent physical reality composed of inert matter is meaningless, as is the notion that inert matter can be identified with, or somehow "emerge" into, consciousness. And of course, so is the notion that there exists some sort of causal connection between mind and body. Again, consciousness connotes that precise set of qualia abstracted away from the set of "things physical," by definition.

This is essentially Berkeley's result, but derived from an epistemic rather than an ontic perspective. Ultimately, the "problem of consciousness" reduces to the following:

---

2 This general property of cognition should not be confused with Bohr's similar-sounding notion of "Complementarity," nor deemed in any manner to validate it.

1. The disparity of qualia associated with the distinction between events deemed physical and those deemed mental, which is connected with the representational aspect of perception and conception. As demonstrated, every conceptual construct, as representation, is necessarily not identical to the experience(s) to which it is applied – just as a map must be different from the territory it represents in order to function as a map. This is the general root of the problem of consciousness, which is a misconception, further confused in the manner of (2) and (3) below.

2. The subject-object dichotomy is a particularly striking form of the figure-ground duality intrinsic to cognitive constructs generally. Neither half of the whole makes sense in isolation; the notion that the foreground can exist independently of background, or that it is possible to identify a boundary that isolates it, is an illusion that is characteristic of every perception and conception. Each representation, taken as a whole and thus referred to as a Gestalt, binds figure and ground in a manner that makes logical inferences based on objectifying the distinction of foreground and background contradictory. And yet objectification is an intrinsic aspect of cognition: the foreground is taken to be the object of representation.

3. The particular image of the mental organ as "a physical thing: brain" – perceived as a soggy gray mass and conceived as an electrochemical machine – confuses, in the manner of naïve realism, perceptual and conceptual representations for the thing-in-itself. There are at least three distinct representations alternating in confusion, the relations among them quite misunderstood. The "thing-in-itself" is understood to be (a) the totality of sensate, "first-person" experience, while two very different representations are applied to it, *viz.*: (b) *perceived* "gray brain matter" and (c) *conceived* "real/physical 'thing-in-itself' electro-chemical brain, which *perception* merely *represents*." The *problem of qualia* – a necessary appurtenance to every representation – is

thus thoroughly confused among these three different ways of viewing what is deemed to be "the same thing."

A physical model of the mind can certainly be imagined and employed – by *analogy* – to represent mental activity. And, to the extent that the analogy can be carried through – which is already known to be quite substantial, for instance via the mathematical concept of information (e.g., the image of the brain as a sort of computer or information processing complex) – it is possible for thinking to be metaphorically broken down into detailed "physical" processes. This does not mean that the qualia of consciousness are somehow "emergent" from physical processes, because physical processes cannot be attributed ontological status. It means rather that an abstract or indirect representation – again, via analogy – can be employed to describe certain relations among experiences; which is all that any representation can ultimately do, including the most sophisticated and successful theories of physics. However, and most importantly, if the analogy can be carried through in great detail, a physical model, because of its potential precision and level of detail – *down*, to the scale of the sub-atomic; *up*, to the scale of cosmos – and thus its susceptibility to mathematical representation, can furnish a means for discovering relations among experiences that otherwise could not even be imagined.

§ 3.4　The Heuristic Significance of Conceptual Constructions

In summary, it has been noted that experience *per se*, irrespective of representational purport, is the proper subject of philosophy and science. But scientific explanation necessarily involves model building and the supposition of *extra-personal order*. Without some means of mathematically representing such order – which is today usually thought of as "a physical world" – theory would be extremely cumbersome and sharply limited. Human beings obviously cannot do without an orienting "overview" or "map" of their experiences, which includes both the idea of a self and the idea that experience in some sense represents a reality that transcends the self, whatever that extra-personal reality might be.

Referring, again, to the illustration afforded by the house of mirrors, though cognizance of the map of the layout of the house, which depicts the geometric relations among the mirrors, is qualitatively different from the first person experience of immersion among those mirrors, the former is yet applicable to the latter; intuitive understanding of the relations represented by the map, and their connection to the immediate objects of perception, furnishes the capacity to successfully navigate the maze.

Moreover, it is just as intuitively evident that the map of the house of mirrors can be connected up with other maps, taken to represent other, continuously navigable regions. And it is by such a linking up or expansion of maps, which seems to be an indefinitely extensible process, that one realizes the orienting intuition of immersion in a three-dimensional manifold. The representational purport of spatial intuition, in this sense, is tied to a comprehension of what the mathematician might call the possibility of translating, rotating and potentially even skewing global geometrical arrangements, while maintaining the integrity, or invariance, of local topological if not metrical relations.

Furthermore, the expansion or generalization of this process – of linking experiences through relations of analogy – can be extended to realms that transcend what is deemed "the physical." Hence study of the biology of the brain – in particular, neurophysiology – has furnished an enormous body of knowledge regarding the activity of the nervous system and the relation of that activity to certain aspects of awareness and cognitive function. And while it is not possible to comprehensively describe a sensation or feeling on the basis of an electrical impulse or chemical reaction, there nevertheless seems to be a delicate, highly elaborate correspondence between models of neural processes and subjective states of awareness. As hereinabove stressed, one must keep in mind that neural processes are nothing more than conceptual constructions – again, *maps* – which should not be attributed ontological significance. However, it seems certain that (a) ever-more detailed information of this type can be obtained as research progresses and, perhaps more to the point, that (b) today's concepts of "physical," "chemical" and "electrical" will be greatly expanded/generalized and maybe, ultimately, superseded, so that an even more comprehensive understanding of such processes might be achieved.

How is all this to be assimilated and understood from an epistemological perspective, given that the significance of such a seemingly fundamental concept as "physical world" cannot be unambiguously established apart from heuristic merit, and hence that the interpretation of experience cannot be grounded on such an idea?

§ 3.5 The Singular Significance of Physical Models

Although the exceedingly abstract mathematical concepts of particles, fields and other entities that currently comprise the primitives of theoretical physics cannot be identified as "really existing things," it is yet true that a much greater comprehension of experience can be gained with their aid than without them. This is fully analogous to the circumstance of everyday life, wherein the perceptions and ideas of naïve realism, though epistemologically unsound, provide an enormous range of practical information. Cognitive maps comprising spatio-physical constructions are extraordinarily useful. Moreover, information codified in accordance with such constructions seems to be more efficiently assimilated and manipulated than any other sort.

This much is not very different from the essential upshot of positivism. The constructions of physics can be viewed as highly refined maps, which must be accompanied by strict guidelines for interpretation and use. On the other hand, from the ontological perspective of classical physics, which uncritically adopts a somewhat naïve form of the subject-object dichotomy, every perception is assumed to reflect or represent an event in time and space; the percept is a "thought thing" that is yet conditioned by that external event, and in important ways corresponds to it. Based on this supposition, it has been possible to construct a fairly comprehensive model of relationships among experiences – again, even to the extent of correlating subjective feelings with physical processes in the brain.

Yet there is meaning to be found in the theories of physics – or rather, in certain types of such theories – that far transcends mere practical utility. Perhaps because of the sense of immediacy and hence credibility associated with physical concepts, physical models of "existence as a whole" – e.g., existence as *Cosmos* – can have a very significant personal, philosophical meaning. Human beings have a powerful desire to understand their existence – really a *need* – which is hardly done justice by the word 'curiosity'. And this need is at least partially satisfied by certain types of theories (more on this aspect of physical theory in section 3.6.2 below).

Existence considered as a whole – insofar as it is imagined to transcend the subject-object dichotomy adopted by modern science – would thus seem to be structured in such a fashion that the concepts of physics, which are connected in an intuitive manner with perceptions of bodily objects, can lead to a very general knowledge of empirical relationships, even among experiences not categorized as physical. Moreover, physical representations admit of a detailed mathematical treatment and provide for a type of causal description that otherwise – for example, through the use of exclusively psychological concepts – cannot be readily achieved. And so the representation of subjective events as neural processes makes possible the description of detailed relationships among the experiences of an individual on the supposition of the existence of a world external to that individual – including that person's own "unconscious" mind (for instance, the relation (a) between the perception of a cube rotating in space and a corresponding image on some graphical display device, (b) between said perception and concomitantly observed patterns of neural processes, and (c) between the neural patterns and further, underlying processes hypothesized to exist in the brain). It is thus possible to describe immediate experiences in relation to those that can only be inferred to exist; whether in memory, subconsciously or even "externally" – for example, in another mind (as when two people communicate).

All manifestations of *discreteness* – from perceptions of "physical objects" to designations of perceptions and conceptions as "elements of cognition" or "individual thoughts" – can be viewed as aspects of some sort of '*system*' or '*totality*' of similar/related things. Such a whole – again, what might be called a "*Sensorium*" of all experiences, occurring 'every-*where*' and 'every-*when*' – is somewhat similar to Berkeley's idea of the mind of God, though neither leavened nor adorned with theological ferments. Moreover, it is possible, on analogy with the model of existence as physical universe, to associate a mathematical structure with such a sensorium, as discussed at length in chapter six.

Each of these two ways of viewing things then – as *sensorium* or as *universe* – can be considered a "symbolic image of existence." Taken separately, however, they have limited merit. So whereas the notion of a sensorium seems to preserve something of the qualitative essence of experience, *relations* among experiences – insofar as the latter are interpreted as comprising the first person phenomenology of mind – cannot be drawn out very comprehensively on this model; especially relations between distinct individual minds. That is, without the notion of a common spatial manifold, it is hard to explain communications "between two minds" – or more specifically, to answer the question: "What is there, in addition to the thoughts of each person, to both (a) distinguish the thoughts of one from those of the other, and (b) to mediate the communication of thoughts between them?" On the other hand, the image of existence as a physical universe is particularly well suited to the modeling of such relations between parts, and between parts and whole, and it furnishes a *comprehensive* yet *compact* image for the notion of the totality. And although representation "as *physical*" gives those parts and that whole a quite abstract, *qualitatively bereft* aspect, it yet furnishes a superb metaphor/model for representation of the place and relation of the "self" to the "totality of other minds" – again, an image without much resolution of detail when viewed from the perspective furnished by the sensorium model.

Taken together, the power of these perspectives is evident, as their utility has already been demonstrated through centuries of practice, albeit in confused contexts. Again, in rough essentials, it is the basic idea at the back of many scientists' and particularly neuroscientists' minds. Its efficacy will only grow as confusion abates and ever more intricate relationships appear. Toward this end, new frameworks are needed, for both the physical and cognitive sciences; frameworks that remove the misconceptions associated with the subject-object dichotomy while employing the best aspects of these (thence) compatible perspectives.

§ 3.6   The Applicability and Intelligibility of Theory

Before moving on to consider the epistemological issues that are specific to the foundations of physics, it will prove advantageous to clarify and review a few subtle aspects of the results obtained above, inasmuch as they beg the question: (1) Given the constraints on cognitive processes, which necessarily limit the logical completeness and consistency of thought in general, must not any analysis, including the hereinabove critique of cognition, be hopelessly confused? Indeed, it may justifiably be asked: (2) To what extent can the present critical analysis have meaning and validity? And again: (3) What is the proper object of scientific theory?

The following remarks are intended to answer these questions in a somewhat tentative, general manner. A more comprehensive treatment of the salient issues will be furnished in chapters five and six, where specific frameworks will be presented for the physical and cognitive sciences, respectively. The reader is thus requested to keep an open mind and not attempt to anticipate the nature of these frameworks on the basis of the following brief remarks, as this may conduce to a prejudiced view of the presentations that follow.

§ 3.6.1   Meaning and Understanding

The previous discussions have focused, for the most part, on certain negative aspects of knowledge and thought. And while recognition of these limitations of cognition has shed light on the origins of certain crucial epistemological problems, it is not yet clear what positive attributes of theoretical knowledge remain – i.e., that survive critical scrutiny and characterize, as it were, "good" scientific theory. A theory of meaning is required to address this matter properly; which is thus fit subject for a separate treatise. But it is not necessary to so expand the scope of this work in order to get a sense of the salient issues involved and arrive at an intuitive understanding of the relevant criteria, if not a hard and fast set of rules for their determination.

In accordance with the figure-ground duality of cognitive processes described above, the negative aspects of these processes imply the positive;

and so knowledge of the limitations of thought to a certain extent involves, already, knowledge of its powers. The information processing functions of perception and thought entail the same properties of Gestalt unity and foreground-background duality that figure into the arguments about logical incompleteness and inconsistency presented above. It was shown, in connection with these properties, how perceptions and conceptions encode complex constructions, and thereby large quantities of information, into simple, relatively small and readily manipulable cognitive constructs, which thus fulfill a representational function.

The "encapsulation" of information content is a general characteristic of percepts and concepts, and there are as many ways to so represent information as there are instances of perception and conception. But among these infinite possible cognitive forms, certain features stand out as distinctive among the modes called "philosophic" and "scientific." Conceptual constructions intended to "explain something" after the manner of these modes share a unique group of qualities. Of course, insofar as there are controversies regarding the relative merits of various theoretical approaches to the empirical content of science, no single set of criteria should be sought as an absolute standard. However, because much that is controversial in this respect is only so because of misconception and confusion, once confusion is quieted much discord can be expected to disappear. With insight into the sources of misconception comes a sharpening of the understanding regarding the positive features of cognition, which thus comprise its powers. And so what follows is a distillation, so to speak, of the conceptual qualities peculiar to "good" scientific theory – where "good" is defined by way of the discussion.

§ 3.6.2   Scientific Understanding: Part A

What does it mean to "understand something" by virtue of scientific explanation? The quotation marks in the previous sentence are intended to indicate that it is of course the meaning of the phrase *understand something* which is the proper subject of the query, inasmuch as to answer the question is to explain the meaning of this phrase in the given context. But again, such an inquiry is well beyond the scope of the present treatise, and so this analysis must pursue the more modest goal of discerning some of the salient modes of thought that characterize successful scientific theorizing. Accordingly, it commences with a quote from Newton, in which the master describes his philosophy of science circa the time of publication of the first edition of the *Principia*. As he wrote in his preface thereto:

"Since the ancients (as we are told by Pappas), esteemed the science of mechanics of greatest importance in the investigation of natural things, and the moderns, rejecting substantial forms and occult qualities, have endeavoured to subject the phænomena of nature to the laws of mathematics, I have in this treatise cultivated mathematics so far as it regards philosophy.

"The ancients considered mechanics in a twofold respect; as rational, which proceeds accurately by demonstration; and practical. To practical mechanics all the manual arts belong, from which mechanics took its name. But as artificers do not work with perfect accuracy, it comes to pass that mechanics is so distinguished from geometry, that what is perfectly accurate is called geometrical; what is less so, is called mechanical.

"However, the errors are not in the art, but in the artificers. He that works with less accuracy is an imperfect mechanic; and if any could work with perfect accuracy, he would be the most perfect mechanic of all, for the description of right lines and circles, upon which geometry is founded, belongs to mechanics. Geometry does not teach us to draw these lines, but requires them to be drawn, for it requires that the learner should first be taught to describe these accurately before he enters upon geometry, then it shows how by these operations problems may be solved. To describe right lines and circles are problems, but not geometrical problems. The solution of these problems is required from mechanics, and by geometry the use of them, when so solved, is shown; and it is the glory of geometry that from those few principles, brought from without, it is able to produce so many things.

"Therefore geometry is founded in mechanical practice, and is nothing but that part of universal mechanics which accurately proposes and demonstrates the art of measuring. But since the manual arts are chiefly employed in the moving of bodies, it happens that geometry is commonly referred to their magnitudes, and mechanics to their motion. In this sense rational mechanics will be the science of motions resulting from any forces whatsoever, and of the forces required to produce any motions, accurately proposed and demonstrated.

"This part of mechanics, as far as it extended to the five powers which relate to manual arts, was cultivated by the ancients, who considered gravity (it not being a manual power) no otherwise than in moving weights by those powers. But I consider philosophy rather than arts and write not concerning manual but natural powers, and consider chiefly those things which relate to gravity, levity, elastic force, the resistance of fluids, and the like forces, whether attractive or impulsive; and therefore I offer this work as the mathematical principles of philosophy, for the whole burden of philosophy seems to consist in this – from the phænomena of motions to investigate the forces of nature, and then from these forces to demonstrate the other phænomena; and to this end the general propositions in the first and second Books are directed.

"In the third Book I give an example of this in the explication of the System of the World; for by the propositions mathematically demonstrated in the former Books, in the third I derive from the celestial phænomena the forces of gravity with which bodies tend to the sun and the several planets. Then from these forces, by other propositions which are also mathematical, I deduce the motions of the planets, the comets, the moon, and the sea.

"*I wish we could derive the rest of the phænomena of Nature by the same kind of reasoning from mechanical principles, for I am induced by many reasons to suspect that they may all depend upon certain forces by which the particles of bodies, by some causes hitherto un-*

known, are either mutually impelled towards one another, and cohere in regular figures, or are repelled and recede from one another. These forces being unknown, philosophers have hitherto attempted the search of Nature in vain; but I hope the principles here laid down will afford some light either to this or some truer method of philosophy.[3,4]

Thus did Newton explain his *modus operandi* – to determine an axiomatic framework from which, by mathematical deduction, the phenomena of nature can be reproduced and thereby predicted. Despite interpretations of some to the contrary, it is evident that Newton considered his philosophy *mechanical.* Although his agency of gravitational force acts at a distance and instantaneously, it is not on that account fundamentally different (and not at all in its mathematical representation) from the concept of contact force. Newton's invocation of force as causative of motion is a mathematical generalization of the idea as it appears in *Statics*, and from ordinary intuition connected with the feeling of physical exertion. It is an anthropomorphism that has survived all ingenious efforts to circumvent it. Notwithstanding much erudite opinion that it has been or will be vanquished, it is yet a keystone concept of modern physical theory.

Indeed, in the physical realm force would seem to be the quintessence of causality. So it is that in the world of Galileo-Newton, absent force, things remain in a state of rest or uniform rectilinear motion – which latter, in Newtonian theory, is the physico-mathematical generalization of rest. Similarly, in the General Theory of Relativity, motion exclusively under the influence of gravity is inertial and so undistinguished from rest. But because Einstein was unable to extend his framework to account for motion generally, the concept of force remains salient. For example, Richard Feynman's sum-over-histories approach to quantum theory is usually understood in connection with the concept of force; which, under Feynman's interpretation, is treated as an exchange of particles, via the diagrams that bear his name.

---

3 Newton, Isaac, Andrew Motte, and John Machin. The mathematical principles of natural philosophy. London: Printed for B. Motte, 1729

4 Emphasis of the last two sentences not in the original. Moreover, the original passage is not divided in this manner into paragraphs. The Author has taken the liberty to so format this extract in order to enhance its readability and emphasize the salient thoughts with respect to the present discussion

So it would appear that at the core of mathematical physics rests an intuition that derives straight from naïve realism. Again, the mathematical definition of force in Newton's system is a generalization of the common sense idea of *cause of motion*, whereas rest denotes the absence or cancelling of such causes. In this sense, the modern concept is rather similar to the Aristotelian, except that Aristotle did not recognize the principles of Galilean inertia and [hence] relativity.

And yet there are circumstances in which intuition calls upon an entirely different mode of explanation, wherein the significance of physical causation seems altogether diminished. For instance, when a psychologist discusses the actions of a sentient being – say a fairly intelligent one, like a rat or a human – its movements are not explained in terms of physical but rather *emotive* causes, which are appropriately called *drives* (literally, '*motive*-ations'). Similarly, when a detective investigates a crime, or a judge hears a case, questions of physical evidence are incidental to the discovery of psychological motive – "e-*motion*" – which is the ultimate object of the inquiry.

So the concept of causality is certainly not limited to physical force. Moreover, the essential meaning of the concept is really more akin to the idea of necessity than force. That is, under the intuition that time/existence is singular, it would seem that things cannot happen other than as they do (i.e., because they only happen once) and, therefore, that reality cannot be changed simply by virtue of how it is looked upon; by any particular mode of dividing up one's perception of the world into time slices. Inasmuch as it appears that every change – which is conceived as a sort of boundary, in time, between perceivably different circumstances – denotes a division imposed by the mind (for time, like space, is intuited to be continuous), like cutting a picture into a jigsaw puzzle, it would seem that regardless of where and how the cuts are made the whole must remain the same. That is, dividing a picture into a puzzle by cutting along the boundaries of recognizable features – clouds, people, etc. – no more changes the whole than cutting it "arbitrarily" where the scene is "featureless" (i.e., where there are no perceivable differences from one place to the next).

In 1813 Arthur Schopenhauer completed his PhD thesis,[5] an extended analysis of the concept of causality under four forms, *viz.*: (a) *physical cause*, as connecting material events – i.e., *force*; (b) *logical reason*, such as the link between deductions in an axiomatic structure; (c) *geometrico-mathematical necessity*, such as the laws of geometry relating the properties of triangles; and (d) *volitional motivation* – i.e., *Will* – which drives the actions of sentient beings. As a form of *Sufficient Reason*, Schopenhauer's categorization of physical causality as a specific type of mental construct reflects the evidently innate requirement of the mind that every change must have an explanation.

As cognitive construct, which again was both Hume's and Kant's understanding of causality (though Hume thought it contingent; purely a habit of experience), the idea of a force acting between physical bodies is clearly an extension of the first person experience of the relation between muscular exertion and the motion it produces; from lifting and pushing to running and jumping (Kant would here differ from Hume by understanding this intuitive sense of force to be an *a priori* property of the mind, which informs and is involved in the very construction of the perceptual experience[6]). Without much thought concerning the nature of the connections involved, the "common sense" of naïve realism combines the intuitive notion of physical body with the similarly self-evident intuition of "being a body oneself," and thus the knowledge of what it is like "from the inside."

That is why the concept of force, which again derives from the first person experience of the relation between exertion and movement, seems somehow empty and confused when applied to relations between inert objects. As a generalization of the subjective sense of force, the mathematico-physical concept is an objectification/representation that differs in meaning/content from the idea it generalizes – i.e., in the "bundle of qualities" which characterizes the set of meanings that comprise the concept. This is to be understood in the sense of the universal constraints of concepts *as representations*, per the epistemological analysis of cognition presented in this and the previous two chapters. Hence the vague feeling of incompleteness and inconsistency that accompanies earnest inquiries

---

5 Schopenhauer, Arthur. On the fourfold root of the principle of sufficient reason. La Salle, Ill.: Open Court, 1974
6 Also Schopenhauer's understanding; that perception is informed by the sense of causality

into the "ultimate meaning" of such ideas; whence derives the rhetorical query, paraphrased elsewhere herein, "Why is it that when a red ball strikes a blue ball motion is communicated but redness is not?"

The confusion associated with this question is not unrelated to that of the putative mind-body problem, inasmuch as the feeling of exertion that accompanies physical work, and the sense of the reaction of the object back upon the musculature, spans the threshold that links mind and body. The feeling, which constitutes the sense of force, is only attributable to the subject, whereas the inert material object is deaf, dumb, blind and insensitive generally. In a literal sense, then, the concept of force is a "notion hovering in a mystic obscurity between abstraction and concrete comprehension..."[7]

This dualistic aspect of the concept also manifests in a surprisingly different context. Early investigations into the nature of visual perception turned up some very interesting analogies between Newton's notion of "levity, elastic force, the resistance of fluids, and the like" and what appeared, to a handful of inspired researchers, to be dynamical properties of the objects of perception (i.e., literally intrinsic thereto). Consider the following excerpts from Wolfgang Köhler's *The Task of Gestalt Psychology*. The first quotation summarizes his findings that the elements of the visual field of perception seem to influence each other, in the same manner as the elements of a dynamical physical system as they tend toward a stable equilibrium state. (It must be borne in mind that this was written over forty years ago, and some of the work to which it refers is decades older still.)

"The Gestalt psychologists suspected that such phenomena are caused by interactions, but they could not yet tell why such interactions occurred, that is, what forces or processes were involved. Perception only showed the effects of such hypothetical causes; it told the observer nothing about the nature of these causes. The forces and processes which underlie such perceptual facts are simply not represented in the phenomenal, the perceptual, world.

"Now, there is only one part of the world the processes of which are in close contact with the perceptual experiences of the human being and which may therefore determine the character of these

7Lange, Friedrich Albert, and Ernest Chester Thomas. The history of materialism and criticism of its present importance ; Volume I. London: Trubner & Co., 1877, 308.

experiences. Hence, the Gestalt psychologists had to assume that the unknown events responsible for such curious interactions in perceptual fields were processes in the corresponding parts of the human brain, mainly in the gray cover of the brain, the cortex. We know from pathological cases, and from other evidence, where in the brain the processes directly related to vision take place, also where the physiological correlates of our hearing, where those of our tactual experiences are located, and so forth. But our main question is, of course, what physiological events occur in these places when human beings have perceptual experiences of one kind or another...

"...But we did not refer to sensory qualities when we began to suspect that certain properties of perceptual fields resemble properties of cortical processes to which they are related. The properties we had in mind were structural properties. If, for instance, under certain conditions, perceptual processes tend to assume particularly regular and simple forms, and if we suspect that, under the same conditions, corresponding processes in the brain show the same tendency, then we refer to what I just called "structural" characteristics. It is only such structural characteristics which, not only in this case but also in many others, perceptual facts and corresponding brain events may have in common.[8]

This theme is elaborated, in some detail, in a subsequent discussion regarding "figural after-effects." It had been discovered, for example, that when a person looks at a segment of a large curve, as time progresses the shape of the curve seems to become flatter. After enough time, if that same gently curved segment is viewed adjacent to a straight line, the latter tangent to the curve at its mid-point, the straight segment will appear bent outward, away from the curve. Likewise, persistent viewing of geometric figures will cause, as after effect in an altogether different scene, an apparent displacement of objects away from each other, from their actual relative positions. And it had been found that similar effects occur in other

---

8 Köhler, Wolfgang. The task of Gestalt psychology. Princeton, N.J.: Princeton University Press, 1969.

perceptual modalities; for example, the kinesthetic sense of the body's motion and position.

Discussing experiments involving reversible patterns, where foreground and background alternate in perception as the vase and faces of figure 1-6, Köhler explains:

> "...It seemed to us quite possible that, in either position, the cortical process corresponding to the figure seen at the time raises a local obstruction to its own continuation, and so weakens itself until, as a consequence, this process suddenly moves from its original location into the other possible area, and so forth.

> "This hypothesis had a far more general meaning than at the time was realized.

> "If the process in question weakened itself (or blocked its own way) in a pattern in which another distribution of this process was possible, the result was a "reversal." But there was no reason why the first part of this sequence, the self-weakening of the process, should occur only in patterns which permitted the process to shift into another part of the pattern. Hence, we had to conclude that the process underlying any visual object would cause a local obstruction in its medium, and thereby cause a change in its own distribution – even when the change could not be a sudden reversal, a transfer to an entirely different location. Wallach and I tested this conclusion in experiments with a great number of visual patterns of all kinds. After a while it became obvious not only that any visual pattern or object is gradually altered when inspected for some time, but also that the change follows certain definite rules. Moreover, we discovered that often, quite apart from the changes of the inspected object itself, other objects later shown in the same place or in its neighborhood were also affected. The most frequently observed effect of this kind was a displacement of such other objects, a displacement away from the area in which the originally inspected object had been located and, more specifically, away from the boundary of that first object.

A striking analogy occurred to Köhler.[9]

"...when physical systems or human perceptions are given time and other opportunities to do so, they change in the direction of greater simplicity or regularity..."

Examining the rules that seemed to govern the interaction of the perceptual elements in the visual field, Köhler speculated on what the corresponding physiological constraints might be:

"1. The cortical process must be such that its occurrence in the brain almost immediately begins to cause an obstruction in the tissue through which it passes.

"2. The process must be stronger at or near the boundary of the object than far in its interior, because test objects shown inside the boundary recede from this boundary into the interior of the first object.

"3. The process cannot be limited to the cortical area corresponding to the visual object itself, because often objects shown at a considerable distance from the first object still recede from it.[10]

On the basis of these considerations, an incredible discovery followed. Köhler continues:

"These rules suffice for identifying the process in terms of physics. At the present time, we know enough about the brain to extract from this knowledge a list of the processes which can possibly occur there. The list is not very long. When it is completed, we compare its items and their functional properties with the rules which I just formulated. The result is simple. One item after another must be eliminated because it does not fit one rule or another. Eventually, only one item is left; but this item satisfies all conditions which follow from the rules. Electric currents which originate and spread in the brain tissue as a continuous or volume conductor are left as the only, and also the most satisfying, possibility.

9 Op. Cit.
10 Op. Cit.

"How could such currents originate? For the sake of brevity, I will mention only the simplest case, that of an object the color of which differs from the color of the environment. The colors may, for instance, be two different degrees of brightness. I will surprise few physiologists if I say that in this situation a direct or quasi-steady current will begin to flow which passes through the object area and the environment area – more or less at right angles to the visual cortex, in one direction within the object area, in the opposite direction in its environment; these two parts of the flow are complemented by the flow which, on one side and the other of the active area, turns around the boundary of object and environment. At the active level, the current, flowing in opposite directions inside and outside the boundary, produces a most radical separation of the object from its environment, which makes the boundary a functional boundary. The behavior of the current thus explains the segregation of the visual object from its background – one of the facts which we could not understand during my first lecture."

"I now return to the rules derived from the examination of figural after-effects. According to the first rule, the process related to a visual object must be able to cause an obstruction in the cortical medium. Will a cortical current cause such an obstruction? Undoubtedly, it will. For about a hundred years, physiologists have known that, when an electric current passes through cell surfaces in the nervous system, it immediately establishes an obstruction where it enters the cells. Naturally, the obstruction is stronger where the current is strong than it is in places where the flow is weak. The obstruction has two components, one physical in the usual sense of the word, the other a biological reaction of the cell to the entering current. We have no time for details, but I must mention that the biological component of the obstruction often survives the duration of the flow for long periods. In neurophysiological discussions, the obstruction goes under the name "electrotonus" (a term once used by Faraday with a somewhat different meaning). Anybody familiar with elementary physics realizes, of course, that the obstruction established by the current will force this current to change its distribution in the tissue; it will be weakened where the obstruction is particularly strong, and will flow with relatively increased intensity through regions where a weaker obstruction has been established. Since part of the obstruction persists when the original current has disappeared, the obstruction will still affect the flow of object currents when the original object is no longer present, but other objects appear about the same region.[11]

In this passage, a remarkable analogy between physical and mental events is revealed. For the writer describes a direct, geometrical analog of the *figure-ground* of perception in the relation of *object-region* to *background-region* of electric currents in the brain. It were as though the three-dimensional distribution of charge and current directly mirrors the shape and structure of the figural object against its background in perception. There is, of course, no *a priori* reason to expect such literal relations to exist, generally, between the phenomenology of experience and the objects of physical theory. However, other such relations have been found in more recent times.

It was noted above that the quotes from Köhler reflect relatively early research. But his hypothesis – that relevant neurological brain processes are dynamic, self-organizing processes tending towards equilibrium – has been borne out in more recent studies. Consider the following passage from Gerhard Werner's 2006 review paper, *Perspectives on the Neuroscience of Cognition and Consciousness*, in which Werner suggests that there is a holistic field property at work – i.e., that the global electromagnetic field generated by the totality of disparate, local synaptic firings seems to be the seat of a coordinating pattern of individual and regional neural activity:

"I will first address the remarkable discovery of stimulus induced oscillatory and synchronous neuronal activity with the predominant frequency in the 40 Hz range. In sensory systems, the phase synchronization of oscillation is thought to functionally "bind" together neuron groups that respond to identical stimulus features, thus forming in some sense a higher-order neuron assembly (Gray

---

11 Op. Cit.

and Singer, 1987, 1989; Gray et al., 1989; Eckhorn et al., 1988). Synchronization can occur across distances of several mm and also across different cortical regions (Roelfsema et al., 1997; Ribary et al., 1991). The functional neuron assemblies thus formed consist of elements which are distributed in space, but 'locked' together by a common signal phase. Physiological data suggest that the synchronous oscillations are mediated by cortico-cortical connections which link preferentially cortical neuron columns responding to related stimulus features and tending to be grouped perceptually (for review: Singer, 1998, 2004). Synchronous cortical oscillations also seem to have a preparatory function in the generation of motor output (MacKay, 1997).

"...Most striking is the virtual simultaneity and the very short latency of onset in the oscillating ensemble, suggesting a process operating on a faster time scale than conduction and synaptic delays would permit

(Singer et al., 1997).

"The unusual properties of this oscillatory activity suggest a process of principled significance for brain theory. Simulation studies of neural assemblies identify the switching phenomenon as a phase transition due to synaptic nonlinearity and fluctuations from peripheral input and/or interaction with other assemblies (Bauer and Pawelzik, 1993). This places the cortical oscillatory activity in the category of Kuramoto's (1984) "self synchronization transitions" in an ensemble of oscillators.

"This phenomenon has been identified in a variety of biological systems as a mode of self-organization. Phase transitions in nonlinear dynamic systems occur as a matter of principle with very short latency. They would then also account for the short onset latency of the assembly oscillations. These considerations identify the oscillations in the 40 Hz range, and their apparent role in perceptual and motor functions, as manifestation of self-organizing nonlinear system properties of neuronal ensembles.[12]

So the intuitive, rather general notion of *force* – connoting a dynamic interaction among entities – has proven extraordinarily fruitful in very diverse contexts. It seems to be the piece of the puzzle that is ever being sought, with respect especially to questions of physical phenomena but even, as is evident from the above example, with respect to certain aspects of cognition as well.

For Newton, although a material connection could not be found between gravitating bodies, it was yet certain that there must exist *some* agency of force, which he was thus compelled to describe purely mathematically. Of course in the final analysis Hume's critique of causality remains valid – as it must, if only on the basis of the hereinabove established, epistemological limitations of cognitive processes generally. And so, in light of its great heuristic value, there is something about the concept of *force* that stands in need of elucidation.

It was mentioned above that psychologists describe emotional motives as *drives*. This is, again, a particular instance of the general notion of force; which is to say that the feeling of being *driven* to an action is related to the inner sense of pressure within the organism generally. Accordingly, psychologists speak of psychological pressure and stress in the same vein as physicists speak of their physical counterparts. These intuitive notions of force – clearly vestiges of naïve realism – are thus deeply ingrained in modern science, inasmuch as they animate the dynamics of both physical and psychological theory alike. This is apparent even in certain theories of Linguistics, which posit *Principles and Parameters* governing the growth and use of language. Rules of syntax are conceived in terms similar to the above-described *Gestalt* interpretation of perception, according to which perceptual objects interact dynamically. Similarly, under certain models of Linguistics *words* dynamically interact in the formation of sentences. Pursuing this line further, logic, arithmetic and geometry seem likewise to comprise rules governing dynamics of thought, and thereby the structure of formal systems.

---

12 Werner, G. "Perspectives on the Neuroscience of Cognition and Consciousness." Biosystems 87, no. 1 (1, 2007): 82-95.

To "understand something" thus means, in the few contexts examined above, to see successive changes linked by a motive cause. "*What's the go o' that?*" and "*Show me how it doos,*" in the words of the three year old James Clerk Maxwell, as quoted by his father.[13] The "drive of curiosity" to understand such connections – here again on the part of the infant Maxwell, "*the hidden course of streams and bell-wires, the way the water gets from the pond through the walls...*" is yet another example of such a motive power; the intellectual need to discover such connections.

This is of course the meaning of Newton's famous disclaimer:

> "It is inconceivable that inanimate brute matter should, without the mediation of something else which is not material, operate upon and affect other matter without mutual contact...That gravity should be innate, inherent, and essential to matter, so that one body may act upon another at a distance through a vacuum, without the mediation of anything else, by and through which their action and force may be conveyed from one to another, is to me so great an absurdity that I believe no man who has in philosophical matters a competent faculty of thinking can ever fall into it."[14]

And indeed, it was in the course of addressing a similar absurdity that a somewhat older James Clerk Maxwell constructed his fluid-dynamic theory of the aether, from which he was able to derive his famous equations for the electromagnetic field, deduce the speed of light, and thus vindicate the mantle he shared with Faraday. And again, such an absurdity is the *raison d'être* of the de Broglie-Bohm *Pilot Wave/Quantum Potential* interpretation of quantum phenomena.

But given the epistemological limitations of conceptual constructions generally, and the vestigial, anthropomorphic blemish of the *force* concept in particular, what is the value of such ideas for the future of science? Before answering this question, another possible objection to such a mode of explanation needs to be examined.

---

13 Campbell, Lewis, and William Garnett. The life of James Clerk Maxwell. With a selection from his correspondence and occasional writings and a sketch of his contributions to science, London: Macmillan, 1882. p 27
14 Bentley, Richard, Alexander Dyce, and Isaac Newton. The works of Richard Bentley. London: F. Macpherson, 1836. pp 212-213

§ 3.6.3   Scientific Understanding: Part B

From an epistemological perspective, it is of great interest that many mathematical relations, taken to represent laws of nature, can be expressed in a multiplicity of quite distinct ways. Such equivalent though different statements of mathematical truths are often referred to as *dualities. Abstract Dynamics* presents a striking range of such dualities, inasmuch as an infinite set of mechanical constructions can be represented by a single dynamical description. Of particular interest are those equivalent descriptions that seem to be mutually incompatible, and therefore paradoxical. Qualitatively distinct ways of looking at a situation that are yet equally valid – like the two perspectives *Vase* versus *Faces* – are thus one of the root difficulties of the erstwhile mind-body problem, in that *both* psychological motives *and* physical forces seem simultaneously valid yet *qualitatively incompatible* ways of describing the actions of a sentient being.

When Maupertuis discovered what he [inappropriately] called "*The Principle of Least Action*," he believed that he had stumbled on a teleological form of causality operating in Nature. He wrote:

"Here then is this principle, so wise, so worthy of the Supreme Being. Whenever any change takes place in Nature, the amount of action expended in this change is always the smallest possible."[15]

It is now known, of course, that Maupertuis' action principle is but one of many possible ways of formulating such [relative] *extremum* laws. Nevertheless, a teleological tinge attends these so-called *stationary principles*, because when a particle moves from one point to another in accordance with such a law, it seems to "know" which path among an infinite set of possibilities to select, without having any apparent way of "scenting out," as it were, the single correct course to its pre-determined destination. The destination seems to determine the path, as though causes were working backward in time, or again the particle had a "purpose" in view. (Somewhat ironically, it is in quantum mechanics that stationary principles make intuitive sense. That is, in the Feynman sum-over-histories approach, a ray of light is the resultant of every possible wave action, constructively inter-

---

15 Maupertuis. Essai de cosmologie. Leiden: s.n., 1751.

fering along the path of the ray while destructively interfering everywhere else. Of course, strictly speaking, the usual interpretation is in terms of 'probability waves' for the photon, not actual electromagnetic waves, but the underlying idea is the same.)

However, in *any* deterministic scheme, where the future is rigidly decided by the past, the past can equally be viewed as determined by the future. This is sometimes expressed by the characterization of the laws of physics as symmetrical with respect to time – that is, they apply equally in either time direction; toward the past or toward the future. This symmetry is intrinsic to Newton's third law of motion, i.e., that *for every action there is an equal and opposite reaction.* A force acts between two objects equally in both directions, from the first to the second and the second to the first. Accordingly, if a motion picture showing billiard balls interacting on a table is played backwards, and the actions of the balls traced out, the forces deemed to propel them in the forward time direction will appear reversed, as acting consistently in the backward time direction. (Of course, the backward playing movie will only be sensible in this manner if all the collisions are elastic. If inelastic collisions occur, then in the forward time direction there will be a dissipation of energy-momentum associated with a permanent change of form of the objects, so that in the backward time direction crunched objects, upon collision, will un-crunch themselves. Such so-called *irreversible processes* involve, again, a dissipation of energy-momentum outside the system comprising the objects under observation; for example, a heating of the molecules of both the objects and the surrounding air. In order for such a process to be reversed, all of the extraordinarily complex motions of the air molecules, as well as those that constitute the objects, would have to be reversed – not impossible to arrange "in principle" but certainly in practice.)

Accordingly, one can describe the behavior of ideal, inert objects, acting exclusively by the motive power of contact force, in terms of a teleological framework – i.e., as though the behavior were goal directed. In this respect, as mentioned above, the mathematical dualities of physical theory are similar to the "big daddy of them all" – the mind-matter dichotomy. Which is to say, they are just as specious vis-à-vis the paradoxes that they seem to connote. Alternate forms of representation can be quite distinct qualitatively while offering just as distinct advantages and disadvantages, all context-dependent (per the *"Regions of Truth"* spheres of figure 2-3). This is of the essence of conceptual representation, inasmuch as every conception creates and brings to experience a novel set of qualities.

The existence of dualities would seem to imply that some sort of abstract order is common to various modes of representation – and hence the notion of an "underlying reality" or "thing-in-itself"; some-*thing* that *instantiates* – literally "stands in" for this order (i.e., something that the mind can "pin" the order to, so to speak, and thereby fix an image to the concept). But again, just as with any abstract idea – which means every conception – the meaning that characterizes what might be called its *set of instances* is not something that can be pointed to. Rather, the meaning of the concept – as it unites instances under the banner of a *general connotation* – does not necessitate some sort of objective existence *independent* of those instances (apart from its existence *as cognitive experience*).

This is a common confusion, from which arises the notion of *Truth* as instantiating independently real, objective meaning, as per the *Platonic Theory of Ideas* – another of the 'quite stubborn' illusions. Consider, in this connection, the thoughts of Roger Penrose, a brilliant mathematician and physicist, who writes:

"But what does true mean, in this context?... Plato made it clear that the mathematical propositions – the things that could be regarded as unassailably true – referred not to actual physical objects (like the approximate squares, triangles, circles, spheres, and cubes that might be constructed from marks in the sand, or from wood or stone) but to certain idealized entities. He envisaged that these ideal entities inhabited a different world, distinct from the physical world. Today, we might refer to this world as the *Platonic world of mathematical forms*. Physical structures, such as squares, circles, or triangles cut from papyrus, or marked on a flat surface, or perhaps cubes, tetrahedra, or spheres carved from marble, might conform to these ideals very closely, but only approximately. The actual *mathematical* squares, cubes, circles, spheres, triangles, etc., would not be part of the physical world, but would be inhabitants of Plato's idealized world of forms.[16, 17]"

A little further along, he discusses the reasons that convince him of the objective existence of mathematical truth, and gives an example that he feels is particularly persuasive. Discussing scientific models, he says:

"If the model itself is to be assigned any kind of 'existence', then this existence is located within the Platonic world of mathematical forms. Of course, one might take a contrary viewpoint: namely that the model is itself to have existence only within our various *minds*, rather than to take Plato's world to be in any sense absolute and 'real'. Yet, there is something important to be gained in regarding mathematical structures as having a reality of their own. For our individual minds are notoriously imprecise, unreliable, and inconsistent in their judgements. The precision, reliability, and consistency that are required by our scientific theories demands something beyond any one of our individual (untrustworthy) minds. In mathematics, we find a far greater robustness than can be located in any particular mind. Does this not point to something outside ourselves, with a reality that lies beyond what each individual can achieve?"

"Let me illustrate this issue by considering one famous example of a mathematical truth, and relate it to the question of 'objectivity'. In 1637, Pierre de Fermat made his famous assertion now known as 'Fermat's Last Theorem'" ...[the conjecture is described]... "Fermat noted: 'I have discovered a marvelous proof of this, which this margin is too narrow to contain.' Fermat's mathematical assertion remained unconfirmed for over 350 years, despite concerted efforts by numerous outstanding mathematicians. A proof was finally published in 1995 by Andrew Wiles..."

"Now, do we take the view that Fermat's assertion was always true, long before Fermat actually made it, or is its validity a purely cul-

16 Penrose, Roger. The road to reality : a complete guide to the laws of the universe. New York: A.A. Knopf, 2005.11-12.
17 Consider in this connection Newton's definition of *Absolute "Mathematical"* space and time...

tural matter, dependent upon whatever might be the subjective standards of the community of human mathematicians? Let us try to suppose that the validity of the Fermat assertion is in fact a subjective matter. Then it would not be an absurdity for some other mathematician X to have come up with an actual and specific counter-example to the Fermat assertion, so long as X had done this before the date of 1995. In such a circumstance, the mathematical community would have to accept the correctness of X's counter-example. From then on, any effort on the part of Wiles to prove the Fermat assertion would have to be fruitless, for the reason that X had got his argument in first and, as a result, the Fermat assertion would now be false! Moreover, we could ask the further question as to whether, consequent upon the correctness of X's forthcoming counter-example, Fermat himself would necessarily have been mistaken in believing in the soundness of his 'truly marvelous proof', at the time that he wrote his marginal note…[18]"

Note the close connection between Penrose's problematic regarding objective truth, and the linguistic problematic of language acquisition. Children evidently have an innate sense of the semantics and syntax underlying every possible natural human language (as opposed to artificially created symbolic systems, such as software languages). In order for a child to learn a word – to pick up its sound, syntactical relations and representational meaning (the concept) on one exposure, as happens in early childhood[19] – he or she must be capable of understanding these things more or less instantly. Moreover, natural languages cannot be learned from grammar books and dictionaries[20] – such works, which can be quite extensive in the sense of documentation, are clearly only useful for someone who "knows a language." Now, none of the concepts of geometry – squares, triangles, circles, spheres, and cubes – are "actual physical objects (like the approximate squares, triangles, circles, spheres, and cubes that might be constructed from marks in the sand, or from wood or stone) but … certain idealized entities…" Just as none of the concepts of natural languages

18 Op Cit, pages 12-14
19 Chomsky, Noam. New horizons in the study of language and mind. Cambridge [England]; New York: Cambridge University Press, 2000.page **6**
20 That is, not as an individual's *first* acquired language.

are "actual physical objects" but rather abstract ideas that human beings innately understand – but not, for instance, rats and monkeys. Does this mean that theories of linguistics must invoke the hypothesis of a Platonic Realm of words/concepts generally? Most linguists – those, at any rate, who believe in the innate faculty of language – seem content with the notion of a genetic endowment, and this would seem equally applicable in the case of mathematics. In other words, whatever the human experience of mathematical truth is, it is certainly that – i.e., a human experience. And as for the perfection of mathematics – considered as a body of work, in comparison with the cognitive experiences of an "untrustworthy" individual mathematician – is this any different from the relation between a comprehensive English dictionary and the vocabulary of an individual English speaker? Without human beings to interpret the collected works of mathematics, would those works have any more meaning than a dictionary without people to reference it (and thus bring to it their innate understanding)?

Just as with the illusory notion of the *referent of representation*, the "thing-in-itself" – whether considered as "physical object" or "mental ideal" or merely "an objective order underlying experience" – is not necessarily anything more than an artifact of cognition. The belief in an abstract order that exists apart from experience, corresponding to the order of theory, has no more ontological validity than the belief in a particular instantiation of such order; for example, a particular type of physical universe. As with the idea of God, that of *order* cannot be tied to any graven image. Notice, in this connection, the similarity between the *Realm of Ideas* of Plato and Penrose, the *Skeletal Structure of the World* of Eddington, the *Thing-in-Itself* of Kant, the *Collective Unconscious* of Jung and the dark side of the moon of Einstein's aphorisms (Einstein more than once alluded to the moon to make his point about objective reality, and what he felt to be the absurd, even mystical aspects of quantum theory – for example, asking if one could really believe that the unseen side is not there at all, or that the moon does not exist unless observed, and that its existence can be conjured by a mouse, simply by looking up, etc.). All of these thinkers are driven by the apparent logical necessity of associating a "substance" – a *substantial* or *independent meaning* – with an object of conception, thus *object*-ifying the representation.

But although extra-personal reality, just as first person experience, cannot be captured in a single representational form, insofar as the existence of dualities enables the concept of *force* to be an effective heuristic device, and given the central role that this concept plays in both everyday experience and the contemplative life of the mind, it is certain that – especially in the latter capacity – this idea, in some form, must remain an object of science indefinitely (i.e., at the very least insofar as science is concerned with the phenomenology of sentience). There is certainly no valid reason to banish it from the conceptual toolset of the theorist. However, there are some rough edges that need to be smoothed. The Newtonian concept of force is certainly not "ready for prime time." This matter is addressed in the following two chapters. With proper attention to detail, an enhanced mathematical concept of force – or rather, a generalization of the concept as currently usually understood – can indeed be quite fruitful, as will hereinafter become apparent.

§ 3.7 Conclusions

While it is evident from the considerations above that ideas about "reality" or "all of existence" cannot be expected to represent their intended referent in an ontological sense, there are yet ways of thinking about such things that are not entirely devoid of significance. In particular, the representation of reality as a physical universe provides a very useful and, if viewed in the proper light, even satisfying image for that which the words "all of existence" are intended to connote. Indeed, it is because of the peculiar credibility and appeal of physical models of reality – deriving from the original sin of *naïve realism* – that theories of physics (with notable exceptions) are so compelling.

If a model of existence as cosmos/universe is taken as an analogy or metaphor for the qualitatively quite different image of existence as a vast "Sensorium" – another conceptual construction with its own limitations, but one that captures many qualities that are usually excluded from physical models – then a vast horizon of possibilities appears. For as the relationships between the two models are refined, those aspects of the world that are directly known – i.e., the variegated *qualia* of experience – may help reveal new types of order and structure, with respect to which exclusively physically-oriented thinking might not be informed (and thus not motivated to explore; which order may yet be describable, via analogy, within the context of the physical model).

If both models are developed in sufficient detail, it may be possible to draw extremely fine-grained correlations between descriptions of physical and mental processes. Correlations of this sort might be elucidated to such an extent that it becomes possible to hypothesize the existence of as yet unrealized patterns of activity, connected with organic brains and possibly even synthetic constructs, and which correlate with new types of experiences – i.e., differing qualitatively from anything any human being has ever experienced – and to thereby extrapolate what those qualities are. In this manner, it might be possible to understand the experiences of other sentient species, in the sense of Nagel's rhetorical question, "What is it like to be a bat?"[21]

It may even be possible to "bridge" the physiological components of a single organism with a corresponding network of sensate relations, which essentially means understanding that certain aspects of experience may be associated with non-neural biological processes as well as the neural, and interpolating the subjective processes that correlate with the various parts of the organism. Still further, perhaps such an extrapolation will be extensible to what is currently considered inanimate matter, recalling, metaphorically, John Locke's conjecture that some form of consciousness might have been "superadded by God" to all matter, in a manner he con-

---

21 Thomas Nagel The Philosophical Review, Vol. 83, No. 4 (Oct., 1974), pp. 435-450 It should be mentioned, in connection with Nagel's discussion on this matter, that he overlooks an important distinction when he compares the understanding of a physical phenomenon such as lightening to the understanding of a psychological phenomenon such as bat sonar perception. In his usage of the terms, the "subjective" nature of bat perception, which is taken to qualify its inaccessibility to human understanding, is the same as the "objective" nature of lightning, an understanding of which he believes can be approached by any intelligent person. But he seems to neglect that a physical process is a hypothetical/ conceptual construct, a representation that does not necessarily have an identifiable referent, whereas bat perception is ostensibly real. This, of course, is the very essence of the mind-body confusion: the thing-in-itself is the experience, while its putative "neural correlate" is a very subjective representation of that thing; i.e., what is generally thought to be subjective is objective, and vice versa, the root of the duality misconception.

sidered analogous to that in which bodies acquire the "intrinsic property" of gravitation. (Of course, Locke considered the primary qualities of physical objects to exist objectively in an extra-personal world, but the ideas are functionally similar).

In any case, it would seem that physical models, that is, theoretical constructs involving the salient intuition of spatial extension, will always be necessary for comprehensive understanding of experience, even experience that is deemed "mental." This should be interpreted in the sense of the house of mirrors discussed in section 1.8.3, with respect to which detailed knowledge of the reflections in the individual mirrors can be acquired without understanding the geometric relations depicted in the map, representing the spatial arrangement of the mirrors. But without the overview of arrangements signaled by the map, there is no comprehensive understanding of the geometric order that exists among the totality of reflections. The sense of orientation is limited to the first person perspective – just as with personal experience generally – and navigation thus depends on rules of serial relations, learned by rote.

The functional value of this complimentary approach, and of spatio-physical models in particular, can be illustrated by the following thought experiment, touched upon above. Whereas there does not seem to be an inherently meaningful connection between the *qualia* of experience (color, sound, etc.) and the so-called "neural correlates of consciousness" (in accordance with the misconceived "mind-body problem") – once the possibility of such a connection is admitted to consideration, possibilities arise that are otherwise inconceivable. For example, on the basis, exclusively, of the phenomenological or first person perspective, it evidently makes no sense to ask: "Is the brain the seat of consciousness, or is it some sort of *transceiver* that allows a 'distant' mind to experience a 'local' world?" Or even: "Are *all* brains in fact transceivers, which link up, ultimately, to a *single* mind?" On the other hand, the possibility of discovering such relationships – usually thought to be the exclusive purview of metaphysics or religion – is within the realm of physical science; or, more precisely, *psychophysics*. That is to say, it is not outside the realm of possibility that the brain might turn out to contain an electromagnetic or other type of transceiver. While this does not seem very likely, nor a credible object of

research, as a thought experiment (i.e., *in principle*) the possibility of un-
covering such connections can yet be conceived – whereas, on the basis
of a solipsist or otherwise exclusively first person view, the thought is essentially meaningless.

The power of physical modelization is the capacity to readily represent
experience in the context of a *manifold*, in the mathematical sense, whereby relations with and among "the extra-personal" can be schematically
traced out. And so a proper understanding of these vastly different but intricately related ways of representing experience is crucial to the progress
of science. The relationship between theories of physical processes and
the experiences that are ordered and predicted with their use has historically been misunderstood as some sort of *"psycho-physical parallelism"* – a
misunderstanding that has appeared in various forms, and has been com-
mented on, from many perspectives, by virtually all philosophers.

In the other extreme, radical empirical approaches to the phenomenology of experience have seemed rather superficial. Consider Mach's proto-
positivism, as he describes what he views as purely habitual and conventional categorizations of sensations, thereby "explaining" the concept of
the physical object as a "nucleus" of colors:

> "…Colours, sounds, and the odours of bodies are evanescent. But
> their tangibility, as a sort of constant nucleus, not readily suscep-
> tible of annihilation, remains behind; appearing as the vehicle of
> the more fugitive properties attached to it. Habit, thus, keeps our
> thought firmly attached to this central nucleus, even when we have
> begun to recognize that seeing, hearing, smelling, and touching are
> intimately akin in character. A further consideration is, that ow-
> ing to the singularly extensive development of mechanical physics
> a kind of higher reality is ascribed to the spatial and to the temporal
> than to colours, sounds, and odours; agreeably to which, the tem-
> poral and spatial links of colours, sounds, and odours appear to
> be more real than the colours, sounds and odours themselves. The
> physiology of the senses, however, demonstrates, that spaces and
> times may just as appropriately be called sensations as colours and
> sounds. But of this later.

"Not only the relation of bodies to the ego, but the ego itself also, gives rise to similar pseudo-problems, the character of which may be briefly indicated as follows:

"Let us denote the above-mentioned elements by the letters A B C . . ., K L M . . ., a b c . . . Let those complexes of colours, sounds, and so forth, commonly called bodies, be denoted, for the sake of clearness, by A B C. . .; the complex, known as our own body, which is a part of the former complexes distinguished by certain peculiarities, may be called K L M . . .; the complex composed of volitions, memory-images, and the rest, we shall represent by a b c . . . Usually, now, the complex a b c . . . K L M. . ., as making up the ego, is opposed to the complex A B C . . ., as making up the world of physical objects; sometimes also, a b c . . . is viewed as ego, and K L M . . . A B C . . . as world of physical objects. Now, at first blush, A B C . . . appears independent of the ego, and opposed to it as a separate existence. But this independence is only relative, and gives way upon closer inspection. Much, it is true, may change in the complex a b c . . . without much perceptible change being induced in A B C . . .; and vice versa. But many changes in a b c . . . do pass, by way of changes in K L M . . ., to A B C . . .; and vice versa. (As, for example, when powerful ideas burst forth into acts, or when our environment induces noticeable changes in our body.) At the same time the group K L M . . . appears to be more intimately connected with a b c . . . and with A B C . . ., than the latter with one another; and their relations find their expression in common thought and speech.

"Precisely viewed, however, it appears that the group A B C . . . is always co-determined by K L M . . . A cube when seen close at hand, looks large; when seen at a distance, small; its appearance to the right eye differs from its appearance to the left; sometimes it appears double; with closed eyes it is invisible. The properties of one and the same body, therefore, appear modified by our own body; they appear conditioned by it. But where, now, is that same body, which appears so different? All that can be said is, that with different K L M . . . different A B C . . . are associated.

"...Ordinarily the complex a b c . . . K L M . . . is contrasted as ego with the complex A B C . . . At first only those elements of A B C . . . that more strongly alter a b c ...., as a prick, a pain, are wont to be thought of as comprised in the ego. Afterwards, however, through observations of the kind just referred to, it appears that the right to annex A B C . . . to the ego nowhere ceases. In conformity with this view the ego can be so extended as ultimately to embrace the entire world. The ego is not sharply marked off, its limits are very indefinite and arbitrarily displaceable only by failing to observe this fact, and by unconsciously narrowing those limits, while at the same time we enlarge them, arise, in the conflict of points of view, the metaphysical difficulties met with in this connexion.

"As soon as we have perceived that the supposed unities "body" and "ego" are only makeshifts, designed for provisional orientation and for definite practical ends (so that we may take hold of bodies, protect ourselves against pain, and so forth), we find ourselves obliged, in many more advanced scientific investigations, to abandon them as insufficient and inappropriate.

"The antithesis between ego and world, between sensation (appearance) and thing, then vanishes, and we have simply to deal with the connexion of the elements a b c . . . A B C . . . K L M . . ., of which this antithesis was only a partially appropriate and imperfect expression. This connexion is nothing more or less than the combination of the above-mentioned elements with other similar elements (time and space). Science has simply to accept this connexion, and to get its bearings in it, without at once wanting to explain its existence.[22]

Now there is certainly much that is meaningful and perspicacious in this analysis, and there are significant points of contact between Mach's perspective and that presented here. However, as already discussed at length, such an extreme, anti-ontological position is hardly conducive to the progress of theoretical science. The similarities with Bohr's radical empiri-

---

22 Mach, Ernst. The analysis of sensations, and the relation of the physical to the psychical. New York: Dover Publications, 1959.

cism are quite evident in this passage, particularly in the last paragraph. What Mach considers an unwarranted attribution of significance to the notion of extra-personal order, which he credits to "the singularly extensive development of mechanical physics," by no means mitigates the need to employ concepts of extra-personal spaces and relations in the theories of science.

§ 3.8 Implications

The most salient upshots of these various considerations are:

(1) Despite the intrinsic limitations of conceptual representation, the range of experience that can be comprehended by its means connotes a significant "truth" content. In particular, when confusion surrounding the subject-object dichotomy is eliminated, the overall ordering function that the paradigm furnishes is invaluable (as briefly hinted at above, and much enlarged upon in chapter six). With the powers and limitations of this dichotomy clearly in mind, an adequate epistemological critique of the foundations of modern science is possible. (Such a critique is offered in chapter four.)

(2) Because of the inherent limitations of conceptual representation, it follows that any "empirical truth" must be emergent. That is to say, inasmuch as the conceptual elements of theory, and the relations among them, are necessarily limited by the cognitive context in which they are embedded, and cannot be deemed, in light of the constraints of conceptual representation generally, to be identical to any "objectively existing referents," any correlation with experience cannot be attributed to "objectively existing" or "fundamental laws of nature" (as represented in theory). So it is clear that any correlations between representations and the experiences to which they are deemed to apply must, at least in this trivial sense, be considered emergent. (More succinctly: There must be a difference between the representation and the object of representation – the representation is not identical to the thing it represents; no model can

model itself. Therefore, the correspondence between the behavior of the model and that of its object must arise from a reason other than identity. QED)

With these thoughts in mind, chapter four examines the foundational concepts of theoretical physics.

Chapter IV

Critique of Physics

§ 4.1 Introduction

THE epistemological renovations in the foundations of physics associated with the advances of relativity and quantum theory mark the beginning of a conceptual revolution that has yet to be completed... A statement much like this is often heard in connection with one or another expression of conventional wisdom, according to which the central tenets of relativity and quantum theory are not merely sacrosanct but, owing to certain intrinsic subtleties and mutual incompatibilities, so enigmatic that it may well require superhuman imagination and mathematical expertise to plumb the depths of their full profundity.

This is not the author's meaning. Rather quite the contrary – i.e., that it should not be deemed a *fait accompli* that the foundations of any future physics must accommodate these precepts and principles just as they are today usually understood. Thus, in the view hereinafter to be developed, the revolutionary import of Lorentz invariance and quantization-of-action is taken to be the common provenance of these [superficially] distinct aspects of modern physics; the deeper order from which they spring and the unification that they thus commend.

The primary heuristic significance of the foundational principles of modern physics is usually taken to be the facilitation of appropriate mathematical constructions, attributed to the constraints these principles impose on the bounds of inquiry – as for instance vis-à-vis the *Uncertainty Principle* and the constant, limit velocity of light. And so it is generally accepted that the notion of local causal action on indefinitely small scales must be excluded from consideration. *Quantization-of-Action* and *Lorentz Invariance* are to be hard-wired, so to speak, into the mathematical foundations of any future framework, *à la* String Theory, rather than treated as emergent properties.

Pursuant to the epistemological investigations of the previous three chapters, it may seem odd to raise an objection against the elimination, from the corpus of theoretical physics, of philosophically outmoded ideas; especially such as have been painstakingly demonstrated to lack objective significance – e.g., physical entities/events with definite demarcations in space and time. However, it is another crucial upshot of those investigations that *all* of the concepts of physics – just as all concepts generally – harbor logical inconsistencies, and are necessarily devoid of meaning in the metaphysical sense. Therefore, in addition to empirical suitability, a judicious balancing of the epistemological pros and cons associated with a given group of conceptions must inform the composition of theory. And while it should be acknowledged that, at some time in the perhaps non-near future, a unification of the theoretical conceptions of physics with those of the cognitive sciences may lead to ideas of a new and altogether different, hitherto undreamt of type, in the meantime it is not advisable to discard any ideas that can be gainfully employed.

It therefore remains a serious drawback that, while some of the most essential concepts of pre-twentieth century physics and philosophy have been expunged from acceptable discourse – ideas that have been central to humankind's sense of existential orientation for centuries – no plausible substitutes have filled the gap. And while, in a narrow sense, it may seem reasonable to dispense with ontologically suspect ideas, there remains a practical need, with respect to the pursuit of the natural sciences as well as the conduct of everyday life, for a guiding ontological vision— albeit accompanied by the hereinabove established disclaimers regarding the limited significance of any such vision as map/guide, and the realization that such maps must necessarily reflect a limited aspect of experience.

Yet the philosophical perspective that has emerged with quantum theory is far too narrow. To prescribe an arbitrarily circumscribed, ambiguous invocation of classical physical concepts for the arrangement and interpretation of "experiment," while leaving the object of experiment undefined – not to mention making experimental outcome contingent upon an interpretation of the concept of *consciousness* – is quite inadequate as an overall philosophy of science. And yet this overly narrow perspective is deemed to capture the best possible reflection of reality available to human beings – in accordance with constraints imposed by Nature herself, as revealed by her prophet from Copenhagen.[1] What follows is an epistemological critique of these concepts, and the roots from which they spring.

§ 4.2 The Rise and Fall of Mechanical Philosophy

Beyond Kant, very little of scientific relevance had been achieved on epistemological issues by the advent of the last century. Ernst Mach had helped bring the problems associated with the Newtonian conception of Absolute Space front and center. Indeed, Mach took up a firm position against any non-observable elements in physical theory. This viewpoint became a precursor to *Logical Positivism*. It had a strong influence on Einstein during the years that he was formulating relativity, and later – especially via the example of Einstein's early success – on the physicists who worked with Bohr on the development of quantum theory.

Mach did not believe in atoms, and rejected Boltzmann's statistical-mechanical treatment of thermodynamics. He believed that sensations *per se* are the only proper object of science, and viewed the notions of physical entities and forces acting between them – even contact forces – purely as utilitarian fictions that facilitate and economize the mind's handling of sensations. While this position is superficially similar to the perspective developed in chapter three, Mach neglected or incorrectly dismissed an essential point; namely, that *all* interpretations of experience – including their categorization as *"sensations"* – are necessarily representational and

1 Although many physicists seem to be intrigued by *"many-worlds"* and *"wave-collapse"* as literal interpretations of the quantum-theoretical superposition of state, the author does not consider such schemes to meet minimal requirements of credibility – i.e., logical economy and explanatory efficacy.

so involve conceptual constructs, the essential role of which cannot, therefore, be eliminated, or even stringently reduced.

Efforts to "stick to observables" must always lead to an arbitrary restriction on the fundamental components of theory, based on one or another philosophical prejudice. The question as to whether a theoretical conception is valid can ultimately only be answered on the basis of how well it engenders the ordering and comprehensibility of experience, not on whether or not it has a direct referent in experience – an illusory characteristic which, as hereinabove established, does not exist. All concepts are representational and yet without direct referents in experience – they carry meaning that is intuitively linked with some *aspect* of experience, but not identical with any-*thing* that can be unambiguously demarcated *in* experience (again, even "sensory" perceptions, taken to be "sensations as such," are yet cognitive constructs that do not reflect identifiable, independent referents, and are subject to the constraints of representational constructions generally).

Mach's primary epistemological contribution was to call attention to the importance of experience and the defects of certain well-worn yet controversial concepts, which were beginning to raise serious empirical problems – such as the Newtonian notion of *Absolute Space*, which he suggested be replaced by the framework established by the overall arrangement of matter in the universe. Mach felt that the so-called "fixed stars" were better suited as a reference frame vis-à-vis the definition of inertia, and that Newton's system was in dire need of clarification.

He was joined by the great mathematician Henri Poincaré who, with penetrating insight, elucidated this and many closely related notions, including key questions associated with the relativity of time and space and even the possible scientific relevance of non-Euclidean geometries. Poincaré illuminated but did not resolve many of the outstanding issues that were addressed decisively by Einstein; who ultimately took a position opposite to that of Poincaré on certain of these key issues – in particular, the relevance of non-Euclidean geometries for physics.[2]

---

2 It should be noted that Einstein most likely did not bring a preference for the conception of dynamic or non-Euclidean geometry to his investigations, and that the idea was evidently suggested by his friend Marcel Grossman, whom Einstein, as is well known, lobbied to help with the extension of relativity and who had studied the subject at school

Prior to the publication of Einstein's first work on relativity, one of the most controversial questions regarding the aether was whether or not it should be spelled with the leading letter *a* – George Fitzgerald, for example, found the use of this character to be irritating and redundant, and Joseph Larmor evidently wrote at least one essay on the subject. In contrast, Mach was among a minority of physicists who, before Einstein, thought the aether redundant as element of physical theory. But if the philosophy of science had not made significant progress after Kant, by the time Einstein appeared on the scene much practical progress had been made in physics – progress that was of great relevance to these epistemological questions.

Contrary to a subtle misconception that lies embedded in the conventional historical narrative, the work of Faraday and Maxwell had, by the late-nineteenth century, brought many physicists back around to a deeply entrenched, mechanically-minded approach to theory; evincing disdain for the concept of action-at-a-distance. Three decades before the publication of Maxwell's famous equations for the electromagnetic field, James MacCullagh had produced an equivalent set of equations for the propagation of light, on the basis of an aether comprising an elastic solid, yet capable of supporting local, rotational elastic stresses. One of the reasons that the relevance of MacCullagh's work was not immediately recognized – beyond the fact that no one at the time, including MacCullagh, was able to imagine how a substance with the qualities of a solid could support such attributes – is that there existed, in important circles, a prejudice against speculative aether constructions, and a preference for the superficial simplicity – and by then familiarity – of "Newtonian" action-at-a-distance.

Maxwell's equations, though employing electromagnetic terms, were also based on a physical model. And though that model was hydrodynamical it was just as mechanical as MacCullagh's. It was thoroughly imbued with the quintessence of Faraday's vision; a space-filling physical agency of mechanical force transmission. However, because Maxwell's construction unified such a wide range of phenomena – from electricity and magnetism to optics – it had to be taken seriously; though acceptance came slowly.

One upshot of this renewal of the aether concept was that, on the basis of such models, one could say that the Newtonian conception of Absolute Space had acquired concrete meaning. In the minds of many, some such notion was deemed necessary for the explanation of centrifugal force; i.e., for the distinction of inertial versus accelerated motion generally. And it was another widely held belief that, as with electromagnetism, gravity might also be explainable on the basis of an aether model; thus eliminating action-at-a-distance from physical theory altogether.

Overall, the nineteenth century was a period of spectacular success for the Mechanical Philosophy. As Einstein put it:

"What made the greatest impression upon the student...was less the technical construction of mechanics or the solution of complicated problems than the achievements of mechanics in areas which apparently had nothing to do with mechanics: the mechanical theory of light, which conceived of light as the wave-motion of a quasi-rigid elastic aether, and above all the kinetic theory of gases: -- the independence of the specific heat of monatomic gases of the atomic weight, the derivation of the equation of state of a gas and its relation to the specific heat, the kinetic theory of the dissociation of gases, and above all the quantitative connection of viscosity, heat-conduction and diffusion of gases, which also furnished the absolute magnitude of the atom. ... Apart from this it was also of profound interest that the statistical theory of classical mechanics was able to deduce the basic laws of thermodynamics, something which was in essence already accomplished by Boltzmann."[3]

Subsequent to these achievements, two new cracks appeared in the foundations of Mechanics, though they were not recognized as such immediately. These were the soon-to-be-infamous problems associated with the blackbody spectrum and the Michelson-Morley experiments. Yet it was not until Einstein's 1905 relativity theory became widely recognized among physicists that Mechanical philosophy was more-or-less clearly on its way out. Of course, even after Einstein, mechanical conceptions continued to be employed by physicists as they groped their way towards new

---

3 Einstein, Albert, and Paul Arthur Schilpp. 1979. Autobiographical notes. La Salle, Ill: Open Court.

models – and still, today, exist in "the back of the mind" as new constructions are grappled with. But from the general acceptance of special relativity to the present it has been taken for granted that Newton's conceptions of space, time and matter are logically and empirically flawed, and, despite their considerable heuristic and practical value, inadequate as foundational concepts of physical theory.

§ 4.3   Newtonian Inconsistencies

To the uninitiated, one of the striking features of special relativity is its incompatibility with the concept of the rigid body. The restriction on how fast information can be transmitted through space imposes far-reaching constraints on the structure of physical objects. That is, because a perfectly rigid object can, in principle, communicate a signal instantaneously across its length, however great, it follows that such rigidity cannot exist in nature. On first thought, this constraint seems implausible – especially if one is unfamiliar with the roots of relativity in constructive theories, such as those of Lorentz and Maxwell. And yet it is an immediate, logical consequence of simple empirical premises. On the other hand, it is not so immediately clear – as evidenced by history – that there exists a similarly inherent incompatibility of the concept of the rigid body with Newton's laws of motion.

It was mentioned in the general introduction to this treatise that Newtonian physics is neither intuitively obvious nor free from logical inconsistencies. Beyond the manifest ambiguities regarding the definitions of *Absolute Space* and *Time* – which make the first law of motion meaningless and incapable of furnishing a context for the second – it can be shown that the second law is logically incompatible with the third, and that the three laws are thus inconsistent. This would seem to suggest that, to the extent they have empirical significance, such significance must be attributable to an order that is *emergent* from *different*, albeit in some way related principles (see *Newton's Laws of Motion* below).

Philosophical Solutions

Newton's Laws of Motions

I. Every body continues in its state of rest, or of uniform motion in a right line, unless it is compelled to change that state by forces impressed upon it.

II. The change of motion is proportional to the motive force impressed; and is made in the direction of the right line in which that force is impressed.

III. To every action there is always opposed an equal reaction: or, the mutual actions of two bodies upon each other are always equal, and directed to contrary parts.[4]

Among the various modalities of physical action comprehended under the Newtonian framework, the transfer of momentum by contact seems the easiest to understand: either through pressure, vis-à-vis objects more-or-less continuously connected, or through impulse, via momentary contact... i.e., collision. And while even this idea has been challenged on general epistemological grounds, to the minds of most the concept of contact force seems acceptable. Yet the constraints that Newton's laws impose on the kinematics of collisions are far from unobjectionable.

The equation "$F = ma$" – i.e., *Force* equals *mass* times *acceleration* – is a mathematical expression of Newton's second law of motion. It restricts the manner in which objects can interact, because changes in velocity must take place smoothly, in finite time. However, a collision between two perfectly rigid objects cannot obey such a law, for the reason illustrated in the following example.

Imagine two identical, perfectly rigid spherical objects approaching each other in an inertial frame, in which both objects have the same speed. Assuming ideal circumstances – that is, ignoring friction, non-contact forces and other dissipative factors (imagine the balls move in an empty Newtonian space absent gravitational forces) – the two objects will collide and make contact at some "point" in space, coincident with a point

---

4 Newton, Isaac, Andrew Motte, and John Machin. The mathematical principles of natural philosophy. London: Printed for B. Motte, 1729

of each ball (the latter, hereinafter, the *"contact-points"* of the two objects, respectively).

Now, because the balls are perfectly rigid – which is to say they cannot suffer any alteration of form – they must behave perfectly elastically, and thus, by conservation of energy, rebound and retreat from their meeting point, in opposite directions from and at the same speeds with which they approached. While it might not seem obvious that perfectly rigid objects must act elastically, in order to do otherwise their pre-collision kinetic energy must be dissipated in some manner that does not involve rebound, and this is precluded by the property of ideal rigidity (by virtue of which even internal heating cannot occur).

From these simple considerations a subtle *non sequitur* arises. For it is impossible for the objects, as they approach collision, to reduce their relative velocity until they actually make contact. But upon contact, it is not possible for these bodies – or, for the sake of clarity, the points of each that are actually in contact (again, the *contact-points*), to undergo a finite acceleration. For as long as they are in contact, they must move together *with the same velocity* – i.e., with the same speed *and* direction. But clearly their velocities must be oppositely directed, so anything other than an *instantaneous* meeting and change of velocity is contradictory.

To see what this means in greater detail, consider the necessary series of events. In order for the contact-point approaching from the right to reduce speed *gradually* upon striking the one approaching from the left it must undergo a sequence of diminishing speeds – for example, using discrete measures for convenience, say moving to the left at x units of distance per second, then moving to the left at $x - 1$ units per second, then $x - 2$ units per second, etc. However, this implies that the contact-point of the other ball must undergo the exact same sequence of velocities. That is, from the moment of contact, it must necessarily move with speed x in the direction opposite to that of its approach; then with speed $x - 1$, then $x - 2$, etc. In other words, it must begin its altered motion at a *higher* speed – in direction *opposite* to that of its approach – while continuously reducing to a *slower* speed, which is of course exactly opposite to what happens. Each ball must instantly reverse direction to match the velocity of the other – and then, as long as contact is maintained, go from a greater speed to a lesser speed in the reverse direction of its approach; again exactly opposite of what is called for by $F = ma$. Therefore, by the inherent symmetry of the situation, it is impossible for the contact-points and the rigid bodies to which they belong to react in the manner that is called for by Newton's law.

By a simple extrapolation from this argument, $F = ma$ cannot apply to collisions generally. For if elastic bodies are conceived to exist on Newtonian terms – i.e., as abstract composite objects ostensibly composed of rigid corpuscular constituents, but deemed either to have a collective elastic quality or to otherwise mimic an ensemble of independent constituents, each possessing an elastic property – then the law $F = ma$ can only be considered approximately valid, and only so on the collective level. So if, as assumed in the example above, only one leading contact-point of each of two objects in collision are at first in contact, followed by others in a cascade, the contradiction must continue to arise with respect to every component pair of contact points, as long as they are assumed to have anything other than an infinite acceleration. That is, $F = ma$ is only conceivable as an approximation, and only then as a collective, emergent property of an ensemble of components. This is because, as long as action is transmitted between rigid objects by contact, and the agents of action (objects in motion) are subject to Newtonian laws, the contradiction described above will always arise on the individual component level.

It is interesting to note how this simple thought experiment contains hints of certain relativistic and quantum properties that were only discovered empirically – i.e., that both continuous action and rigid objects are incompatible with the laws of physics. For centuries, rigid bodies have been regarded as idealized objects of Newtonian dynamics, yet their behavior must contradict the laws of Newton. In order to overcome this problem, and maintain Newton's second law as a literal [as opposed to approximate, emergent] truth, continuous action between solid rigid objects, in finite time, must be rejected. This can be accomplished, for example, by (a) assuming objects to have spaces inside them, and that – during what appear to be collisions – the spaces of each object line up with the masses of the other, so that they can interpenetrate without any head-on collision, and then – by some friction-like, inertial absorption mechanism that can capture, store and redirect energy – gradually reduce their relative motions and ultimately reverse them, or by (b) conjuring some special agency of action, such as a Newtonian gravitational field; i.e., an abstract, purely mathematical force that operates on objects continuously and non-mechanically. In any event, the simple case of perfectly rigid, solid objects in collision cannot satisfy Newton's laws. It seems that a prescient Newtonian physicist, noting these circumstances – or, perhaps, merely becoming cognizant of the problematic nature of the concept of discrete objects generally, as per the epistemological arguments of chapter one – could have anticipated some of the later developments of physics; i.e., Einstein's discovery that rigid objects cannot exist.

But, as argued above, the universal epistemological limitations of a given concept does not necessarily negate that concept's unique *practical* (heuristic) utility. And $F = ma$ is an empirical truth, inasmuch as it can be used to accurately represent a broad range of experiences. Therefore, insofar as it is true, it must either be deemed a fundamental feature of reality – something that "just is" on the "classical level of experience" – or it must be considered derivative; i.e., an emergent, collective property. On this note, it is helpful to remember that quantum mechanics was founded on the principle that, on the macroscopic scale, the consequences of the laws of quantum theory must correspond with those of classical mechanics (Bohr's original *Correspondence Principle*). So, at least in the trivial sense that Newtonian law does not apply in the quantum and relativistic regimes, it is necessarily not fundamental but rather emergent.

It is also instructive to note that, whereas Newton's first and third laws of motion are considered to be the consequences of a general symmetry – i.e., that of translation in space – his second law is usually accepted simply as the *definition* of force. But there is another, far more salient sense in which Newton's second law can be understood; a perspective from which classical physics as a whole seems more intelligible, and the concept of physical law, in general, more logical and satisfactory. It will be seen, in chapter five, just how penetrating and wide-ranging the view from this perspective can be.

§ 4.4 From Constructive Mechanical Theory to Mathematical Theory of Principle

Einstein more than once expressed the view that, although Maxwell was tireless in his efforts to depict the electromagnetic field mechanically, he was yet primarily responsible for the ultimate ascendance of the field concept and the demise of the mechanical paradigm. But, as previously noted, this attribution does not accurately reflect the circumstances. Again, it was not until the widescale acceptance of special relativity – well into the twentieth century – that physicists at large came to renounce the possibility of basing constructive theories on mechanical conceptions. In other words, it was *Einstein* who was primarily responsible for the demise of the mechanistic view and ascendance of the field concept as representation of a new physical primitive.

Maxwell conceived the electromagnetic field to be a manifestation of dynamical processes in a material substrate – the *aether*. And, while he did not believe that the hydrodynamical model from which he derived his field equations was a perfect representation of these processes, he was nevertheless convinced that such dynamical processes existed. Moreover, he understood that a potentially infinite set of mechanical models could be mathematically isomorphic to any single such dynamical formulation, in the sense of the *Abstract Dynamics* that had recently become a powerful tool of mid-nineteenth century physics. Dynamical approaches based on *action principles* were then being popularized, in particular by Maxwell's close colleagues Peter Guthrie Tait and Lord Kelvin (*Né* William Thompson *circa* this period).

It is only in this sense that Maxwell did not feel compelled to complete the construction of a perfect mechanical model in order to be convinced that material processes are responsible for electromagnetic phenomena. Maxwell remained a strong proponent of the use of mechanical models, for heuristic and pedagogical purposes, to the end of his tragically short life,[5] and he went to great lengths to make his mechanical constructions as detailed as possible, in order to convince himself that his mathematical treatment was sound.

---

5 He died in 1879 at the age of 49, the year Einstein was born.

It is also often said that Maxwell discovered the connection between electrodynamics and optics as a consequence of his equations, from which he was able to derive light's velocity. But this story is also misleading. Maxwell long suspected – as had Faraday before him and other inspired thinkers – that light is an electromagnetic phenomenon, and one of the primary purposes of his hydrodynamical model was to furnish a plausible basis upon which he could understand the connection between electricity and light.

§ 4.5   Faraday and Maxwell versus Weber *et al*

In 1820, Hans Christian Oersted discovered that an electric current can turn the needle of a nearby magnetic compass. Over the course of the next forty-four years, detailed study of these apparently different but related phenomena – electricity and magnetism – culminated in a grand theoretical synthesis. This unification became a source of great inspiration for succeeding generations of physicists, and it generated the seeds of relativity and modern field theory on which that inspiration, in turn, acted.

The Oersted effect was, at first, difficult for mathematically trained physicists of the early nineteenth century to understand, because it did not lend itself to facile interpretation on the basis of Newtonian forces acting in straight lines between the central points of objects deemed to be their sources. The kinematics of the observable effects seemed all wrong for an interpretation along such lines. But to Faraday, who was self taught and largely ignorant of advanced mathematical methods, an intuitive, visualizable model came naturally. It seemed obvious to him that something was acting in the space between the sources of electricity and magnetism, and he modeled that action after the curved lines traced out by iron filings, spread over a piece of paper held near a magnet. According to Newtonian physics, such curved lines of force did not make any obvious sense.

The induction of electric current by a moving magnet, discovery of which followed quickly on the heels of Oersted's finding, was likewise interpreted by Faraday. He conceived the phenomenon to depend on the relative motion of conductor and magnet, and for him it was natural to understand the current in terms of changes in position, relative to his imagined lines of force, as the latter cut across the moving conductor. Again, these phenomena were not immediately understandable on the basis of Newtonian ideas because of the "sideways" actions.

Ampere, however, was able to devise a Newtonian explanation on the basis of mathematical elements of continuous current (i.e., arbitrarily divided, purely abstract entities), the forces between which varied not only with distance but also with their relative velocity. On this model, and interpreting rotational motion of these current elements as sources of magnetism, Ampere could account for turning forces in current-carrying conductors. His theory thus linked electricity and magnetism in a new unity, inasmuch as permanent magnetism came to be imagined as an effect of perpetually circulating electric currents.

Ampere's construction was quite complex and, from an ontological perspective, unsatisfying – in this regard reminiscent of Ptolemaic epicycles. But by the time Maxwell published his first paper on Faraday's experiments, Wilhelm Weber, in Germany, had developed a single, unifying equation summarizing Ampere's ideas, employing the concept of discrete charged particles in place of Ampere's abstract current elements. So Maxwell's first, incomplete work appeared on a stage already set with a more or less complete, mathematically unified account of electromagnetism – based on the well accepted paradigm of Newtonian forces acting at a distance, and nicely analogous to Newton's inverse square law of gravitation.

Weber's model, which was elaborated and extended by other German physicists, indeed seemed correct. It could, in principle, explain the known electromagnetic phenomena on the basis of observable quantities and Newton's laws of motion. Any attempt to reach beyond such a well-founded and empirically confirmed theory conformable to Newton's framework – especially on the basis of a speculative, Cartesian-like aether model – would surely be met with scorn by the cognoscenti of the times. At best such a model might be deemed 'metaphysical' and at worst counterproductive: precisely as *"Hidden-Variable"* interpretations of quantum mechanics tend to be viewed today. What, then, could the justification for the Faraday-Maxwell program have been, and how could it have come to surpass the approach of Weber *et al*?

Maxwell described his own thoughts on this matter in some detail, in three landmark papers elucidating and expanding on Faraday's work. In these papers, the production of which spanned a decade, Maxwell developed his mathematical theory of electromagnetism and the equations that bear his name. Below are some relevant excerpts from these monographs. The first quotation is from the introduction to his first paper, "*On Faraday's Lines of Force*," read to the Cambridge Philosophical Society on December 10, 1855:

"The present state of electrical science seems peculiarly unfavorable to speculation. ... [T]o appreciate the requirements of the science, the student must make himself familiar with a considerable body of most intricate mathematics, a mere retention of which in the memory materially interferes with further progress. The first process therefore in the effectual study of science, must be one of simplification and reduction of the results of previous investigation to a form in which the mind can grasp them. Results of this simplification may take the form of a purely mathematical formula or of physical hypothesis. In the first case we entirely lose sight of the phenomena to be explained; and though we may trace out the consequences of given laws, we can never obtain more extended views of the connections of the subject. If, on the other hand, we adopt a physical hypothesis, we see the phenomena only through a medium, and are liable to that blindness to facts and rashness in assumption which a partial explanation encourages. We must therefore discover some method of investigation which allows the mind at every step to lay hold of a clear physical conception, without being committed to any theory founded on the physical science from which that conception is borrowed, so that it is neither drawn aside from the subject in pursuit of analytical subtleties, nor carried beyond the truth by a favourite hypothesis.

"In order to obtain physical ideas without adopting a physical theory we must make ourselves familiar with the existence of physical analogies. By a physical analogy I mean that partial similarity between the laws of one science and those of another which makes each of them illustrate the other. Thus all the mathematical sciences are founded on relations between physical laws and laws of numbers, so that the aim of exact science is to reduce the problems of nature to the determination of quantities by operations with numbers. Passing from the most universal of all analogies to a very partial one, we find the same resemblance in mathematical form between two different phenomena giving rise to a physical theory of light.

"The changes of direction which light undergoes in passing from one medium to another, are identical with the deviations of the path of a particle in moving through a narrow space in which intense forces act. This analogy, which extends only to the direction, and not to the velocity of motion, was long believed to be the true explanation of the refraction of light; and we still find it useful in the solution of certain problems, in which we employ it without danger, as an artificial method. The other analogy, between light and the vibrations of an elastic medium, extends much farther, but, though its importance and fruitfulness cannot be overestimated, we must recollect that it is founded only on a resemblance in form between the laws of light and those vibrations. By stripping it of its physical dress and reducing it to a theory of "transverse alternations," we might obtain a system of truth strictly founded on observation, but probably deficient both in the vividness of its conceptions and the fertility of its method. I have said thus much on the disputed questions of Optics, as a preparation for the discussion of the almost universally admitted theory of attraction at a distance.

The next quotation is from the introduction to Maxwell's "*On Physical Lines of Force*," first published in 1861. In this passage, Maxwell is somewhat less ambiguous about the value of physical analogies:

"In all phenomena involving attractions or repulsions, or any forces depending on the relative position of bodies, we have to determine the magnitude and direction of the force which would act on a given body, if placed in a given position.

"In the case of a body acted on by the gravitation of a sphere, this force is inversely as the square of the distance, and in a straight line to the center of the sphere. In the case of two attracting spheres, or of a body not spherical, the magnitude and direction of the force vary according to more complicated laws. In electric and magnetic phenomena, the magnitude and direction of the resultant force at any point is the main subject of the investigation. Suppose that the direction of the force at any point is known, then, if we draw a line so that in every part of its course it coincides in direction with the force at that point, this line may be called a line of force and indicate the direction of the force in every part of its course.

"By drawing a sufficient number of lines of force, we may indicate the direction of the force in every part of the space in which it acts.

"Thus if we strew iron filings on paper near a magnet, each filing will be magnetized by induction, and the consecutive filings will unite by their opposite poles, so as to form fibers, and these fibers will indicate the direction of the lines of force. The beautiful illustration of the presence of magnetic force afforded by this experiment, naturally tends to make us think of the lines of force as something real, and as indicating something more than the mere resultant of two forces, whose seat of action is at a distance, and which do not exist there at all until a magnet is placed in that part of the field. We are dissatisfied with the explanation found on the hypothesis of attractive and repulsive forces directed towards the magnetic poles, even though we may have satisfied ourselves that the phenomenon is in strict accordance with that hypothesis, and we cannot help thinking that in every place where we find the lines of force, some physical state or action must exist in sufficient energy to produce the actual phenomena.

"My object in this paper is to clear the way for speculation in this direction, by investigating the mechanical results of certain states of tension and motion in a medium, and comparing these with the observed phenomena of magnetism and electricity. By pointing out the mechanical consequences of such hypotheses, I hope to be of some use to those who consider the phenomena as due to the action of a medium, but are in doubt as to the relation of this hypothesis to the experimental laws already established, which have generally been expressed in the language of other hypotheses.

The following, final passage is from the introduction to Maxwell's "*A Dynamical Theory of the Electromagnetic Field*," in which he more or less completed the development that he initiated with his 1855 paper. By this time he is no longer speaking ambiguously of analogies, and seems quite definite regarding the *aether* – having succumbed, evidently, to "a favourite hypothesis." This paper was first read on December 8, 1864:

"The most obvious mechanical phenomenon in electrical and mechanical experiments is the mutual action by which bodies in certain states set each other in motion while still at a sensible distance from each other. The first step, therefore, in reducing these phenomena to scientific form, is to ascertain the magnitude and direction of the force acting between the bodies, and when it is found that the force depends in a certain way upon the relative position of the bodies and on their electric or magnetic condition, it seems at first sight natural to explain the facts by assuming the existence of something either at rest or in motion in each body, constituting its electric or magnetic state, and capable of acting at a distance according to mathematical laws.

"In this way mathematical theories of statical electricity, magnetism, and the mechanical action between conductors carrying currents, and of the induction of currents have been formed. In these theories the force acting between the two bodies is treated with reference only to the condition of the bodies and their relative position, and without any express consideration of the surrounding medium.

"These theories assume, more or less explicitly, the existence of substances the particles of which have the property of acting on one another at a distance by attraction or repulsion. The most complete development of a theory of this kind is that of M. W. Weber, who has made the same theory include electrostatic and electromagnetic phenomena.

"In doing so, however, he has found it necessary to assume that the force between two electric particles depends on their relative velocity, as well as on their distance.

"This theory, as developed by MM. W. Weber and C. Neumann, is exceedingly ingenious, and wonderfully comprehensive in its application to the phenomena of statical electricity, electromagnetic attractions, induction of currents, and diamagnetic phenomena; and it comes to us with the more authority, as it has served to guide the speculations of one who has made so great an advance in the practical part of electric science, both by introducing a consistent system of units in electrical measurement, and by actually determining electrical quantities with an accuracy hitherto unknown.

"The mechanical difficulties, however, which are involved in the assumption of particles acting at a distance with forces which depend on their velocities are such as to prevent me from considering this theory as an ultimate one, though it may have been, and may yet be useful in leading to the coordination of phenomena.

"I have therefore preferred to seek an explanation of the facts in another direction, by supposing them to be produced by actions which go on in the surrounding medium as well as in the excited bodies, and endeavoring to explain the action between distant bodies without assuming the existence of forces capable of acting directly at sensible distances.

"The theory I propose may therefore be called a theory of the Electromagnetic Field, because it has to do with the space in the neighborhood of the electric or magnetic bodies, and it may be called Dynamical Theory, because it assumes that in that space there is matter in motion, by which the observed electromagnetic phenomena are produced.

"The electromagnetic field is that part of space which contains and surrounds bodies in electric or magnetic conditions.

"It may be filled with any kind of matter, or we may endeavor to render it empty of all gross matter, as in the case of Geissler's tubes and other so-called vacua.

"There is always, however, enough of matter left to receive and transmit the undulations of light and heat, and it is because the transmission of these radiations is not greatly altered when transparent bodies of measurable density are submitted to the so-called vacuum, that we are obliged to admit that the undulations are those of an aethereal substance, and not of the gross matter, the presence of which merely modifies in some way the motion of the other.

"We have therefore some reason to believe, from the phenomena of light and heat, that there is an aethereal medium filling space and permeating bodies, capable of being set in motion and of transmitting that motion from one part to another, and of communicating that motion to gross matter so as to heat it and affect it in various ways.

"Now the energy communicated to the body in heating it must have formerly existed in the moving medium, for the undulations had left the source of heat sometime before they reached the body, and during that time the energy must have been half in the form of motion of the medium and half in the form of elastic resilience. From these considerations Professor W. Thompson has argued, that the medium must have a density capable of comparison with that of gross matter, and has even assigned an inferior limit to that density.

"We may therefore receive, as a datum derived from a branch of science independent of that with which we have to deal, the existence of a pervading medium, of small but real density, capable of being set in motion, and of transmitting motion from one part to another with great, but not infinite, velocity.

"Hence the parts of this medium must be so connected that the motion of one part depends in some way on the motion of the rest; and at the same time these connections must be capable of a certain kind of elastic yielding, since the communication of motion is not instantaneous, but occupies time.

"The medium is therefore capable of receiving and storing up two kinds of energy, namely, the "actual" energy depending on the motion of its parts, and "potential" energy, consisting of the work which the medium will do in recovering from displacement in virtue of its elasticity.

"The propagation of undulations consists in the continual transformation of one of these forms of energy into the other alternately, and at any instant the amount of energy in the whole medium is equally divided, so that half is energy of motion, and half is elastic resilience.

"A medium having such a constitution may be capable of other kinds of motion and displacement than those which produce the phenomena of light and heat, and some of these may be of such a kind that they may be evidenced to our senses by the phenomena they produce.

In these excerpts, Maxwell identifies two broad reasons for pursuing an aether model, one pedagogical and the other epistemological. From the epistemological standpoint, he evidently understands physics to be an application of mechanical ideas to natural phenomena. In this sense, not only is action-at-a-distance implausible, but so too the sorts of causal connections that seem to be required by the variance of force with velocity as well as distance. Overall, the complexity and implausibility of the Weber scheme, in comparison to Faraday's, seems to augur against it, in the same sense that the Copernican vision diminishes the Ptolemaic. Moreover, the value of a greatly simplified, efficacious model is of the utmost pedagogical value, as Maxwell describes.

But if a single salient feature of the Faraday-Maxwell framework need be identified as decisive with respect to its ultimate ascendency, that feature would likely be the power and ease of application of its methods with respect to the solution of real world problems. Linearly superposable vector quantities, reflecting the same attribute of forces and wave amplitudes, makes calculating the combined electromagnetic effects of multiple-charges much more tractable than under Weber's Newtonian scheme. And so the bottom line would seem to be that the field model is simply much more useful.

Of course, this property is closely related to the pedagogical factor, but it is not identical. For if there existed an extremely powerful means of solving important physical problems, which was yet vastly more difficult to learn than a much inferior method, even if that inferior method provided a "better understanding" of the underlying ideas, students would yet be forced to learn the more efficacious procedure.

From the epistemological standpoint, however, it is crucial to remember that Faraday and Maxwell were both driven by a keen desire to understand nature. Although this was perhaps more transparently the case with Faraday, while Maxwell clothed his curiosity in epistemologically more sophisticated terms, they clearly shared a similar vision. This is crucially important because, regardless of the practical reasons for the ascendency of Maxwell's model (in the face of Weber's more-or-less adequate, pre-existing theory), it is not evident that the Faraday-Maxwell framework could ever have gotten off the ground absent great faith in the comprehensibility of nature and a deep longing to understand.

Therefore, explanatory adequacy must be considered a key factor in theory construction, if only because, historically, it has played a crucial role in the development of the best theories, especially when, from an empirical or positivistic point of view, the line pursued is not necessarily called for, or even "appropriate." This aspect of theory construction is often paid lip surface, but is not consistently respected, especially at present.

In summary, the moral of Maxwell's devotion to the vision he shared with Faraday is this: Although physical theories need not, from a strictly logical standpoint, be reducible to visualizable models – and although, ultimately, the effort to push any such model to completion must hit a wall – the mathematical theories that result from visualizable structures in three space dimensions plus time, with causality mediated by local action, have an historical record of extraordinary success – not only direct success, with regard to the explanation and prediction of phenomena, but also indirect success, vis-à-vis the discovery of new directions of exploration and the precipitation of new paradigms.

§ 4.6 Lorentz versus Einstein

Lorentz extended the work of Maxwell, further elaborating the concept of the electron, or particle of electricity, which Maxwell had made a key mechanical component of his aether model (one of its "gears and wheels," as Richard Feynman sardonically referred to the purported workings of such mechanical artifices). In Lorentz's theory, the electron is an independent unit of electric charge that moves through a static material aether. And so the aether establishes, in Lorentz's model, a fixed frame of reference with respect to which he was able to deduce the coordinate transformations that form the basis of the *Special Theory of Relativity*.

In its most mature incarnation, the Lorentz aether lacked most of the physical characteristics that Maxwell and earlier physicists had attributed to it, and seemed largely (and in the view of some astute critics, *only*) to serve the purpose of defining a unique Galilean reference frame, absolutely at rest – thus fulfilling Newton's requirement by establishing an absolute inertial framework. Einstein noted this weakness, and it was with his novel interpretation of the Lorentz equations that epistemological issues first became a central concern of twentieth century physics.

Lorentz had shown how it can come to pass that a moving electron – which is simultaneously both the source of an electromagnetic field and its captive – is the seat of a self-action that effectively increases the electron's inertia, and shortens, in the direction of motion, the shape of its surrounding field. Lorentz assumed that molecules of matter are held together primarily by electrical forces, in which case the equilibrium distance between molecules must be determined by the shape and strength of the ambient electromagnetic field. Accordingly, the acceleration of an electron must impose both a drag on its own translatory motion (a reaction to the changing velocity, per *Lenz's law*) and a shortening, in the direction of motion, of the length of any collection of particles (i.e., composite object) to which the electron belongs. Such a drag is effectively identical to a velocity-augmented mass, and so inertial mass must appear to increase with velocity.

Again, assuming that matter is composed of electrons, detailed calculations of these effects conform to the experimental finding that material objects behave as though their entire mass increases in accordance with the law for the increase of electromagnetic inertia. This inertial modification manifests as a reduction both in the magnitude of the particle's translatory acceleration and in any transverse or "internal" acceleration – such as oscillations about the mean path, i.e., *vibrations*; hence the clock slowing effect.

As a consequence, it could be theorized that all of the mass of an object is electromagnetic in origin. A number of physicists took this as their point of departure, and sought to reinterpret mechanics on the basis of electromagnetism. Of course, it could simply be accepted that all matter, regardless of composition, is augmented with velocity, without regard to underlying causes. For though the idea of the electromagnetic origin of inertia was compelling, experiment could not arbitrate the relative merit of these positions. This is the essential meaning of Lorentz invariance; the putative *properties of the aether* conspire to make it impossible to determine their identity as such. Thus the velocity of light must always be measured to have the same value, regardless of the state of motion of the observer.

Einstein recognized these inherent symmetries of Lorentz invariance, and saw that it is not only impossible, in practice, to give meaning to the notion of Absolute Space – because motion relative to the carrier of the field could never be observed, thereby precluding the possibility of distinguishing one particular Galilean reference frame from another – but that it is also impossible, in practice, to coordinate clocks in such a fashion that meaning can be given to the notion of absolute simultaneity, except when events are immediately proximate. This gives rise to a subtle difficulty with respect to the synchronization of time-keeping devices, inasmuch as it undermines the effective meaning of distant simultaneity. If two clocks are synchronized at the same location and then gradually separated, they will just as gradually become out of phase and no longer be synchronous.

Moreover, any effort to compensate for this phase shift must employ detailed knowledge of the "true" velocity of the clocks relative to the aether. But because velocity relative to the aether cannot be measured, this is an ambiguous notion, and so the synchronization cannot be effected. Regardless of how the clock coordination scheme is contrived, "absolute" simultaneity cannot be established.[6]

Einstein's fundamental and novel insight was to view these relations as the upshot of *any* viable aether model, regardless of details – i.e., that the constancy of the velocity of light and hence the Lorentz transformations follow necessarily from the electrodynamic facts. The constancy of the velocity of light, in particular, was well established experimentally. Moreover, because various efforts to explain the Lorentz relations on the basis of detailed models did not produce, directly or simply, the exact results that could be derived on the single premise that the velocity of light is invariant, and because the aether, like distant simultaneity, is in principle unobservable, Einstein felt compelled to simply attribute the invariance of the velocity of light to the property of relativity itself. In other words, if the velocity of light reflects a constant quantitative relationship between the electric and magnetic magnitudes of the field, as Maxwell established, and if that relationship is invariant under inertial frame changes, then the invariance of the velocity of light can be viewed as a law of nature, which, as with any law of nature, should not be expected to vary by virtue of a Galilean frame change. Galilean relativity need thus only be modified and extended with respect to the addition of velocities – per the Lorentz transformations – to accommodate this invariance.

In order to complete this line of thought, it is necessary to find a way to deal with clock coordination and space-time measurements generally. Under the circumstances, the only practical option for establishing measurable simultaneity of non-spatially-proximate events is to specify, by definition, some operational procedure that gives the concept of simultaneity a non-ambiguous meaning in all cases. And the most practical way to define simultaneity in a simple, meaningful way is by the stipulation that along any light path all times are "simultaneous." Again, this means

---

6 This brief description of the problems attendant to distant clock synchronization is not intended to exhaust the logical possibilities, but rather merely to characterize the way that the problem evidently appeared to Einstein in the spring of 1905, and the manner in which the possibilities seem to have entered into his reasoning at the time.

accepting that light signals travel at a constant speed, always and in all directions – but this is a reasonable accommodation of convention to a large body of well-established empirical facts.

In the mathematics and phraseology that grew out of these considerations, light paths are *null geodesics*; time is deemed to "stand still" along them. With respect to the foundations of physics, this turned out to be a crucial convention, with many consequences. Based on the understanding of physics at the turn of the last century, and with electromagnetic signaling the only (realistic) imaginable means by which a clock-coordinating scheme could be implemented, it implied the necessity of abandoning altogether the *very idea* of "absolute simultaneity."

It should be noted that it is precisely at this juncture that a fundamental break is made with previous philosophies of science. For it is not yet impossible, in principle, to imagine that absolute simultaneity might have a physical meaning. For example, if gravitational action were deemed to act instantly, as physicists usually assumed at the time, one could – in principle – employ the motions of matter to signal across any distance, thereby determining simultaneity to an arbitrary degree of precision. Or, if matter were able to take arbitrarily rigid forms, one might simply use highly rigid, mechanical connections between far-flung places to determine simultaneity. Perhaps most importantly, one might hypothesize that there are other, as yet unknown phenomena, not necessarily limited to the speed of light, by use of which one might be able to establish distant simultaneity in a manner different from that based on known means. So an enormous nexus of logical and physical relations hinges on the assumption that the velocity of light is an absolute limit, including the seemingly self-evident notion that matter can exist in perfectly solid forms.

Einstein was well aware of this nexus, and felt that he had uncovered a deep property of nature. Thus, gravity cannot act instantly; perfectly rigid bodies – just as perfect clocks – cannot exist; etc. By the time Einstein had begun thinking about the relativity of simultaneity, he had more or less abandoned his earlier efforts to understand the electromagnetic field in terms of mechanical constructions. In his mid-teens, he had written an essay on the possibility of measuring what he called the 'magnetic states of the aether' and, at about the same time, was struck by the paradoxes that seemed to ensue from the possibility of material objects moving with the velocity of light. He was also disturbed by asymmetries in the descriptions of electric and magnetic fields, inasmuch as, on the basis of the prevailing view, these fields could be theoretically distinguished by the absolute state of motion of their sources, whereas the motion of the sources relative to each other was the only observable action.

By 1905, with thermodynamics as his model, Einstein was looking for an overall ordering principle that could guide him toward eliminating the artifacts of electromagnetic theory, which he believed were obscuring the hidden meaning buried within the Maxwell-Lorentz framework. He found this principle in the constancy of the velocity of light.

In the century that has elapsed since the publication of Einstein's first relativity paper, it has been well established that Lorentz invariance holds in a wide variety of circumstances, and for all forms of matter; including those which have been observed to move with velocities virtually equal to that of light. Lorentz invariance thus extends beyond electrodynamics, inasmuch as the dynamics of all known "particles" – regardless of their charge, composition or structure – conform to the Lorentz equations. Hence the attribution of the properties of Lorentz invariance to "space" and "time" *per se*, and the unity of "spacetime." That is, these properties do not seem dependent on any particular attributes of matter; just as acceleration in a gravitational field is deemed independent of the material composition of the accelerating body – an invariant property that Einstein utilized vis-à-vis the extension of relativity to encompass gravitation (i.e., he treated the equivalence of gravitation and uniform acceleration in the same manner in which he treated the invariance of the velocity of light, and the properties of gravity likewise became properties of space and time *per se*).

§ 4.7 Einstein's "Positivism"

As noted above, Einstein said that he was looking to thermodynamics as a guiding model during his work on special relativity. He was seeking universal principles, like conservation of energy, to develop an effective or phenomenological as opposed to a constructive theory. This decision seems to have followed on a decade of failed efforts to construct a physical model. When he despaired of this pursuit, which he felt was condemned by certain asymmetries inherent to mechanistic constructions, whereas the mechanisms must in turn be obscured by the emergent symmetries of such models, he adopted a somewhat positivistic approach. That is, he found his way forward via an operationalist methodology, with strict attention to phenomenological facts – an approach he would ultimately characterize as a *Theory of Principle*. And of course he found his principle in the constant, limiting velocity of light, together with the universal validity of [Lorentzian modified] Galilean relativity. In particular, he came to a firm belief in the principle of relativity as something truly fundamental – almost transcendent.

He also came to favor a theory of principle because it facilitated a significant reduction of the number of assumptions required to explain the phenomena – in this case no more than two, and in an important sense really one, *viz.: relativity*. This aspect of relativity theory – that many physical consequences seem to follow from a single, simple *symmetry* principle – has exerted a strong influence on succeeding generations of physicists. Also significant is the fact that the resulting theory greatly simplified the mathematical treatment of optical and electrodynamic phenomena, which were otherwise not so tractable in situations involving relative motion. In this connection – the application of the new, simplified viewpoint to technical problems of computation – the advantages of the operationalist paradigm became immediately obvious (for example, as applied to the treatment of optical phenomena in currents of water).

This latter benefit has historical precedents. As noted, one of the most important reasons why Maxwell's construction became favored over Weber's was that it facilitated the calculation and solution of real world problems – even though, in contradistinction to special relativity, it entailed a turn from a phenomenological theory to a speculative construction. This alternation of theory type has happened more than once throughout the development of modern physics, and the heuristic value of both types of theory – and, more to the point, their mutual interdependence with re-spect to the evolution of science – cannot be emphasized too strongly.

It is also important to remember that, during the early years of the twen-tieth century, many physicists thought that electromagnetism might pro-vide a theoretical basis for the explanation of all natural phenomena. So, if electromagnetic effects were deemed fundamental – that is to say, if the electromagnetic field were *the* primary physical primitive, replacing mate-rial elements – it would of course make no sense to attempt to explain the properties of light in terms of a derivative mechanical construct. In such a case, nothing could be more elemental than the field – it would be as though matter were *made of light*, which of course would explain all rela-tivistic effects. And if all physical observations and measurements were limited by the properties of light – from the time it takes to become aware of events to the delicacy with which they can be resolved – then redefining the nature of time and space in accordance with the limitations imposed by the nature of light would not be such a farfetched proposition.

On the other hand, because relativity dispenses with the requirement that space and time dilation have a physical explanation in terms of me-chanical models, and because of its focus on operational factors, Einstein's interpretation of Lorentz invariance can be described as simply a modifi-cation of the prescription for how space and time measurements are to be performed and understood. In this sense, and in Einstein's own words, his 1905 relativity paper "employs a modification of the theory of space and time."

Also, and perhaps most significantly, the notion that time and space measurements are a function of circumstance, rather than absolute and independent of context, has been imbued in the collective psyche of phys-ics. Today, while some might argue that Lorentz's interpretation of relativ-ity has certain merits that recommend it over Einstein's, even if only as a pedagogical utility, no one will take a position against the fact that time and space measurements vary with motion and gravity. The core idea that

the local rhythm and rate of physical processes *is* time has been thoroughly incorporated into the fabric of physics.

It is in this sense that the philosophy of science was first significantly
altered in the twentieth century, albeit, unfortunately, without adequate
understanding of what this alteration meant. For example, as will become
clear in the sequel, it is a key but insufficiently appreciated underlying
aspect of this development that "instants of time" have no physical meaning – and so, of course, neither do "infinite velocities," although it is really only the latter point (and its implications) that was at all clarified by
Einstein's analysis. All physical considerations must involve the context of
finite time – this is the ultimate, general meaning behind the realization
that distance and time measurements are necessarily coupled, which is a
corollary of that more general meaning. In this sense, Minkowski's unification of spacetime is merely a mathematical artifice of limited, context
dependent utility (again, this will be clarified in subsequent sections).

In summary, it was really only with the ascendance of special relativity that mechanical philosophy became marginalized by the physics community at large. According to some historians, this philosophy had been
purged from physics by Newton himself, at the very birth of the science he
created, by invoking the expedient of action-at-a-distance. And Newton
apparently came to believe that, absent a mechanical explanation for gravity, some sort of spiritual substratum of the universe, which he assumed
on theological grounds must exist in any case, would ultimately justify
his concept of Absolute Space, another questionable concept that he was
compelled to employ, like action-at-a-distance, in order to make sense of
accelerated motion and gravity. But it is not accurate to say that Newton
thus demolished the mechanistic underpinnings of physical theory. Rather, the revival of the popularity of mechanical models by Maxwell merely
re-vitalized a somewhat dormant tendency.

There were also other significant issues, besides mechanical action *per se*,
involved in the change of view. Einstein was very familiar with the works
of Hume and Kant, and of more modern skeptical thinkers such as Mach
and Poincaré, and he had been giving much thought to Newton's concept
of Absolute Space, particularly with regard to Mach's critique, which condemned the invocation of non-observables in theory – especially New-

ton's establishment of empty space as an inertial frame of reference. He had also been thinking about the practical technicalities of coordinating far-flung networks of clocks, by virtue of his day-job at the patent office. In that context, he had examined many schemes for clock synchronization, which was a major technical issue of the day. The question of how such co-ordination can be accomplished – absent a method for determining an ab-solute spatial-temporal reference framework and without frame-invariant measuring devices for local distance and time determinations – is a non-trivial problem, the solution to which is the special theory of relativity.[7] In this sense, it is again understandable that Einstein would refer to special relativity as employing "a modification of the theory of space and time."

Before Einstein, physicists did not often refer to the conceptions of space and time as "theories." The reference to "the theory of space and time" can be interpreted to mean Newton's definitions of Absolute Space and Abso-lute Time in the *Principia*, and Kant's subsequent validation of the con-cepts as *a priori* givens. So Einstein was challenging two notions. First, he was objecting to the assumption that space and time, as elements of physical theory, can be defined in the manner that Newton had prescribed. Second, while possibly accepting, from a philosophical standpoint, that space and time are *a priori* subjective constructs – i.e., *given, general forms of perception* per Kant – he was yet rejecting the premise that those forms, even if *a priori*, are for that reason adequate for the description of physical phenomena. While Newton had given space and time objective meaning, because his framework of absolute space and time is necessary to make sense of the concept of acceleration, Einstein – at least with respect to the development of special relativity – was not necessarily seeking to redefine space and time as objective aspects of the world; rather, he wanted to re-move the ambiguities that arise from the conjunction of Newtonian ideas with the fact of the constancy of the velocity of light.

Rather than signalling an electrodynamic theory of mat-ter – i.e., attempting to explain the problems of the day on the basis of the electromagnetic field – Einstein's focus on the concepts of space and time reflects the circumstance that the formulation of a dynamical model of the field seemed to be a hopeless task. Not only were the empirical data and the mathematical tools grossly insufficient, but the symmetries of Lorentz

7 That is to say: *in principle, in accordance with the conventional understanding of the topic*.

invariance seemed to bode against a confirmable success. While the symmetries would seem to be emergent they conceal their origins, with the consequence that emergence is evidently impossible to verify.

This imposes another noteworthy epistemological constraint on theory, supplemental to the general limitations discussed in chapters one through three. As will become manifest in what follows, this type of limitation – which moreover is an intrinsic *aspect of emergence* – is a pervasive characteristic of theoretical physics; from Newtonian mechanics to relativity and quantum theory. But it is a two-sided property. Thus, while emergent symmetries limit the ability to confirm their origins, they also advance knowledge by yielding *Order Parameters* – e.g., "constants of nature" – which in turn enable phenomenological (*'Principle'*) theories to be constructed absent exact knowledge of the reasons for those constants.

Again, with regard to special relativity, Einstein recognized that an effective or operational theory was needed as a stopgap – a method for establishing the mathematical transformations between reference frames in uniform relative motion – and that a solution was at hand merely by accepting the velocity of light as a constant of nature. And this is exactly what special relativity brought to physics – a pragmatic way to deal with the electrodynamics of moving bodies. By "the theory of space and time" Einstein need not have meant anything more than "a prescription for carrying out distance and duration measurements and understanding their inter-relations." Such a relational view of physics was in general accord with the philosophies of Hume and Mach. And such relational or operational theories, employed judiciously, are just as crucial to the development of science as are constructive models. The two types are complimentary, and mutually catalytic. The gravest danger of either viewpoint is that it be taken too seriously, while its compliment is considered meaningless.

For example, while relativity provides a pragmatic, operational prescription for time and distance measurements, it does not preclude thinking in terms of Newtonian or intuitive notions of space and time. That is, the "common sense" concepts of space and time establish a *cognitive context* in which relativistic space-time can be understood. This is the salient aspect of arguments made by those who favor what is sometimes called *Lorentzian relativity*. In order to *understand* how clock rates vary with context, the mundane conception of time is required, in the sense of the figure-ground duality intrinsic to cognition. The variability of clock rate only makes sense in the context of something in the background that doesn't change – i.e., *"Time."* One can *say* that the existence of other clocks and the relations among them establish all the background that is mathematically necessary, but *cognitively* the context is the same. And of course this also applies to the notion of "Absolute Space." In order to fix the idea of a dynamic metric, *cognitively* there need be some background against which change occurs, even if the idea is ultimately inconsistent. Per the arguments of chapters one through three, the fact that absolute space and time have no objective referents is only in a limited sense a discovery of physics. In a more general sense it is an epistemological fact, inasmuch as it is a feature of all concepts. And as will be shown below, there still exists considerable confusion regarding the concepts of space and time in physics.

§ 4.8.a   The Pros and Cons of Positivism: Part A.

Many of the putatively positivistic interpretations of physics that have followed on the success of special relativity tend to overemphasize the virtues of the *"Principle"* type of theory. Again, the reason that it made good sense to put aside the aether concept in 1905 was that there was no empirical basis on which to establish its properties. This is the same reason why it made sense for Newton to avoid hypothesizing arbitrarily about gravity. There is, however, an important role in science for the sort of approach that Lorentz pursued. So, for example, after establishing the special theory of relativity – on relatively positivistic lines – it yet made sense for Einstein to seek a more elucidated description of gravity than that which had been possible in Newton's time. This is a quintessential instance of the historic alternation between the two broad types of approach to theory construction; an alternation that – as will be seen – played out in the development of quantum mechanics as well.

Recapitulating the developments of special relativity: Einstein's interpretation of the electrodynamics of moving bodies arose naturally from the consideration that, if one takes as a working assumption the existence of a more or less static Lorentzian aether, it follows that it is not possible to establish the inertial reference frame of that aether. The concept was self-defeating in this plain and simple sense. The symmetries of Lorentz invariance make it impossible to synchronize clocks in such a manner that one can determine absolute simultaneity between arbitrarily distant points, and therefore no basis would seem to exist for realizing the idea of a rigid, and thereby unitary "extended thing" in nature.

But a way forward can be forged without building upon such ideas. By accepting the constant, limit velocity of light as a law of nature – beyond the need of explanation *at least for the time being* – an enormous simplification can be made in the theoretical interpretation of electrodynamics. None of the details of a mechanical aether need be worried about; the whole concept becomes moot – at least, that is, at this stage of the theoretical development, within the technologically accessible regime of time and distance measurements.

And the non-trivial issue of accounting for the effects of changing frames of reference can be dealt with in a simple, straightforward way, without worrying about whether the Lorentz transformations are idealized approximations to some physical modification of clocks and rods due to electrodynamic effects. Again, in 1905 and for some time thereafter, there was good reason to believe that all of physics might be reducible to electrodynamics. If this were achieved, relativistic effects would have become universal principles of physics at any rate, because the world would thus be conceived as "made of light" – and Lorentz invariance would be absolute.

But by taking the Lorentz transformations at face value a subtle, new development is precipitated. Einstein essentially rewrites Newtonian law, because the Lorentz transformations apply to mechanics as well as to electrodynamics. The symmetries that these transformations embody thus become *revealed truth* – and thus a new focus of attention in their own right, like a *Platonic Ideal*. Indeed, these symmetries are particularly perspicuous as a result of Minkowski's formal unification of *spacetime*. By way of this treatment, the far-reaching consequences of Lorentz invariance can be addressed in a straightforward and mathematically elegant way, via Minkowski diagrams. Whereas Einstein's use of the phrase "a modification of the theory of space and time" to describe special relativity is understandable in the operational sense described above, as a result of the Minkowski formalism the unification of space and time begins to take on an ideological significance of its own.

Absent a physical interpretation of Lorentz invariance and a basis for defining accelerated motion, it is a natural trend of thought to objectify the symmetries together with the overall Minkowski framework – attributing independently existing properties to them; metaphysical or *Ideal*, again in the Platonic sense. Minkowski's unified spacetime, with its invariant interval, tends to be treated as something objective. From this standpoint, it seems sensible to regard the erstwhile "naturalness" of the outmoded Galilean transformations as "a prejudice of common sense" – a holdover of "classical thought." But it is instructive in this context to note that, whereas familiarity with the Minkowski perspective, and acceptance of it as "natural," is indeed an instance of what is sometimes derogatorily referred to as "habit and custom," understanding of the Galilean transformations, in contradistinction, is merely an application of the rules of arithmetic (i.e., the linear addition of velocities) to the *a priori* forms of perception in the sense of Kant (in Kant's terminology, it is synthetic, *a priori* knowledge). It is a realization of the symmetries of space and time as given to intuition, and is not 'merely habitual or conventional,' albeit certainly *commonly sensible* if not instinctive. Although certain writers, such as Poincaré, have speculated that the spatial intuition associated with Euclidean geometry is based entirely on experience, it would appear from modern neuroscience (as noted in chapter six) that the Euclidean sensibility of spatial relations is more-or-less hard-wired into the brain, *genetically*, just as the faculty of natural language is also now believed to be. (Pursuant to the researches of Noam Chomsky and in accordance with those who share his views, all natural human languages are thought to be essentially the same, sharing a genetically determined core structure. As with any organic structure connected with cognition, such as the visual system, very young children do not learn a language so much as they *grow* it – acquiring their vocabularies, at times of peak development, on only a single exposure to individual words in real time.[8])

---

8 cf. Chomsky, Noam. New horizons in the study of language and mind. Cambridge [England]; New York: Cambridge University Press, 2000.(Op. Cit. page 6)

Taking the next step – that is, beyond the tentative establishment of a new prescription for space and time measurements – and defining away the possibility of *ever* employing a concept along the lines of *aether* (i.e., where deviations from Euclidean metrics and time dilation have a physical underpinning) is quite sterile in terms of the future development of theory. From an epistemological standpoint, this was just as unwarranted *circa* 1905 as it is today, despite arguments that the symmetries of mathematical physics have an objective significance that is deeper than (and even contradictory to) any possible physical model. Proclaiming *Lorentz Invariance* to be an absolute property of nature that cannot be violated, on any scale, under any circumstances, is quite arbitrary.

Regarding the existence of the aether, within a decade or so Einstein had revised his position, as it had become clear to him that General Relativity had actually re-introduced the concept into physics in covariant form. Moreover, independently of considerations concerning General Relativity, a similar change of thinking later occurred among many physicists, particularly in connection with Dirac's work on relativistic quantum theory. Today, *the "Vacuum" of space* is conjectured to contain approximately 95% of the substance of the known universe. So there is in fact good reason to believe that theoretical descriptions of a deeper level of spatial structure might eventually come to dominate the models of science, and that Lorentz was not wrong, but only premature, in attempting to effect such a development.

However, the manner in which the concept of *spacetime* emerged from special relativity, and subsequently became the object of general relativity, left the ideas of space, time and especially *spacetime* in an epistemologically unclear state, inasmuch as geometry *per se* came to be viewed as an objective, physical property rather than a conventional, subjective construct. Absent adequate epistemological analysis, the concept of the aether became taboo, while the confused concepts of space and time became "re-objectified" – though in a form that makes Newton's original objectification of these concepts seem quite tame in comparison. Certain properties that had previously been ascribed to the aether became attributed directly to *space* and *time*, which thence became dynamic objects of physics on a par with matter. There is nothing wrong with this attribution from a purely logical point of view, but the confusion and misconceptions that surround these ideas, as they are today "understood" by many physicists, have become genuine obstacles to progress.

§ 4.8.b   The Pros and Cons of Positivism: Part B.

As noted, the incarnation of dynamical spacetime hatched with General Relativity soon came to be regarded by Einstein as a sort of aether. Although in principle the theory does not necessitate preferred coordinate frames, it yet deems regions of spacetime that are empty and sufficiently far removed from matter/energy to be "flat" – i.e., geometric measures in such regions are expected to conform to Euclidean geometry. And gravitational energy, like any energy, is associated with the curvature of spacetime. Accordingly, while the gravitational field contains energy and so in itself is a source of gravitation – which is to say that curved spacetime is a source of gravitation – flat spacetime is not. Moreover, such induced or self-gravitation conforms to conservation of energy – i.e., gravitation may be nonlinear but not indefinitely self-reinforcing: a black hole cannot contain more energy than is put into it – just as the concentration of energy corresponds to the degree of curvature generally. This means that spacetime can be thought of as having an inherent resistance to curvature. In other words, while technically there are no preferred reference frames, the mere fact that any deviation of geometry from a Euclidean structure requires the presence of and in turn stores energy, implies that ostensibly empty space has some characteristics that are at least partly analogous to those of an elastic substance. And Euclidean structure is associated with the lowest, default energy state of that substance. However, while spacetime is indeed dynamic, it is yet covariant and so considered to have no identifiable "parts."

The theory of relativity does not address the finer points and conflicting aspects of these issues. That is, *epistemologically,* space and time are considered to be nothing more than conceptual constructs for the ordering of experience; which, in turn, is taken to indirectly represent an abstract order of some underlying "world at large." But the spacetime of general relativity does not represent a purely relational order; rather, it references a kind of independent existence that is much like that associated with Newton's *Absolute Space*, as well as the physical objects of classical physics generally; which carry energy, etc. And so the foundational concepts of relativity, while not as problematic as those of quantum theory, are yet in need of improvement.

Einstein's pursuit of general covariance was motivated by the seemingly self-evident notion that in "pure," i.e., "truly empty" space, there are no features by which it is possible to establish Galilean reference frames with respect to which accelerated motion can have meaning. In this sense, general relativity starts from the assumption that there are no elements of space beyond those thought to exist *circa* 1915 – that is, shortly after Einstein banished the aether – and so no seat of physical activity, at a sufficiently elemental scale, to give meaning to the definition of acceleration. This is the same consideration at the heart of special relativity – i.e., if there is no observable thing to instantiate a mechanism that might be responsible for space and time dilatation, Lorentz invariance "simply is."

This perspective is described very clearly and simply by Einstein himself, writing for a lay readership in his 1916 popular exposition of the theory of relativity. He begins by describing a thought experiment, in which the surface of a large marble table is intended as a metaphor for a two-dimensional Euclidean continuum. He explains how it is possible on this surface – using a number of small identical metal rods, all rigid and straight – to construct a grid resembling a Cartesian coordinate system; i.e., a matrix of rods, laid out on the marble slab in such a manner that four rods meet at a point, and every corner so formed is a right angle. The arrangement is then altered, in accordance with the following quotation:

> "By making use of the following modification of this abstract experiment, we recognize that there must also be cases in which the experiment would be unsuccessful. We shall suppose that the rods "expand" by an amount proportional to the increase of temperature. We heat the central part of the marble slab, but not the periphery, in which case two of our little rods can still be brought into coincidence at every position on the table. But our construction of squares must necessarily come into disorder during the heating,

because the little rods on the central region of the table expand, whereas those on the outer part do not.

"With reference to our little rods – defined as a unit – the marble slab is no longer a Euclidean continuum, and we are also no longer in the position of defining Cartesian coordinates directly with their aid, since the above construction can no longer be carried out. But since there are other things which are not influenced in a similar manner to the little rods (or perhaps not at all) by the temperature of the table, it is possible quite naturally to maintain the point of view that the marble slab is a "Euclidian continuum." This can be done in a satisfactory manner by making a more subtle stipulation about the measurement or the comparison of lengths.

"But if rods of every kind (i.e., of every material) were to behave in the same way as regards the influence of temperature when they are on the variably heated marble slab, and if we had no other means of detecting the effect of temperature than the geometrical behavior of our rods in experiments analogous to the one described above, then our best plan would be to assign the distance one to two points on the slab, provided that the ends of one of our rods could be made to coincide with these two points; for how else should we define the distance without our proceeding being in the highest measure grossly arbitrary? The method of Cartesian coordinates must then be discarded, and replaced by another which does not assume the validity of Euclidean geometry for rigid bodies. The reader will notice that the situation depicted here corresponds to the one brought about by the general postulate of relativity..." [9,10]

Under such circumstances, the notion that the structure of Euclidean geometry is an arbitrary imposition on the description of phenomena might seem appropriate. But again, General Relativity does not merely replace Euclidean geometry with a substitute system of measurement. Rather, spacetime becomes 'a something'; its metric a *dynamic* something that

---

9 Einstein, Albert. Relativity: the special and the general theory; a popular exposition. New York: Crown Publishers, 1961
10 Italics/emphasis, quotation marks and parenthetical remarks all in the original.

changes with the activity of matter-energy. The geometrical dynamics, in turn, carry energy, and so in a sense are like elastic deformations of an underlying Euclidean structure. But such structure is explicitly excluded, and so the spacetime of General Relativity is posited to be a fundamental element of reality – a *refinement* of but *not solution* to Newton's inadequate conception of *Absolute Space*.

It should be noted in this connection that Mach did not accept General Relativity as an appropriate solution to the problem of Absolute Space; somewhat to the dismay of Einstein, who believed his scheme followed *Mach's Principle* – i.e., that acceleration should be measured with respect to observable reference points; again, the "fixed stars." A schematic representation of the development of General Relativity makes the reason for Mach's rejection clear. In order to equate Galilean systems with arbitrarily moving (accelerating) systems, the device of gravity is required. But gravity, in Einstein's thinking, does not act at a distance, and so must reside locally throughout space, rather than only in distant matter. And matter affects space – *literally*: space dynamically reacts to matter – to the same extent that space affects matter. Therefore, because space can store energy, the notion of space as a substance remains; and the problem associated with Newton's conception of Absolute Space is only partially solved.[11]

An important epistemological lesson emerges here. For Mach, it was imperative to avoid hypothesizing about things that cannot be observed, such as a substantive spacetime. He believed that inertia and gravity arise together in a phenomenological relation between individual observable objects and the total ponderable matter of the cosmos – which again, for him meant mainly the visible stars. Einstein, on the other hand, felt it was crucial to avoid hypothesizing about occult causes, such as action-at-a-distance. While he shared Mach's opinion that the so-called fixed stars account for inertia and gravity (Einstein too equated inertia with gravity), he would not accept that the influence of the stars could be felt instantly across empty space, without an intermediating physical connection.

Note how arbitrary the implementation of positivistic philosophy must be. In Mach's view, the idea of action-at-a-distance is acceptable, because nothing is observed to exist in the space between gravitating objects. Only

---

11 The confusion and epistemological problems associated with these ideas are discussed, in greater detail, in the section below entitled *The Ehrenfest Paradox*.

objects are observed, so only objects exist. From Einstein's perspective, the idea of action-at-a-distance is "spooky," because a force is hypothesized to be active throughout empty regions of space, i.e., where nothing is observed to exist, and therefore space cannot be truly empty. Ultimately, the line is drawn in accordance with *philosophical prejudice*, which is to say personal, ideological bias.

However, some preferences make more sense than others. And though physical theories need not be limited by general, wide ranging philosophical concerns, some yet have relevant implications. For example it is clear, on the basis of general epistemological considerations, that an observer-independent model of the cosmos is a desirable objective. Moreover, this preference is reinforced by the reasonable prognostication that (ultimately) unification of the physical sciences with the psychological will be a relevant issue.

All of this must be understood in the context of the arguments presented in chapter three, in accordance with which it makes sense to consider the relevance of such a possible unification along two general lines. First and foremost, there is the concern that physical theory be established on a firm, observer-independent basis. Neither the physical nor the cognitive sciences have reached a stage of maturity sufficient to contemplate a near-term unification of these two fields in any substantial sense. But in the meantime physics must provide its *"mapping"* function as clearly as possible, and this evidently demands a much more extensive development of physical models than hitherto achieved.

Second, given the plausibility of unification as a viable long-term goal, the concepts of science must be rich enough to comprise the basis for some sort of combinatorial model of subjective experience; providing, in some fashion, conceptual elements that represent sensations, emotions, semantic meanings – the *qualia* of subjective experience generally – that can relate in an *intra-*, *inter-* and *extra-personal* fashion or *format*; one which is not, moreover, necessarily *space-like*. On the other hand, the existence of *space-* and *time-based* experience must somehow arise from such theoretical elements; or at least within a context that supports them. And none of this can be premised on the basis of a circular foundation that presupposes the existence of a "conscious observer."

Of course, it is possible that such a scenario might never be attainable. But it will not be possible to advance towards such a scenario without continued development of sound, objectively anchored physics. The need for refined physical maps will therefore not diminish. And to fulfill that need, theoretical physics must in any case be observer-independent.

It is important at this juncture to warn against a possible misconception that may arise in conjunction with these considerations. The belief that physics and psychology should ultimately be unified can be misleading. In particular, when this belief is based upon unclear ideas regarding the subject-object dichotomy, and is not guided by a sound epistemological understanding of the underlying relations, it is easy to fall prey to the misconception that the concepts of theoretical physics must ultimately diverge from ideas characterized by material, spatial or mechanistic qualities. This prejudice, which confuses the subject-object dichotomy as hereinabove elaborated, is based on the view that the concepts of physics will ultimately need to be coordinated along entirely novel, *non-spatial* lines of order, which can thus somehow directly accommodate the phenomenology of subjective experience – a view which tends to draw credence from the putative "*non-local*" properties of modern physics.

Such a concern seems to have influenced David Bohm's tendency away from a physical interpretation of quantum mechanics and toward more abstract conceptions of order – thus transcending what he termed the "Cartesian" order of spatial organization. He felt that such extra-spatial ideas are in better accord with the properties of consciousness. Roger Penrose has expressed similar concerns, albeit in a different context. Penrose believes that it is a corollary of *Gödel Incompleteness* that the contents of cognition are beyond algorithmic description, and evidently that the non-deterministic aspects of quantum theory – e.g., collapse of a really-existing *superposition of states* corresponding to the wave function – is precisely what is needed to account for physical brain processes correlated with consciousness (more or less along the lines of the quantum computer premise; i.e., the notion that a superposition of states can represent infinite parallel computations working to solve a problem, and that the "collapse" of superposition in a measurement process corresponds to a solution.)

Again, both of these views confuse the subject-object dichotomy in the manner of psycho-physical parallelism. Bohm's position seems weak in the following way: that it does not explicitly recognize the subjective character of theoretical conceptions concerning "physical space"; that such concepts must be interpreted as heuristic tools *only*, and certainly not as referencing something objective that stands in need of explanation. On the other hand, from the *psychological* perspective, spatial structure *is* objectively real and in need of description within an appropriate (*psychological*) context. But to attempt to describe spatial order as something "physical" which emerges from an "objective" albeit *non-spatial "order"* is to fall on the wrong side of the fence, so to speak, and thus confuse the two sides of the *subject-object dichotomy*. It makes no sense to attempt to account for space in terms of non-spatial concepts unless the goal is to understand how *spatial intuition* arises. From such a point of view, spatial order is a proper object of representation – i.e., attempting to explain it from this perspective will, at the very least, not confuse matters gratuitously.

This is a crucial but subtle point, which is easily forgotten. While it may, of course, be conjectured that extra-personal reality is of a non-spatial nature, with space a purely subjective construct – indeed, this would seem to be a plausible ontological position – to attempt to transcend the subject-object dichotomy directly is to leave the ground without an aerial map. As discussed, the best approach is to develop the physical and psychological models separately, as far as possible under the subject-object dichotomy, and then, after a sufficient body of meaningful data (regarding the correlations of phenomena across the divide) is obtained, to attempt some convergence of these models *as the data permits*. But because of the overwhelming heuristic value of spatial representation, it is likely that – even under circumstances that permit unification – spatial models will yet enhance understanding.

Penrose's view seems to stem from insufficient appreciation of the fact that, per the analysis of chapter two, *Gödel Incompleteness* emerges from the very nature of the concept-forming process itself, and that the inadequacy of algorithmic description merely reflects the fact that *no* representation can be perfectly complete and consistent. Note that Penrose's view is precisely opposite to this – i.e., he believes that axiomatic construction, despite its great utility, is yet not up to the task of capturing the essence of human thought because the latter is preternaturally transcendent. The correct answer to Penrose's dilemma is that *algorithms* – i.e., axiomatically structured conceptual constructs intended to represent a set of meanings (i.e., propositions) – cannot perfectly represent human thought because, again, *no* representation can perfectly reflect its professed referent. More-over, the interpretation of quantum mechanics as "non-local" and "non-causal" does not furnish a positive basis upon which a special link with cognitive processes can be forged. The "collapse of the wave function" re-flects precisely that aspect of the phenomena that is not understood under quantum theory via its usual interpretation (and largely merely reflects the distinction between *analog* and *discrete* processes). To equate this aspect of quantum phenomenology with some aspect of thought which seems similarly enigmatic is simply to say that *neither* is understood.[12]

§ 4.9   Origins of Relativity, Misunderstood, Inspire Quantum Mechanics

Most of the theorists who participated in the development of quantum mechanics believed that Einstein's work on relativity had been motivated by the conviction that only observables should enter into theoretical de-scriptions of natural phenomena. The reasoning that Einstein employed in his 1905 relativity paper is indeed based on an analysis of the operative meaning of the concepts of space and time – that is, how measurements of distance and duration are to be carried out and interpreted given the re-strictions that nature imposes on observation. But that analysis was aimed at bringing out the implications that are inherent to the Maxwell-Lorentz theory, and eliminating the artifacts of theory that were obscuring these matters – separating the wheat from the chaff, so to speak. Although Ein-stein originally interpreted his analysis to signify the redundancy of the aether concept, and that it should be eliminated from theory, he not long

---

12 While there are aspects of quantum phenomena that appear to be analogous to certain interesting, fairly recently discovered features of neural processes, these analogies are better understood in terms of the framework proposed in chapter five than that of current quantum theory, and are discussed in a more appropriate context elsewhere herein.

thereafter recanted, as he came to realize that some such concept is needed, whether referred to as *field, vacuum, spacetime geometry* or something else.

But the Founding Fathers of quantum mechanics adopted a more strictly positivistic credo, inasmuch as they explicitly sought to employ none but observable magnitudes – although, ironically, they considered themselves opposed to the philosophy of *Logical Positivism*. Bohr's constructs – the *Correspondence Principle* and *Complementarity* – are extreme examples of this philosophical confusion. In Bohr's vision, classical concepts are required for orientation in the world, and for carrying out and describing the results of experiments and observations. However, classical ideas are to be strictly limited to the "macroscopic" realm, to which, and only to which, they are suited. Therefore, theories employed to describe the results of quantum experiments must correspond with the predictions of classical physics on the macro level. On scales where Planck's constant-of-action is significant there does not exist a "real world" with objective significance, independent of observation. When experiments are carried out on such levels all that can be discussed are probabilities of outcomes – purely on the basis of arrangements of equipment and other macroscopic, classically describable things, such as instrument dial readings and images on photographic plates or display devices.

Notwithstanding Bohr's detailed and methodical elaboration of his Correspondence and Complementarity principles, he does not, unfortunately, say anything epistemologically significant about what any of it means. With Kant, he does not object to the thesis that the "thing-in-itself" is not describable in physical terms. Yet he maintains that classical physical concepts are valid on the level of experience and must be retained in physics for practical reasons – again, orientation and the carrying out of experiments and observations. But the concepts of classical physics clearly reflect a distinct ontological vision, a world manufactured by the mind that is conceived to be meaningful on all levels of scale, spatial and temporal. So Bohr's philosophy is only partially consistent – to the extent that it is viewed as operative and perhaps merely tentative – inasmuch it says nothing at all about the meaning of his prescriptions. However, he evidently did not view his philosophy as merely tentative.

This much is similar to Einstein's initial relativistic interpretation of space-time measurements. And so the same sort of confusion creeps into Bohr's interpretation as that which is found in connection with the objectification of space and time via the framework of General Relativity. While it is inconsistent to attribute objective reality to macro-physical entities and yet deny it on the microscopic scale, this is the upshot, in the minds of most physicists, of Bohr's philosophy. The only consistent position is to consider physical concepts on all scales as nothing more than heuristic devices. While Bohr does not seem to disagree on this, he is yet not clear on what he does mean, which is why his interpretation leads to confusion.

Physicists are not solipsists. As is the case with most other human beings, they employ ontological models for their understanding of experience. And a convincing model cannot be built on the basis of a division of reality into "large" and "small." To most physicists, Bohr seems to associate the Kantian thing-in-itself with an unknowable micro-reality, from which there somehow emerges a more knowable macro-realm, susceptible to representation on the basis of the concepts of classical physics. This is a misconception; again, to be consistent, Bohr's position can only be that even on the "large" scale the concepts of physics have nothing but metaphorical meaning, which physicists must nevertheless employ for orientation and communication because *they work* sufficiently well.

To be consistent, it should be no more taboo to speak of "things" on the micro-scale than of "things" on the macro-scale, as long as the heuristic context is always kept in mind. But again, while Bohr is not clear in this respect, he *is* clear about his belief that he has followed in Einstein's footsteps, viz.:

"The necessity, in atomic physics, of a renewed examination of the foundation for the unambiguous use of elementary physical ideas recalls in some way the situation that led Einstein to his original revision on the basis of all application of space-time concepts which, by its emphasis on the primordial importance of the observational problem, has lent such unity to our world picture. Notwithstanding all novelty of approach, causal description is upheld in relativity theory within any given frame of reference, but in quantum theory the uncontrollable interaction between the objects and the measuring instruments forces us to a renunciation even in such respect. This recognition, however, in no way points to any limitation of the scope of the quantum-mechanical description, and the trend of the whole argumentation presented in the Como lecture was to show that the viewpoint of Complementarity may be regarded as a rational generalization of the very ideal of causality."[13]

"...in quantum mechanics, we are not dealing with an arbitrary re-nunciation of a more detailed analysis of atomic phenomena, but with a recognition that such an analysis is in principle excluded. The peculiar individuality of the quantum effects presents us, as regards the comprehension of well-defined evidence, with a novel situation unforeseen in classical physics and irreconcilable with conventional ideas suited for our orientation and adjustment to ordinary experience. It is in this respect that quantum theory has called for a renewed revision of the foundation for the unambiguous use of elementary concepts, as a further step in the development which, since the advent of relativity theory, has been so characteristic of modern science."[14]

Indeed, Bohr attributes the break with classical ideas directly to special relativity, and interprets quantum mechanics as a continuation of this trend.

"Even the formalisms, which in both theories within their scope offer adequate means of comprehending all conceivable experience, exhibit deep-going analogies. In fact, the astounding simplicity of the generalization of classical physical theories, which are obtained by the use of multidimensional geometry and non-commutative algebra, respectively, rests in both cases essentially on the introduction of the conventional symbol $\sqrt{-1}$.[15] The abstract character of the formalisms concerned is indeed, on closer examination, as typical of relativity theory as it is of quantum mechanics, and it is

13 Einstein & Schilpp, Op. Cit. pages 210-211
14 Einstein & Schilpp, Op. Cit. page 235
15 Author's note: By invoking the concept of imaginary numbers Bohr is of course alluding to the Minkowskian treatment of special relativity, which the Author maintains is a primary reason for the idealization of space-time (in the Platonic sense of Penrose).

in this respect purely a matter of tradition if the former theory is considered as a completion of classical physics rather than as a first fundamental step in the thoroughgoing revision of our conceptual means of comparing observations, which the modern development of physics has forced upon us."[16]

Physicists who had difficulty understanding Einstein's rejection of quantum mechanics occasionally raised the same rebuttal – i.e., that Einstein himself had employed the same abstract, positivistic lines of thought in the development of special relativity. When Heisenberg raised this point, during his first conversation with Einstein, Einstein replied: "Perhaps I did use such reasoning, but it is nonsense all the same." And on more than one occasion, when queried on the same issue, he replied: "A good joke should not be repeated too often."

A more modern recognition of the positivistic influence of special relativity can be found in the writings of Jeffrey Bub:

"I argue that quantum mechanics is fundamentally a theory about the representation and manipulation of information, not a theory about the mechanics of nonclassical waves or particles. The notion of quantum information is to be understood as a new physical primitive — just as, following Einstein's special theory of relativity, a field is no longer regarded as the physical manifestation of vibrations in a mechanical medium, but recognized as a new physical entity in its own right."[17]

§ 4.10 Philosophical Prejudices of the Founding Fathers

Both Bohr and Born confessed, early on in the development of quantum theory, to having vociferous, anti-deterministic predilections. And both Heisenberg and Pauli came to share similar views – against visualizable physical constructs obeying causal laws – well before de Broglie and Schrödinger appeared on the scene. Von Neumann seemed to have sealed

---

16 Einstein & Schilpp, Op. Cit. pages 238-9
17 Bub, Jeffrey. "Quantum Mechanics is About Quantum Information." Foundations of Physics 35, no. 4 (4, 2005): 541-560.

the inevitability of this anti-realistic approach to microphysics with the formal veneer of a (now known to be false) mathematical "proof" ostensibly demonstrating the impossibility of accounting for quantum phenomena on the basis of causal theoretical constructions, or representations of independent, objectively existing micro entities or events – the latter often referred to as "*Hidden Variables*."

Accordingly, Heisenberg's Uncertainty Principle came to be viewed as posing an absolute limit on the level of detail that physical theory can in principle ever describe. Just as Lorentz invariance seemed to impose a limit on what can be known about any physical structure underlying the electromagnetic field, the uncertainty relations were understood to mark a sharp boundary between the world of experience and the eternally unknowable "thing-in-itself."

In both cases, the "thing-in-itself" is more or less equated with "things that are very small." This is another example of pushing the meaning of ideas – in this case, those of a rather positivistic philosophy – well beyond a reasonable limit. Positivism or extreme empiricism can only serve as a critical reagent against misconceptions that might arise from artifacts of theory. It cannot be interpreted to mean, literally, that only observable things can have a place in theory – for, as detailed analysis reveals, no objects of theory *nor referents of ideas generally* are "observable" in the strict epistemological sense.

Such radical positivistic philosophy – or rather, misuse thereof – is quite in tune with the basic attitudes of the physicists who worked closely with Bohr, especially Heisenberg, Pauli and Born. These 'four-fathers' indicate in their various writings that they have been influenced by the philosophy of Hegel. The following Hegelian commentary regarding Kant is illustrative in this respect:

> "According to Kant, ... thought has a natural tendency to issue in contradictions or antinomies, whenever it seeks to apprehend the infinite. But Kant ... never penetrated to the discovery of what the antinomies really and positively mean. The true and positive meaning of the antinomies is this: that every actual thing involves a coexistence of opposed elements. ... The old metaphysic, ... when it studied the object of which it sought metaphysical knowledge,

went to work by applying the categories abstractly and to the exclu-
sion of their opposites. Kant, on the other hand, tried to prove that
the statements issuing through this method could be met by other
statements of contrary import with equal warrant and necessity."[18]

Superficially, this quote from Hegel might seem to reflect certain conclusions of the epistemological arguments of chapters one through three
above. But Hegel clearly misunderstood the crux of Kant's problematic.
There is a major misconception at the foundation of Hegel's paradigm,
which guides his attribution of objective significance to just those attributes of the concept that are the quintessence of its subjective, representational limitations. On the other hand, this superficial Hegelian interpretation of Kant's analysis does capture the essence of Bohr's *Complementarity*
quite aptly.

Another attitude shared by these four denizens of Copenhagen is, as
mentioned, a strong anti-deterministic bias. Consider Max Born's feelings
on this topic, as expressed in a letter to Einstein:

> "To me a deterministic world is quite abhorrent. Maybe you are
> right, and the world is that way, as you say. But at the moment it
> does not really look like it in physics – and even less so in the rest
> of the world."[19]

Werner Heisenberg, in a collection of memoirs called *Physics and Beyond: Encounters and Conversations*, discussed his anti-physical bias at
length. Recalling his first conversation with Einstein, which occurred immediately after a 1926 seminar at which Heisenberg spelled out the spe-
cifics of the new matrix mechanics he was developing, he recounts the
following transaction:

> "We cannot observe electron orbits inside the atom ... but the radi-
> ation which an atom emits during discharges enables us to deduce
> the frequencies and corresponding wave numbers and amplitudes
> ... since a good theory must be based on directly observable magni-

18 Hegel, Georg Wilhelm Friedrich, and William Wallace. The logic of Hegel, translated from the
Encyclopaedia of the philosophical sciences, Oxford: Clarendon Press, 1892, p 99.
19 Born, Max. The Born-Einstein letters : correspondence between Albert Einstein and Max and Hed-
wig Born from 1916 to 1955 / with commentaries by Max Born. New York: Walker and Company,
1971.

tudes, I thought it more fitting to restrict myself to these, treating them, as it were, as representatives of the electron orbits."

to which Einstein responded:

"But you don't seriously believe ... that none but observable magnitudes must go into a physical theory?"

Heisenberg parried:

"Isn't that precisely what you have done with relativity? ... After all, you did stress the fact that it is impermissible to speak of absolute time, simply because absolute time cannot be observed; that only clock readings, be it in the moving reference system or the system at rest, are relevant to the determination of time."

to which Einstein answered, characteristically:

"Possibly I did use this kind of reasoning ... but it is nonsense all the same."[20]

About a year later, at another seminar, Erwin Schrödinger described his recently completed wave mechanical formulation of quantum mechanics. Heisenberg claims to have been delighted by the fact that a new mathematical formalism, easier to use than his own matrix mechanics, had thus confirmed – by an entirely different route – his own approach to the quantum problem. However, he also admits to being quite put-off by Schrödinger's intention to interpret his wave equation realistically, as representing a physical wave. Here is an excerpt from his own account of this seminar:

"Unfortunately, however, the physical interpretation of the mathematical scheme presented us with grave problems. Schrödinger believed that, by associating particles with material waves, he had found a way of clearing the obstacles that had so long blocked the path of quantum theory. According to him, these material waves were fully comparable to such processes in space and time as elec-

20 Heisenberg, Werner. Physics and beyond; encounters and conversations. New York: Harper & Row, 1971.

tromagnetic or sound waves. Such obscure ideas as quantum jumps would completely disappear. I had no faith in a theory that ran completely counter to our Copenhagen conception and was disturbed to see that so many physicists greeted precisely this part of Schrödinger's doctrine with a sense of liberation. The many talks I had had with Niels Bohr, Wolfgang Pauli and many others over the years had convinced me that it was impossible to build up a descriptive time-space model of inter-atomic processes – the discontinuous element Einstein had mentioned to me in Berlin as a characteristic feature of atomic phenomena saw to that. Admittedly, this was no more than a negative feature, and we were still a long way from a complete physical interpretation of quantum mechanics, yet we were certain that we must get away from the idea of objective processes in time and space."[21]

This passage should be stunning to all open-minded scientists. Nothing can be more revealing of the prejudices at the root of modern quantum philosophy than these words: "…talks I had had with Niels Bohr, Wolfgang Pauli and many others over the years had convinced me that it was impossible to build up a descriptive time-space model of inter-atomic processes." Heisenberg reports he was extremely upset that so many important physicists greeted Schrödinger's formulation with a sense of relief and joy – "I went home rather sadly," is how he put it. Accordingly, he immediately contacted Bohr, who had not been present at the event, but who just as quickly, in turn, contacted Schrödinger and invited him to Copenhagen. The sense of urgency is palpable in Heisenberg's account. He reports that, upon the acceptance of the invitation by Schrödinger, "I, too, sped back to Denmark." But his own fervor was apparently far exceeded by Bohr's. Heisenberg reports how Bohr waylaid the hapless Schrödinger in an effort to dissuade him from his deviant beliefs:

"Bohr's discussions with Schrödinger began at the railway station and were continued daily from early morning until late at night. Schrödinger stayed at Bohr's house so that nothing would interrupt the conversations. And although Bohr was normally most consid-

---

21 Op. Cit

erate and friendly in his dealings with people, he now struck me as an almost remorseless fanatic, one who was not prepared to make the least concession or grant that he could ever be mistaken. It is hardly possible to convey just how passionate the discussions were, just how deeply rooted the convictions of each, a fact that marked their every utterance. All I can hope to do here is to produce a very pale copy of conversations in which two men were fighting for their particular interpretation of the new mathematical scheme with all the powers at their command."[22]

This remorseless fanaticism, which Heisenberg, from well within the Bohr camp, projects so well, was much more characteristic of the Copenhagen crowd than of their opponents. Moreover, Bohr was in a position to influence policy within powerful academic circles, and apparently did not hesitate to do so, whereas opponents of Copenhagen, for the most part, could do little more than return home to their private enclaves and continue working on their own ideas.

Einstein – the one person who could have made powerful counter efforts in official circles, because of his great prestige – also preferred, for the most part, to simply work on his own ideas. He believed that quantum mechanics should be attributed a purely statistical interpretation – i.e., that it should be regarded as applicable to ensembles of entities and events only, not individuals – and that he would soon develop a successor theory based on relativistic but otherwise classical field concepts. One imagines that if he had taken up the cause of countering Copenhagen, by organizing like minded physicists and lobbying in academic circles with the same zeal as Bohr, he would have been a much more formidable opponent. And so, by the power of his indomitable will – or perhaps by simply "caring more" – Bohr shaped the course of theoretical physics, and made it conform to his views.

It is interesting to consider that Bohr, Born, Heisenberg and Pauli – again, four of the Founding Fathers most closely associated with the Copenhagen Interpretation of quantum theory – did not think of themselves as positivists. In fact, they all expressed disdain for the strictures of Logical Positivism, and rejected its rigid interpretation of truth as a function of

---

22 Op. Cit.

language – particularly as expressed in Wittgenstein's credo: "what can be said at all can be said clearly, and what we cannot talk about we must pass over in silence." Rather, the personal philosophical views of these four men shared a strong, common strand – antithetical to such purely formal doctrines – which can best be characterized as *mystical*. They believed that human language is inadequate for the representation of reality, which can only be approached mathematically. Natural language, they felt, must be used as an adjunct, poetically or metaphorically, to hint at what is mystically sensed to be "the truth" behind the mathematical description. Therefore, Bohr's Complementarity was an acceptable formulation for his collaborators – it did not undermine but rather seemed to embrace the mystical philosophies of nature that they endorsed. However, to the extent that they understood quantum mechanics as a set of operational prescriptions for the handling of "observables" they had much in common with the Logical Positivists; as did Karl Popper, who was yet also a critic of the creed (Popper was a critic not only of Logical Positivism – despite the fact that he is often referred to as a fellow traveler – he was also a staunch critic of Copenhagen. His views were actually more in line with those of Einstein).

The radical or mystical ontological positions adopted by many twentieth century theoretical physicists have been attributed, by John Bell, to a sort of romanticization of theory, and he cites this as a motivating factor in the adoption of such extreme positions. The remark often attributed to Arthur Eddington and J. B. S. Haldane, that the universe is "not only stranger than we imagine, it is stranger than we can imagine," has been taken to heart by many physicists. For example, Richard Feynman described his own fascination with the counterintuitive aspects of quantum theory thus:

> "The peculiar answers that we get from calculating probabilities in this manner match perfectly the results of experiment. I'm rather delighted that we must resort to such peculiar rules and strange reasoning in order to understand Nature, and I enjoy telling people about it. There are no "wheels and gears" beneath this analysis of

Nature; if you want to understand Her, this is what you have to take."[23,24]

The ideological aspect of this trend is apparent in the writings of David Deutsch. His ideas are exemplary of the tendency, absent a plausible ontological orientation, for the positivistic conceptions of quantum and relativity theory to morph into metaphysics. His preference for "parallel-universe" (in his terminology, *Multiverse*) interpretations of quantum phenomena over so called "hidden-variable" approaches is described in his book *The Fabric of Reality*. He begins by recounting a childhood memory:

"I remember being told, when I was a small child, that in ancient times it was still possible for a very learned person to know everything that was known. I was also told that nowadays so much is known that no one could conceivably learn more than a tiny fraction [of that knowledge...]" [25]

Further along he describes the realization that the existence and study of the physics of the Multiverse will lead to a:

"moment when the scope of our understanding begins to be fully universal. Up to now, all our understanding has been about some aspect of reality, untypical of the whole." [26]

His ideas build on a speculative concept called *The Omega-Point*. This idea, developed by Pierre Teilhard de Chardin, a Jesuit priest and philosopher, and physicist Frank Tipler, is a conjecture that existence as a whole is evolving toward a point of ultimate complexity, omniscience and omnipotence – i.e., God. This is the end point, and purpose, of the evolution of the physical universe (and, along with it, a fantastically evolved, human-sourced consciousness) or, in the view of Tipler and Deutsch, the Multiverse. Tipler describes it thus:

23 Feynman, Richard Phillips. QED : the strange theory of light and matter. Alix G. Mautner memorial lectures. Princeton, N.J.: Princeton University Press, 1985.
24 Capitalization of Nature and Her in the original.
25 Deutsch, David. The fabric of reality : the science of parallel universes-- and its implications. New York: Allen Lane, 1997. p. 1
26 Op Cit. p. 29

"He/She is not part of the physical universe of spacetime or matter. The Omega Point is the future c-boundary --- the future singularity --- which is not part of spacetime, but is instead the "limit" of spacetime..."[27]

Deutsch believes that the best theories currently available – in physics, epistemology, cosmology and computation – imply that this evolution is necessarily taking place:

"What is or is not an 'extrapolation' depends on which theory one starts with. If one starts with some vague but parochial concept of what is 'normal' about the possibilities of computation, a concept uninformed by the best available explanations in that subject, then one will regard any application of the theory outside familiar circumstances as 'unjustified extrapolation'. But if one starts with explanations from the best available fundamental theory, then one will consider the very idea that some nebulous 'normalcy' holds in extreme situations to be an unjustified extrapolation. To understand our best theories, we must take them seriously as explanations of reality, and not regard them as mere summaries of existing observations."[28]

And, of course, the implications of these theories – i.e., precisely those that must be taken to be accurate reflections of reality *per se* – happen to conform to Deutsch's personal bias:

"...it seems clear to me that the present trend in our overall understanding of reality is just as I, as a child, hoped it would be."[29]

This sort of ideological motivation is also well exemplified in the popular writings of the physicist Brian Greene, who has argued that the conceptions of theoretical physics have a significance that far transcends their pragmatic or heuristic value. Quoting from *The Myth of Sisyphus*, Greene writes about his adolescent reaction to Camus' off-hand dismissal of the importance of ontological questions:

27 http://www.math.tulane.edu/~tipler/physicist.html
28 Op. Cit.- Deutsch does not put quotation marks about the words *explanations of reality* but does place them around *extrapolation* and *unjustified extrapolation*.
29 Op. Cit. p 366

"There is but one truly philosophical problem, and that is suicide," the text began. I winced. "Whether or not the world has three dimensions or the mind nine or twelve categories," it continued, "comes afterward", such questions, the text explained, were part of the game humanity played, but they deserved attention only after the one true issue had been settled."[30]

To which Greene follows up, later in the same chapter:

"If superstring theory is proven correct, we will be forced to accept that the reality we have known is but a delicate chiffon draped over a thick and richly textured cosmic fabric. Camus' declaration notwithstanding, determining the number of space dimensions – and, in particular, finding that there aren't just three – would provide far more than a scientifically interesting but ultimately inconsequential detail. The discovery of extra dimensions would show that the entirety of human experience had left us completely unaware of a basic and essential aspect of the universe. It would forcefully argue that even those features of the cosmos that we have thought to be readily accessible to human senses need not be."[31, 32]

Greene goes on to elaborate, quite optimistically, the transcendent significance of these theoretical conceptions, explicitly rejecting Camus' allegation that they are irrelevant to the great existential question. This is a stunning example of the tendency to objectify physical concepts. It also makes one wonder if the counselors who staff suicide hotlines are adequately versed in *String Theory*. More to the point, perhaps it is the desire to avoid suicide – i.e., to find meaning, satisfaction and even gainful employment via one's work – that motivates the ontological belief in certain ideas. That is, philosophical prejudice can be understood, at least to some extent, in terms of the personal gratification it affords, in stark repudiation of the inherent limitations of ideology. Not that intellectual satisfaction is a negative feature; quite the contrary – but it is an aspect of motivation

---

30 Greene, B. The fabric of the cosmos : space, time, and the texture of reality. New York: A.A. Knopf, 2004
31 Op. Cit..
32 No quotation marks about the words *proven* and *reality* or any qualification whatsoever about being "forced to accept" the nature of reality in the original.

that demands scrutiny, as scientists must be vigilant against counterproductive bias (in the *extra-personal* sense, that is, of scientific investigation; everyone is naturally free to believe what they will).

Ultimately, in any case, the scientific significance of theoretical constructs is decided pragmatically, on the basis of correspondence with experience, i.e., predictive efficacy. And so mundane theoretical constructions based on local causal action in three spatial dimensions are not of lesser philosophical significance than more "romantic" speculations based on radical or surreal conceptions. As with all models, local deterministic theories are necessarily incomplete, and are only meant to reflect limited *aspects* of experience on the basis of relations that can be discovered *within* experience (where *discovered* should be interpreted as "creatively read into") – they are not intended to reflect *"reality in all its glory,"* whatever, as previously discussed at length, that might be. Moreover, models that reduce to unambiguous causal connections in the minimum number of dimensions required to describe physical experience (three space plus one time) are decidedly simpler than models with topologically more complex relations. And, perhaps most significantly, topological connections in three spatial dimensions are quite different from those in spaces of more than three dimensions (they do not map in one-to-one correspondence between spaces of such differing dimensionality). Furthermore, topological connections (and corresponding mathematical transformations) are more constrained within continua than within discrete systems. So the *a priori* uniqueness and simplicity of local causal connections in three-space-plus-one-time continua bodes powerfully in their favor. And there is strong reason to believe that such constructions are the most appropriate *a posteriori*; for example, on the basis of computer models of cosmic evolution employing powerful numerical techniques (such as *Causal Dynamical Triangulations*[33]). While radical and surreal ideas might furnish serviceable results, unless simpler constructions are unable to do the same job in a convincing fashion they should not be superseded.

Another errant predilection is the above-noted tendency to transform favored hypotheses into inviolable axioms. As discussed, given the knowledge and technology available at the turn of the last century, it was sensible

33 Loll, R. "The emergence of spacetime or quantum gravity on your desktop." Classical and Quantum Gravity 25, no. 11 (6, 2008): 114006.

to conclude that any microscopic structure of space is unobservable and hence need not be included in natural theories of the day. But this pragmatic and reasonable position became transfigured into the dogmatic law that a dynamic underpinning of "*The Vacuum*" does not and cannot exist. In General Relativity, a similar concept led to the position that light paths must be attributed to "the geometry of spacetime as such" – as opposed to any underlying field or constitutive structure of space. In quantum mechanics, the *Uncertainty Principle* – first deduced from the premise that all micro-action is discrete, thereby limiting the resolution of observational detail via the Heisenberg relations – subsequently became the tenet that *no micro reality exists*, and cannot exist *in principle*.

Regardless of the subtlety and cleverness of arguments that support such doctrinaire proclamations, they should seem dubious to any lucid mind. One need merely recall Bohr's famous remonstrance to Einstein: "Stop telling God what to do." Scientific dogma, which is generated so promiscuously under the auspices of positivistic philosophy, is quite contrary to the spirit of empiricism. It is a symptom of ideological perverseness, which invokes and elevates to the status of *The Absolute* the very quintessence of arbitrary convention – *viz.*, the belief in fixed objective boundaries. And so the reasonable working assumption that reality is non-denumerably infinite is displaced by rigid demarcations – e.g., "The Planck Scale," below which the laws of physics break down; and "the beginning of time," before which the laws of physics did not exist, etc.

The notion that there are objective boundaries of scale is *a priori* much less sensible than the conjecture that the constants and laws of physics merely reflect, imperfectly, some sort of order, which can be represented by emergent patterns in theoretical models. That is, it seems most probable that the manifestation of natural constants in physical theory reflects a conjunction of two aspects of the circumstances, viz.: (1) the existence of objective processes that are independent of, and cannot be [fully] captured by, constructive models, and (2) the emergence of patterns or orders within constructive models – comprising "concepts" and "images" – which in turn reflect collective organizational principles (thus "cognitive" in nature) and that [weakly] represent "by analogy" the objective processes. On the other hand, it seems somewhat improbable that the constants and laws of physics reflect absolute, fixed bounds or fissures running through "reality" or "all of existence," and which somehow limit that existence in ways that happen to be convenient for human thought and understanding – "cutting it down to size" so to speak.

Moreover, irrespective of such plausibility considerations, there are salient pragmatic issues at stake. While limits of scale must necessarily arise in physics, the distinction between the two aforementioned ways of viewing such limits is not hair splitting, inasmuch as these two views have quite different operational meanings and the distinction between them is crucial. One implies that limits are absolute barriers to knowledge – e.g., making the Uncertainty Principle a fixed boundary – while the other implies that such limits are heuristic constructs, and should not be assumed to be absolute or final. In other words, the difference is between a closed and an open mind.

In this vein, it makes no sense to argue that a single physical theory must be capable of explaining the evolution of a unique cosmological configuration or "existence as a whole" (i.e., that "boundary conditions" are deducible from first principles), or that physical laws cannot be considered, at least in part, emergent properties of such configurations – in a sense, environmental. Moreover, in accordance with all of the previous considerations, it is epistemologically problematic to invoke boundaries as anything other than tentative guidelines for the study of limited aspects of experience.

This is the proper interpretation of *Empiricism* and *Positivism*, which should not be taken to mean: "That which cannot be observed does not exist, and has no place in theory." Rather, the operative intent should be to guard against problems that might stem from overreaching in any given go at theoretical description. Thus, efforts to find mathematical forms that constrain not only the dynamics of the vacuum but also its creation, evolution and boundary conditions – in order to justify fixed "ultimate building blocks" that possess unique characteristics connected with those constraints (as in the approach of *String Theory*) – would seem, necessarily, to produce a correspondingly too flexible "landscape" of possible solutions; the diversity of which is inversely related to the degree of detail and overall extent of information content attributed to the nature and behavior of those ultimate elements (as encoded in their definitions). This is the ultimate root of the invocation of the so-called *Anthropic Principle:* to justify unlikely boundary conditions – i.e., that the world is as it is because otherwise human beings would not be around to question its existence (or, more perniciously, that the existence of humans *now* is in some sense determinative, retroactively, of boundary conditions *then*). Needless to say, this move is one of the most unfortunate upshots of recent research programs in theoretical physics. The specious nature of the arguments is a manifest sign that artifacts of theory have become confused with fact.[34]

In the case of String Theory, not only is the deductive chain between premises and observables immense and complex, but the premises are convoluted as well – largely because the foundational principles of relativity and quantum theory are built-in, as opposed to being derived from them. In this regard, the foundational assumptions of String Theory seem similar to those of early Cartesian aether theories intended to account for all aspects of the world, before the electromagnetic attributes of light were even known. However, in the case of String Theory natural law is extraordinarily sensitive to intricate cosmological boundary conditions, which thus determine an enormous "landscape" of physical universes and laws. This is the antithesis of an emergentist approach, whereby the evolution of natural law appears to have an intrinsic aspect of inevitability.

The idea of a deterministically evolving universe is not inherently unsound. However, if it is to be sound, and not artificially constrained, it will likely be resistant to the imposition of absolute boundary conditions; to being captured – *in Toto* – via a single algorithmic formulation, especially if the elements of that formulation are complex and rigidly specified. This much is compatible with the hereinabove interpretation of Gödel Incompleteness. In light of the epistemological limitations arising from the general inconsistency and incompleteness of concepts, it seems rather dubious to complicate theory with the adoption of complex foundational elements, the properties of which are deemed to be absolute and invariable.

---

34 Arguments supportive of and defining the role of the *Anthropic Principle* in physics have been examined and deconstructed, in non-technical parlance suitable to a popular work, by Lee Smolin in: The Trouble With Physics The Rise of String Theory, the Fall of a Science, and What Comes Next. Paw Prints, 161-169, 2008.

The heuristic value of employing complex constructions – even if mathematically required and constrained, as deemed to be the case vis-à-vis the extra dimensions of String Theory – is overshadowed by the corollaries of that complexity. In String Theory, these corollaries include not only the difficulties attendant to extra dimensions but also the intrinsically conflicting properties of an unnecessarily intricate set of conceptual elements. "Wiring-in" Lorentz invariance and quantization-of-action makes no sense in a theory aimed at being fundamental. Just as General Relativity demands background independence – that the geometry of spacetime not be pre-determined – so too must Lorentz invariance and quantization-of-action be emergent properties in a meaningful unification of relativity and quantum theory. (Should the reader believe that this last statement is merely an expression of personal prejudice, without justification, the author respectfully requests that judgment be suspended until the end of chapter five.)

The epistemological problems that Bohr sought to remove from quantum theory – and those surrounding Newtonian mechanics and Maxwell's electrodynamics, which motivated Einstein's work on relativity – have not been extinguished from the foundations of physics. Rather, they have morphed into new, more intransigent forms, and continue to impede progress.

The essence of Bohr's position is that the mathematical formalism of quantum theory, understood in the context of his *Correspondence Principle*, is an effective algorithm for predicting the statistical results of experiments. Complementary but apparently conflicting qualities – such as particle and wave – and non-commutative quantities under the Heisenberg uncertainty relations, such as position and momentum – are not interpreted to have independent, objective meaning on the "micro" scale whereas, on the macroscopic scale, all the concepts of classical physics are to be thought of in the usual way.

If the Copenhagen message ended here – as nothing more than a pragmatic stipulation for the application of mathematics to micro-experiments – it would be more-or-less uncontroversial. But (by implication if not explicitly) the prescription for thinking is pushed much further, so that the Uncertainty Principle becomes more than a practical guide. It becomes, rather, a postulate regarding the nature of reality; an absolute proclamation to the effect that, if an objective reality corresponding to the "micro-physical" scale be taken to exist, it is not possible to represent that reality via visualizable concepts of events in space and time. Furthermore, aside from the ontological question of existence on such a scale, it is asserted that experimental outcome cannot be predicted on the basis of *any theory* that employs visualizable physical concepts. In other words, if one thinks in terms of "things" that can be pictured in time and space, and makes deductions/calculations based on the (visualizable) behaviors of such things, then the predicted outcomes cannot correspond with those of actual experiments.

This extraordinary conclusion is premised on the Uncertainty Principle. Just as thinking through the implications of naïve realism leads to the overturning of naïve realism and the conviction that things in the world cannot be as they appear – and just as considerations on the electrodynamics of moving bodies make it evident that motion relative to the aether cannot be measured – so do the uncertainty relations imply that any small-scale goings-on are beyond direct observation. Accordingly, it is logically permissible to doubt the existence of such goings-on. But the stipulation that *nothing* [physically describable] beyond the "Heisenberg Horizon," so-to-speak, can *in principle* exist, is ultimately neither a necessary conclusion nor even a reasonable scientific hypothesis, but rather an ideological belief, posited by fiat as a matter of philosophic bias. It is only demanded by a peculiar (albeit common) interpretation of the quantum mechanical formalism, and is in no way required by experience. It is similar to postulating that nothing can exist beyond the horizon of a black hole, even though something is seen to fall behind it, because there is no direct way to observe a thing after it crosses the horizon. Out of sight means not merely out of mind but *non-existent*, even if that were a contradiction (i.e., if other reasonable assumptions cannot be upheld in conjunction with it). As with most extreme positivist/empiricist positions it is, ironically, extremely anti-positivistic/anti-empirical: It puts forth a definite, unqualified statement about what should be a *strictly empirical* matter; and does so in the name of *Empiricism*.

It should be noted that Bohr's interpretation is not strictly adhered to by most physicists, who seem to share a less rigorously operationalist approach – in many ways closer to that of von Neumann, in which quantum states have a more or less objective, albeit ill-defined significance. In the latter view, the wave function is considered to represent some potential, developing state of indefinite nature – or rather, a superposition of such potentialities – which somehow "collapses" into a well-defined state when an "observation" occurs.

In contradistinction to the strictly orthodox formulation – which, while narrow, is at least moderately well-pruned epistemologically – this interpretation cannot even furnish an unambiguous definition of the word *observation*, and from it springs the infamous puzzles of quantum theory: *Schrödinger's Cat, Wigner's Friend, EPR* and *Delayed Choice Paradoxes, Counterfactuals*, etc. *Consciousness* of an experimental outcome, as opposed to the registration of a result by some macroscopic physical device, plays a decisive role in the process. And so Cartesian *mind-body dualism* is resurrected in a new and surprising form; right at the heart of physics. Moreover, it cannot be unambiguously decided *whose* consciousness is determinative – this is the paradox of *Wigner's Friend* – because each observer-object pair can be made the object of yet another observer, in an infinite chain that can apparently only end with God.

This is so confusing that many physicists consider the face-value meaning of the so-called *many-worlds* scenario, as usually associated with Hugh Everett's *relative state* proposal, to be a convincing and compelling alternative to the von Neumann interpretation. The purport of this "realistic" formulation is that with every quantum transaction – i.e., which in the sense of von Neumann signifies a collapse of the wave function – all components of the superposition-of-states (considered to have existed prior to collapse as a real condition) are objectively realized – by observers spanning (and in some interpretations *spawning*) an infinite array of parallel realities (in the original proposal, an infinite superposition of mental states, perhaps configurations of brain states, akin to a superposition of dead and alive Schrödinger cats). That this interpretation has gained currency among a not-insignificant number of physicists – in particular, the notion that every quantum "measurement" spawns an infinity of branching universes – is at least partly symptomatic of the subtlety of von Neumann's arguments, which made the flaw in his non-existence proof of hidden-variable theories difficult to detect. But when the defect was exposed, and viable hidden-variable theories proposed, the consensus of the physics community did not change – a remarkable sociological phenomenon that continues to this day.

While a few physicists, David Deutsch notable among them, explicitly reject *hidden-variable* approaches in favor of what may be called *hidden-worlds* – on openly ontological grounds – in such cases the philosophical prejudice is perfectly clear. The very assertion that the hidden-*worlds* proposal is *more parsimonious* [and symmetry-preserving] than hidden-*variables* (as Deutsch asserts) speaks for itself. For in place of *each variable* is substituted an *entire world*; each, in turn, containing a non-denumerably infinite set of [*necessarily hidden*] variables – thus saving theory from redundant metaphysical assumptions [and the dreaded loss of symmetry].

Everett's original proposal – based on the proposition that a quantum state corresponds to an infinite superposition of information-processor configurations (e.g., of brains) – was an effort to confer ontological plausibility on von Neumann's *quantum state* via a deterministic scheme. But as noted in chapter three, if the observer is brought into physical theory as an essential element, the subject-object dichotomy becomes quite confused. Again, in trivial illustration, if an observer is required to precipitate collapse of the wave function, then it is impossible *in principle* to construct a meaningful theory of the physical world – such as *cosmology*, and *physiology* – because observers are irreducible elements of the construction, and therefore only definable in circular fashion. In this sense alone, the usual interpretations of quantum theory are much too incomplete.

It is all too easy, without extra complications, to confuse representational constructs with their ostensible objects, thus multiplying the intrinsic complications and inevitable inconsistencies of theory beyond what is minimally unavoidable. Because physics furnishes the basis for all the natural sciences, from astronomy to biology, it is evident that such avoidable confusions and inconsistencies must be eliminated, or mitigated as much as possible.

Figure 4-1 In an infinite superposition of universes *everything* conceivable is occurring. All joking aside, given the supposition of such a *Multiverse*, the scenario depicted above is hardly farfetched – in fact, it is quite tame, inasmuch as *String Theory* seems to demand something on the order of 10^{500} *universes* in order for the existence of the one that is observed to be deemed plausible.

§ 4.11 Invalidation of Objections to "Locally Causal" Quantum Theories

The popularity of such fantastic interpretations of quantum theory as *many-worlds* is surely symptomatic of the absence of a satisfactory philosophy of science. And the reason that no satisfactory philosophy has emerged – or, perhaps more accurately, at least one crucial reason why no such philosophy has gained consensus status – is that confusion and bias have tended to block acceptance of ideas that are potentially fruitful. This supposition need not be accepted unchallenged, as vivid illustrations of how this suppression occurs are close to hand.

As alluded to above, while virtually all of quantum theory has been successfully formulated in terms of the de Broglie–Bohm deterministic framework, there exists a general recalcitrance among physicists to even learn about the extent to which this program has succeeded. Moreover, misunderstanding abounds regarding the potential that this program has to precipitate new, transcendent ideas. Thus, the chilling effect of this sort of suppression is more profound than might be immediately apparent. It prevents criticism of alternate ideas from being adequately addressed, and in this way has a self-catalyzing, reinforcing effect.

The primary reason that the deterministic program has not received sufficient recognition, or that there is such hesitancy to recognize what it has accomplished, is that to the extent the accomplishments of the program are known they appear, superficially, to undermine the reasons for having undertaken it in the first place. For example, the putatively "non-local" aspects of quantum phenomena stand out in strong relief, so that, as with the Lorentzian aether program, any success would seem pyrrhic. Moreover, it does not seem to most physicists that this approach opens any new vistas for the development of a quantum theory of gravity.

But the seemingly self-defeating aspects of the scheme are, indeed, only superficial. The alleged non-locality, almost universally accepted without question – together with the belief that the *pilot-wave* or *quantum potential* is not sufficient to physically determine the motion of a momentum-carrying entity – are entirely fallacious. Even David Bohm and John Bell succumbed to these misconceptions, while contributing, especially Bohm, so much to the progress of the program. But notwithstanding the fact that many champions of determinism believe non-locality to be a necessary appurtenance to any formulation of quantum theory, this is not the case. The roots of the widespread misconception regarding non-locality are revealed in the following section.

§ 4.11.1 Non-Locality not a Quantum Fact

It is generally believed that one of the permanent upshots of quantum theory is that the phenomena of nature are inherently correlated – i.e., *entangled*, in the technical, quantum mechanical sense of the term – in a manner that cannot be accounted for on the basis of local action transmitted at a finite velocity. In fact, belief in non-locality has been one of the primary upshots of efforts to interpret quantum theory on a deterministic basis. It is almost universally accepted that any model of quantum phenomena premised on causal action must involve non-local or instantaneous connections operative across any distance – similar, in this sense, to Newtonian gravitation – but unlike Newtonian gravitation, without attenuation of the efficacy of the causal connection with distance.

This is a fallacy. All of the apparently non-local properties of quantum theory, without exception, are misconstrued features of the formalism. They are not at all supported by empirical fact, despite widescale opinion to the contrary. The non-locality fallacy derives its persuasiveness largely from the following four sources of confusion.

(1) *The many-body Schrödinger equation.* The many-body equation describes an ensemble of objects in terms of their instantaneous positions in configuration space, the number of dimensions of which is equal to the total degrees of freedom of the ensemble, which is taken to be three real space directions per particle – a total of 3N dimensions for an ensemble of N particles. Because the description is in configuration space, the wave-field has values at a "multiple infinity" of points – that is to say, mathematically, the field fully occupies the configuration space, although only one point in the space is required to represent the simultaneous positions of all the particles. This means that a manifold of "3N times infinity" field points exist per particle. Again, whereas the actual ensemble of objects or particles occupies, at any given time, a single point in 3N dimensions, the field has a value at every point in 3N dimensions, at all times. This occupation of every point of configuration space, at every instant of time, leads to certain complications and contradictions, as demonstrated in the next sub-section.

On the other hand, it is an unambiguous and usually uncontroversial aspect of this representation (and an inherent feature of the Hamiltonian formalism on which Schrödinger based his approach) that particles represented by a single point in configuration space are, necessarily, instantaneously connected – like the parts of a perfectly rigid object or ideal machine (i.e., all N particles share the same time coordinate). Thus, the motions of the particles are simultaneously determined by the 3N dimensional wave equation, in such a manner that the particle motions cannot be fully accounted for on the basis of the mutual relations between any set of only two particles, extrapolated pair-wise to the whole. So, for example, it is not possible, in the general case, to calculate the force on a given particle by superposing on it, linearly, the force from every other object considered independently, as per the simple vector treatment applicable to classical electromagnetic interactions. Rather, in the general case, in strict accordance with the formalism, predictions can be made only on the basis of a law that describes the overall, multi-dimensional set of relations comprising the entire configuration of particles and waves at any given instant of time. This aspect of the formalism, which is examined in the next section, comes to the fore particularly clearly in pilot wave or quantum potential models of quantum mechanics.

(2) *Fourier Analysis.* The Schrödinger equation evolved from the Hamiltonian analogy between wave and particle incarnations of the action principle, but the mathematical handling of waves, in general, is steeped in the methods of Fourier analysis. And the mathematical apparatus of the latter is just as inherently non-local as that of the former.

The embryonic idea of the Hamiltonian dynamics of multi-component bodies involves the notion of fixed constraints among the components; i.e., the concept of the perfectly rigid, or perfectly rigidly connected, multi-component object. This is a limit ideal for quasi-rigid bodies, taking the force propagation time between components to be so negligible as to be effectively zero. Indeed, it was based on this ideal that Heinrich Hertz sought to re-establish Newtonian mechanics on a geometrical basis, by eliminating entirely the idea of force and replacing it with that of fixed connections; again, in the sense of the perfectly rigid body (this was pre-relativity, but not by many years). Of course, in general, the geometry of such constructions must be both non-Euclidean and greater than three-dimensional.

Fourier analysis is not only non-local but *a-temporal* as well. Fourier discovered that a large class of functions can be represented as an infinite sum of sine and cosine functions. For example, via his method, the analysis or construction of a wave packet involves an infinite superposition of plane waves, each infinitely extended in space *and* time. At those places and times where the packet appears, there is a constructive interference of every infinitely extended wave. In order for a packet (which is a pattern of constructive interference) not to repeat – i.e., for the group to be singular – the number of constituent waves must be infinite. So a packet is represented as occupying a finite portion of space, at a given moment of time, by the appropriate reinforcement and cancellation – i.e., constructive and destructive interference at those appropriate points of space and moments in time where the parts of the packet either should or should not exist, respectively – of an infinite, linear superposition of waves. Accordingly, in order to represent motion of the packet, the composition of the underlying waves must change as a function of time, in such a way that the points of constructive interference comprising the packet occupies an appropriately changing region of space with time. However, while the infinite set of plane waves that comprise the packet change, they are yet taken to exist continuously, at all places and at all times. Thus, by integrating the Fourier formalism into quantum mechanics, the non-local and a-temporal aspects of the underlying mathematics become (tacitly) integrated as well.

(3) *The Bell Inequalities.* While the Hamiltonian and Fourier origins of Schrödinger's wave mechanics are inherently non-local, it does not follow that the phenomena are. More to the point, it does not follow that only a theory based on non-local and/or a-temporal mathematical methods can account for the observed phenomena. In this regard, Bell's theorem has had a striking influence on the evolution of physics over the past 45 years, helping to convince many died-in-the-wool "realists" that locally causal theories are not viable.

The Bell Inequalities, and the experiments that have been conducted to test them, are widely believed to empirically confirm the existence of non-local connections – as well as the belief that such connections are likely operative over unlimited space-like intervals. However, this belief is incorrect, and the core "theorem" itself is flawed and invalid. While experiments performed on so-called entangled systems are usually interpreted to confirm the predictions of quantum mechanics, a proper understanding of these experiments and the Bell inequalities does not demand the invocation of non-local action. Opinions to the contrary arise from a major oversight in Bell's reasoning.

As with most fallacies of such stature and longevity, this one is based on a subtle, tacit assumption, the implications of which find their way into theory absent awareness of the tautological connection between the assumption and conclusions drawn from it. As a consequence, the relevance of the so-called Bell Inequalities to the issue of non-locality is entirely illusory. Unfortunately, to paraphrase Einstein in a different context, *the illusion is a stubborn one.*

(4) *The Quantum Potential versus Momentum.* As mentioned above, David Bohm felt that the concept of the quantum potential could not be understood in a classical dynamical sense, as representing a "real force," one that literally pushes or pulls the particle whose motion it guides. Rather, he believed that this new potential should be understood in terms of a concept he called *"Active Information."*

Bohm proposed that the wave field conveys information (in the sense that a radar signal provides information to a missile guidance system) about the conditions of the space around the particle, and the direction that it must take, but that the particle moves, for the most part, under its own power. He came to this belief by virtue of the fact that the influence of the quantum potential does not appear to diminish with distance, and because he felt that the mere form of the field could not convey sufficient momentum to jostle and direct a massive particle in the classical dynamic sense.

In addition, there is a long-standing misunderstanding regarding the relationship between the structure of the quantum probability amplitude distribution/field (comprising the quantum potential or pilot wave in the de Broglie-Bohm interpretation) and the particle motion it ostensibly guides. This has to do with the belief that particles should tend to move towards the nodes, and away from the antinodes of the field; if, indeed, their motion is shaped by the physical dynamics of the field.

This belief is premised on an analogy with pressure waves, with respect to which nodes correspond to regions of low pressure and antinodes regions of high pressure. However, this analogy is inappropriate to the quantum case. Moreover, the view herein to be developed draws a quite different picture, the details of which are revealed in the next chapter.

These several objections to local causal theories – representation in configuration space and the Hamiltonian framework, Fourier analysis, Bell's Inequalities and the issue of the Quantum Potential versus Momentum – are addressed in more detail in the following four subsections: 4.11.2 through 4.11.5.

§ 4.11.2 Configuration Space and Non-Locality

In an unpublished 1927 paper entitled: *Does Schrödinger's Wave Mechanics Determine the Motion of a System Completely or Only in the Sense of Statistics?*, Einstein, elaborating on this theme – i.e., attempting to interpret Schrödinger's configuration-space wave along the lines of Hertz's geometrical reconstruction of mechanics – unwittingly uncovered the tacitly imposed, instantaneous connections that are a consequence of the formalism. Using methods of tensor analysis, Einstein's paper reveals unique, definite trajectories corresponding to each of N particles in a 3N-dimensional configuration space – demonstrating that the Schrödinger equation does in fact contain complete information about the state of each particle. The reason Einstein did not publish the paper is that he realized, several weeks after giving his first reading of it and just before it was to go to press, that the trajectories of the particles are indeed entangled, and he did not believe that this could be consistent with what he called "a general condition that must be placed upon a general law of motion of systems." (The same reason that, over twenty years later, Einstein did not take a liking to Bohm's revitalization of the de Broglie pilot wave theory.)

Interestingly, Schrödinger had previously developed, but also not published, a tensor-based approach to Hertz's mechanics, which might have been the indirect inspiration for Einstein's paper, inasmuch as Schrödinger had alluded to a metric formulation of his wave mechanics early on, in the second of his four original papers on the subject. The salient point is that entanglement is an automatic consequence of the mathematical backbone of this sort of dynamical treatment, in which space-like connections are an inherent, tacitly posited feature of the configuration.

Another key point is that superimposing a wave function on a 3N-dimensional space effectively turns the representation into a "3N-times-infinity" configuration, as noted above. This is the source of the (seemingly extra) information that leads to the complete determination of the trajectories of every particle – i.e., which ties the boundary conditions to the paths and inter-relations of the particles. The information comes from the continuous wave field superimposed on the configuration space, and determines the particle motions.

However, there is an inherent problem of interpretation with the mathematical description of such a field in configuration space. To clearly see this, consider a simplified configuration space of two dimensions, which represents the motion of two particles, each in one-dimensional space (i.e., a line – see figure 4-2).

The position of a particle is given, at any moment in time, by one of the two coordinates; each of which represents the momentary position of an individual particle in its respective one-dimensional space (the coordinate system is identical to the ordinary two-dimensional representation of a single particle in the plane). In order to describe a field in a configuration space of two dimensions, an infinite set of points is required, to allow for the specification of the value of the field at every point in that space. And it is the wave field that determines the motion of the particles – by the probability-density gradient in the orthodox interpretation, or the corresponding gradient of the quantum potential in the de Broglie-Bohm causal model.

But because of the symmetries of configuration space, the values of the multi-dimensional field are highly constrained, as shown in the diagram. Because the field value at a single point in the configuration space is projected onto two coordinate axes, it must be constant along the entire length of each respective projection – i.e., the two straight lines through the point

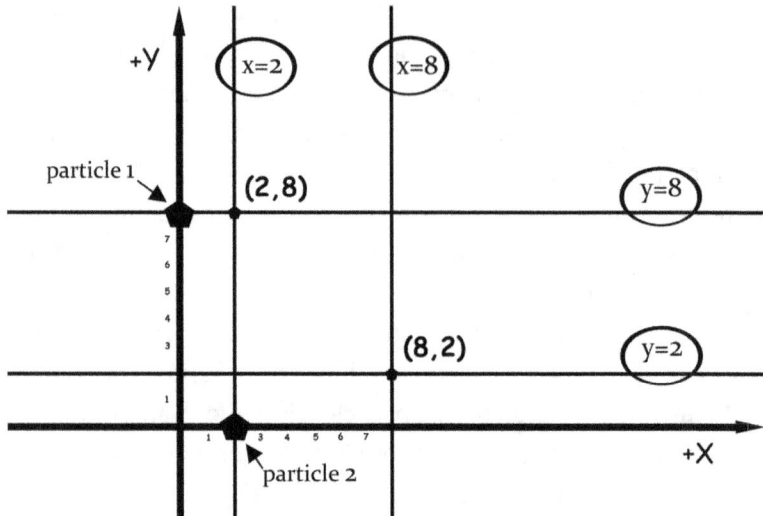

Figure 4-2 In order for the field to have unambiguous values at the particle positions – here at 2 and 8 on the *x* and *y* axes, respectively – those values must be constant at all points along the lines *x*=2 and *y*=8. But this of course implies that the field values must be the same everywhere in configuration space, because the line *x*=2 intersects the line *y*=8, which in turn intersects every other line perpendicular to the *x* axis.

in configuration space that are perpendicular to and intersect the x and y axes at the points where the particles are situated.

Accordingly, the field value associated with a given point in the configuration space, say (x_i, y_j), cannot be unique, because it must be the same at every value of y where x has the value x_i. But this in turn means that for every value of the x-coordinate the field must also have a single, unique value, because every projection of the y-axis crosses a projection of the x-axis (again, see the figure and its caption). To avoid this, one could stipulate that all of the values of the field other than at the point that corresponds to the positions of the particles are meaningless, and that only the value at the point in multi-dimensional space that represents the momentary configuration of the system is valid, which of course negates the premise of a field in configuration space. Therefore, to say that the value of the field at each point in the configuration space is a function of both of the particles' positions is to obscure the fact that the field in configuration space is everywhere the same at any given time, with unique, moment-to-moment values corresponding only to those places where particles are actually situated.

Only particle trajectories in configuration space, which are uniquely determined by Schrödinger's equation, are directly resolvable to [individual trajectories in] real space. In fact, this is the only aspect of the model that has unambiguous meaning in real space. There cannot exist unique field values at every point in configuration space because such points do not resolve in one-to-one correspondence with any points in real space. Only points along actual trajectories have meaning in this sense, and it is the correlation of the changes in particle positions with the changes in field values – along actual trajectories – that can be resolved to three-space vector components. Only these correlations have a direct meaning in three-space. Indeed, these are what might be called the "over-determined" trajectories that Einstein discovered and dismissed in 1927, and which Bohm later found to arise under his quantum potential treatment.

Of course, the interpretation of this rather messy circumstance under the protocols of quantum mechanics is that the representation in configuration space is not one of actual values but rather of potential values (probability amplitudes) and that it is only a *superposition*, i.e., *a function of those values* along each projection axis that has meaning in the real space. Therefore, the field can have a different amplitude at each point in configuration space because it is only a superposition (function) of those various amplitudes that is "projected" on each particle location, not the single-valued amplitude at the unique point in configuration space that corresponds to the particle positions.

But regardless of interpretation, the formalism ensures that the field values at all points are simultaneously determined. Thus Schrödinger, by extending to configuration spaces of arbitrary dimension the method that he conceived in three dimensions (for the modeling of the hydrogen atom), swept into his formalism the premise of the rigidly connected system – that is, a multi-component entity with instantaneous causal bridges between the parts and simultaneously determined, mutually constrained motions.

Here is a salient quote from Schrödinger, taken from a December 1926 article in *The Physical Review*:

"Now how are these conceptions to be generalized to the case of more than one, say of N, electrons? Here Heisenberg's formal theory has proved most valuable. It tells us though less by physical reasoning than by its compact formal structure that Eq. (29) giving a rectangular component of total electric moment has to be maintained with the only differences that (1) the integrals are 3N-fold instead of three fold, extending over the whole coordinate space; (2) z has to be replaced by the sum $\Sigma e_i z_i$ i.e., by the z-component of the total electrical moment which the point-charge model would have in the configuration $(x_1 y_1 z_1; x_2 y_2 z_2; \ldots x_N y_N z_N)$ that relates to the element $dx_1 \ldots dz_N$ of the integration.

"But this amounts to making the following hypothesis as to the physical meaning of ψ which of course reduces to our former hypothesis in the case of one electron only: the real continuous partition of the charge is a sort of mean of the continuous multitude of all possible configurations of the corresponding point-charge model, the mean being taken with the quantity $\psi \psi^*$ as a sort of weight-function in the configuration space.[35]

Note Schrödinger's reference to Heisenberg's matrix formulation. He explicitly expresses the fact that purely formal considerations *based on Heisenberg's extant theory* (and not physical intuition) suggest the generalization to configuration space, which also, of course, is suggested by the Hamiltonian formalism itself. In a lecture on the development of his wave mechanics, Schrödinger describes this explicitly. After explaining the connection between what he calls the "Hamiltonian-Maupertuis" principle and the Fermat principle for wave propagation – i.e., that action is stationary for both the path of a ray of light and the motion of a mechanical particle – he shows that the general expression for the kinetic energy, T, is:

" ... not of the simple form $\frac{m}{2}(\frac{ds}{dt})^2$ but

$$2T = \sum_l \sum_k b_{lk} \dot{q}_l \dot{q}_k,, \quad \ldots \ldots \ldots \quad (41)$$

where the b_{lk}'s are functions of the generalized coordinates q_l. We now define a line-element ds in the generalized q-space by

$$2T = \sum_l \sum_k b_{lk} \dot{q}_l \dot{q}_k = (\frac{ds}{dt})^2 ,$$

35 E. Schrödinger, Physical Review, **28/6, 1049 (1926)**

or $ds^2 = \sum_l \sum_k b_{lk} dq_l dq_k$ . . . . (42)

"The generalized non-Euclidean geometry, which is defined by the latter formula, is exactly the one which Heinrich Hertz used in his famous mechanics and which allowed him to treat the motion of an arbitrary system formally as the motion of a single mass-point (in a non-Euclidean, many-dimensional space). Introducing this geometry here, we easily see that all the considerations of the first lecture which led us to the fundamental wave equation may be transferred, even with a slight formal simplification, viz. that we have to put m = 1. ..."[36]

In the case of three-dimensional systems, Schrödinger's treatment can be interpreted directly in terms of visualizable vibrations of a sphere in three-dimensional space. And in the case of relatively simple systems, such as the hydrogen atom and a select set of others, his framework provides an accurate [non-relativistic] description of the circumstances. On the other hand, in the case of more complex systems in configuration space, analysis is generally so difficult that approximation methods must be employed. Therefore, it is not possible to say that the configuration space formalism has been experimentally confirmed; nor is it confirmable – not simply because of the need for approximation schemes to deal with multi-component phenomena, but because Schrödinger's treatment is explicitly non-relativistic. And so the errors of calculation that might follow from the assumption of instantaneous connections among the components can be washed out and fall below the radar.

More to the point, such representations in configuration space do not elude description in three dimensions. Indeed, actual approximation schemes used for calculation, such as the Hartree equations, treat linear equations in configuration space as sets of coupled nonlinear equations among points in individual particle space. In other words, the many-body problem in configuration space can be treated as a collection of individual entities in ordinary three dimensional space and, moreover, it is by such approaches that the configuration-space model is deemed to be "confirmed."

36 "Four Lectures on Wave Mechanics, delivered at the Royal Institution, London on 5th, 7th, 12th and 14th March, 1928 | London, Glasgow: Blackie and Son. [1928]. Third Lecture, 27-42

§ 4.11.3   Fourier Analysis and Non-Locality

David Bohm and his collaborator B. J. Hiley have called attention to a point of similarity between Einstein's approach to the description of material particles under his unified field program and the various non-local aspects of particles under quantum mechanics.[37] In particular, Bohm describes the Einsteinian point particle as a *Soliton* – i.e., a small region of space in which field values are particularly intense and stable, surrounded by exponentially decreasing values. Figure 4-3 shows two linear wavepackets resulting from a superposition of five component waves – technically not solitons, but the essential idea is similar (note that in order for the superposition to result in only two packets – i.e., in order for the pattern to be non-repetitive – there must be in an infinite series of appropriate component waves, rather than only the five pictured here).

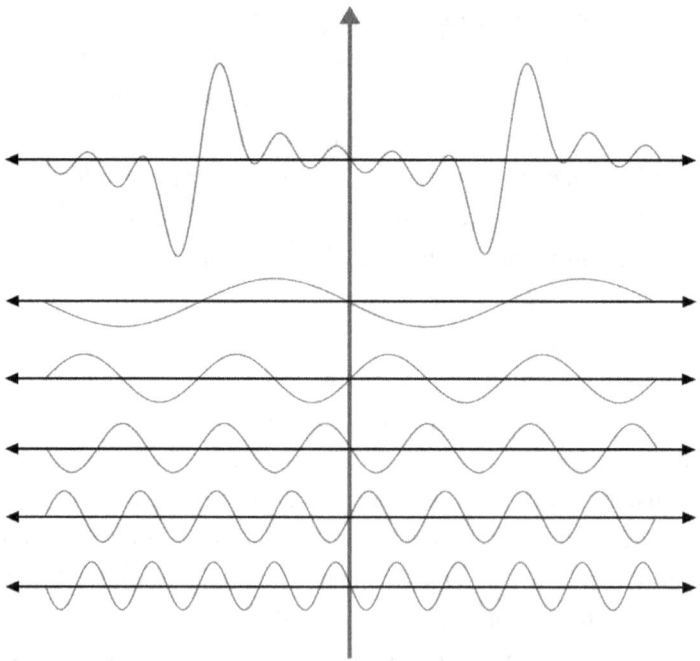

Figure 4-3   The two pulses along the top axis are a simple, linear superposition of the five wave components shown below them.

37 Bohm, D., and B. J. Hiley. "Nonlocality in quantum theory understood in terms of Einstein's nonlinear field approach." Foundations of Physics 11, no. 7-8 (8, 1981): 529-546.

It was known to Einstein that stable, pulse-like wave phenomena could be accounted for on the basis of nonlinear wave equations, although the tools for dealing with such equations was much less adequate in Einstein's time than today. Bohm's primary point was that a wave representation of particles is intrinsically non-local, because there is no sharp boundary at which the particle ends and the extended field begins. And so where there are two such wave-groups, as in figure 4-3, there is no clear demarcation between them – in the sense of a Fourier treatment, they are both described by the same set of overlapping waves.

In broad strokes, the representation of a Soliton via a non-linear construction of waves is not altogether different from the representation of a wave packet via a linear superposition of plane waves. In both cases, an extended wave field, which exists simultaneously everywhere in space and all throughout time, interferes to produce a phenomenon that is relatively localized in space and time. But Einstein, as is well known, did not believe in non-locality, and it was his purpose in attempting to represent local phenomena via the field to circumvent this aspect of quantum theory.

Was it naïve of Einstein to think that such a wave construction could be physically localized? Not at all. It is merely that the mathematical formalism has non-local properties. Consider again the two-particle construction of figure 4-3. Both entities are represented by the same sum of plane waves; they interfere constructively and destructively at different places in space. In order to represent a change in position of one of these wave pulses, it is necessary to change the waves that constitute both of them. In this sense, they are intrinsically connected, instantaneously, in a way that defies description in physical terms.

This intrinsic entanglement is represented in figure 4-4[38]. Signal A, ostensibly comprised of an infinite sum of Fourier components (the pictured waveform is actually the sum of a small number of waves), has zero amplitude prior to time t, indicated by the amplitude axis. If one of the components of the signal, C, is removed from the group, it will no longer have zero amplitude before t. In accordance with the formalism, in order to compensate for the removal of C there must be a shift in the phase of all other components. The requirement for this phase shift reflects the *disper-*

---

38 Roughly similar but not identical to a diagram in J Toll's paper cited below

sion relation, as there is a formal equivalence between the dispersion relation and strict causality.[39]

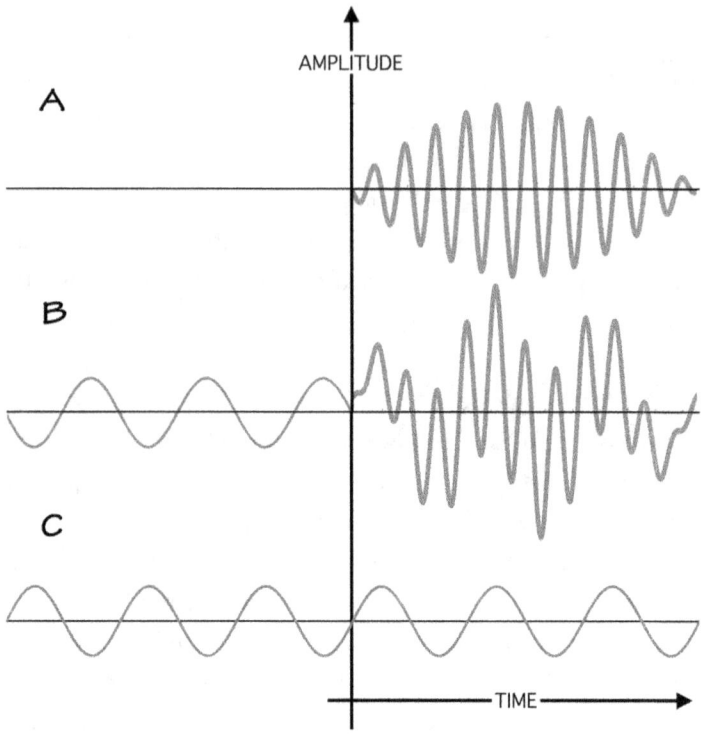

Figure 4-4   Wave packet A is a finite pulse, which has a definite amplitude at every moment in time – here zero at all times prior to that indicated by the amplitude axis. However, under the Fourier formalism, a singularity is treated as a group consisting of an infinite sum of waves, each extended infinitely in time. Wave C represents one component of group A, which, when removed, results in waveform B.

Because of this aspect of quantum theory, entanglement appears to be intrinsic to quantum phenomena. But again, this is an artifact of the formalism. It is certainly possible to find alternate mathematical methods for the description of waves; which do not demand infinitely extended constructions in space and time. Indeed, such methods are routinely used to compress digital images in a wide range of software applications – *viz.*, by the treatment known as *Wavelet Analysis*.

---

39 apparently first noted in: Toll, John. "Causality and the Dispersion Relation: Logical Foundations." Physical Review 104, no. 6 (12, 1956): 1760-1770.

This is the salient point with respect to all of these non-local aspects of quantum mechanics; i.e., they are artifacts of the mathematical formalism – the empirical evidence does not demand them. In the case of the Schrödinger equation, this is obvious on the basis of one simple consideration: It is not merely "inapplicable" to the relativistic case, but rather is entirely devoid of meaning in a relativistic setting. Therefore, it can only be an approximation.

It also cannot be used to calculate anything directly except in the simplest circumstances. If, in the many body case, instantaneous connections could be replaced with connections limited by a finite velocity – say the velocity of light – then, in all settings that can be experimentally addressed, any changes introduced by the [ostensible] velocity lag could not be distinguished against the overall backdrop of errors that must creep into the calculations via the use of approximation techniques (the belief that the Bell Inequalities belie this statement is false, as will be demonstrated below). Because the non-local aspects of the many-body scenario make a non-measurable contribution to the situation, there currently exists no empirical basis upon which it is possible to prove quantum phenomena to be non-local (again, contrary to the usual understanding of the significance of the Bell Inequalities). Moreover, there is no compelling theoretical reason to believe that non-local phenomena exist.

§ 4.11.4 Bell's Inequalities Redacted

Many if not most physicists seem to believe that the Bell Inequalities, and the apparent confirmation of their violation by experiment, establish non-locality beyond the reach of any reasonable doubt. This misunderstanding is eerily similar to the one that prevailed for so long vis-à-vis von Neumann's theorem (falsely demonstrating "hidden variables" to be impossible), which continued to be almost universally accepted as valid even after the publication of Bohm's existence proof to the contrary (and both his and John Bell's demonstration of the flaw in von Neumann's reasoning). Ironically, a similar scenario is playing out with respect to the arguments developed by Bell, who was inspired by Bohm's work and, as noted, published his own deconstruction of von Neumann's theorem. Bell was skeptical about such impossibility proofs generally – a skepticism that was well founded, and which he could have heeded a bit more carefully.

Bell's theorem (or, perhaps more aptly, collection of demonstrations) is indirectly concerned with the so-called *EPR Paradox*, attributed to a thought experiment proposed by Albert Einstein and presented in a 1935 paper, which Einstein published in collaboration with his colleagues Boris Podolsky and Nathan Rosen, entitled "*Can Quantum-Mechanical Description of Physical Reality be Considered Complete?*" As the title suggests, the essay intends to demonstrate that quantum theory is incomplete with regard to the description of individual events and entities (as opposed to the statistical description of multiple experimental outcomes). It examines hypothetical circumstances configured in imagination to reveal the existence of information that – according to the orthodox interpretation of quantum mechanics – should not exist. The broad strokes of the argument, by now so well-known as to scarcely need elaboration, are as follows.

In accordance with the dictums of quantum mechanics, non-commuting variables – such as position and momentum – cannot be simultaneously determined with a joint precision exceeding that fixed by Planck's constant via the Uncertainty relations. By virtue of the prohibition against simultaneously measuring such non-commuting observables with indefinite precision, it would seem impossible to become cognizant of the value of more than one such observable at a time. Einstein sought to demonstrate that this feature of quantum mechanics merely illustrates its incompleteness.

In his thought experiment, he imagines two physical systems created in a single event, and thus defined by a single, non-factorable equation. In broad strokes, Einstein argues that determining one of the observables, of one half of the system – such as the position of one of two entangled particles – does not preclude determining the complementary observable of the other half – i.e., the momentum of the second particle.

Most modern renditions of the experiment obviate Einstein's full intention by invoking directions of spin rather than attributes that are truly non-commuting under quantum mechanics; i.e., strictly on the basis of Heisenberg uncertainty. One system decays into a pair of identical particles, which head off in opposite directions with opposite spins, each with a different but correlated orientation. In accordance with the standard interpretation of quantum mechanics, a particle known to have "spin up" with respect to one [say x] axis of a reference system does not have a predetermined spin direction along the other [say y and z] axes. But because the system is correlated, a determination of the value of one of the observables, of one of the particles, should imply the value of the counterpart observable of the other particle. So if one particle is found to have "spin up" along the x-axis, then the counterpart particle must have "spin down" along that axis, because the total angular momentum of the two must sum to zero in each such orientation. Thus, it seems possible to know the spin of one of the particles without directly measuring it. In addition, it seems that it also should be possible to measure another of the observables of the yet un-measured particle, say spin along the y-axis – and thereby indirectly reveal the y-spin of the first measured particle. By this means, the outcome of the two measurements can evidently furnish both observables for both particles. Moreover, because, in accordance with theory, all of these measured attributes are *non-predetermined*, it would seem that the act of measuring one observable, of one particle, instantaneously causes the complementary observable of its [space-like separated] partner to assume the requisite value – the property that Einstein referred to as "spooky action at-a-distance."

Although this scenario – just as spin-offs, so to speak, based on electromagnetic polarization – does not directly contradict the quantum mechanics prohibition against the simultaneous knowledge of more than one of two Heisenberg non-commuting variables, it would yet seem to capture the essence of entanglement because, regardless of the distance between the particles, the two observables are deemed to be more strongly correlated than can be accounted for by past interaction, and/or present interaction based on signals traveling at light speed. (Again, in the original EPR argument, the case is made that it is possible to know the momentum of one particle by measuring the momentum of a second, and to know the position of the second by measuring the position of the first, thus circumventing Heisenberg's restriction and revealing that quantum mechanics does not account for all the knowable facts.)

Bohr responded to the original EPR argument in a manner that many physicists took to be decisive, although it actually left the possibilities wide open. In 1964, almost thirty years later, John Bell attempted to establish the correctness of the EPR position. Bell was sympathetic to Einstein's viewpoint and, as noted, quite unsympathetic to "impossibility proofs" such as von Neumann's. He constructed a simple hypothetical scenario and a demonstration based upon it, by which he believed he could elucidate and settle the matter beyond a reasonable doubt. To his surprise, and contrary to his personal position, he felt compelled by his own analysis to conclude that local, hidden variables are not possible in quantum mechanics – i.e., that any deterministic scheme must involve non-local action. And to this day – just as most physicists accepted Bohr's counterargument to EPR – most agree with Bell, even though he was wrong.

Bell's fundamental error – and he made more than one – was to assume that the restriction of local determinism necessitated the mathematical treatment of correlated events as statistically independent, even though probability theory is quite clear with regard to the fact that conditional probabilities cannot be treated as independent; that is, as the product of two or more independent probabilities. This mathematical truism has nothing to do with locality or non-locality, but only with the fact that conditional probabilities are different from those that are independent, and cannot be calculated as such, regardless of whether or not there is any question of a causal connection involved. And so, because the quantum expectation for the correlations under consideration are determined by the cosine of an angle (the dot product of two vectors), which cannot be factored, the Bell inequalities must be violated.

Although Bell made a belated effort to address this issue in response to criticism, he evidently was unable to overcome his initial confusion. For example, in a paper entitled: *"Bertlmann's socks and the nature of reality,"* he writes:

"To explain the 'inexplicable' we explain 'explicable'. For example, the statistics of heart attacks in Lille and Lyons show strong correlations. The probability of M cases in Lyons and N in Lille, on a randomly chosen day, does not separate:

$$P(M,N) \neq P1(M)P2(N)$$

"In fact when M is above average N also tends to be above average. You might shrug your shoulders and say 'coincidences happen all the time', or 'that's life'. Such an attitude is indeed sometimes advocated by otherwise serious people in the context of quantum philosophy. But outside that peculiar context, such an attitude would be dismissed as unscientific. The scientific attitude is that correlations cry out for explanation. And of course in the given example explanations are soon found. The weather is much the same in the two towns, and hot days are bad for heart attacks. The day of the week is exactly the same in the two towns, and Sundays are especially bad because of family quarrels and too much to eat. And so on. It seems reasonable to expect that if sufficiently many such causal factors can be identified and held fixed, the residual fluctuations will be independent, i.e.,

$$P(M,N|\alpha,\beta,\lambda) = P1(M|\alpha,\lambda)P2(N|\beta,\lambda) \dots\dots\dots \quad (10)$$

"Where α and β are temperatures in Lyons and Lille respectively, λ denotes any number of other variables that might be relevant, and $P(M,N|\alpha,\beta,\lambda)$ is the conditional probability of M cases in Lyons and N in Lille for given (α,β,λ). Note well that we already incorporate in (10) a hypothesis of 'local causality' or 'no action at a distance'. For we do not allow the first factor to depend on β, nor the second on α. That is, we do not admit the temperature in Lyons as a causal influence in Lille, and vice versa.[40]

"Note well" Bell's "residual" confusion here. He is unable to fully accept that, by virtue of the correlated information that underlies conditional probabilities, it is not possible to factor them into independent terms. He believes that "such causal factors can be identified and held fixed," by virtue of giving them a name, and the relations can thus be treated as independent. He fails to note the salient fact, necessarily true in the quantum case, that α and β are *always related*, by a constant factor, in every experi-

---

[40] Bell, J. S. "BERTLMANN'S SOCKS AND THE NATURE OF REALITY." Le Journal de Physique Colloques 42, no. C2 (3, 1981): C2-41-C2-62 (widely available in *Speakable and Unspeakable in Quantum Mechanics*, cited below).

ment: $\alpha = \beta -$ [a Constant] and $\beta = \alpha +$ [the same Constant] in each experimental run, so that the variables are (α) and ($\alpha + C$). More to the point, *it is not the statistics of single events that are being measured* (e.g., "temperature α when event x happens" or "temperature β when event y happens") *but rather statistical correlations exclusively,* which *irreducibly* depend on the relation between α and β (such as the average temperature of α and β).

Because of the subtlety of the arguments involved, it is easy to become confused. In this case, confusion arises from the fact that, under such interpretations of the quantum state as von Neumann's, it would seem as though "the collapse of the wave function" demands an instantaneous connection among correlated systems, and that it should therefore be possible to arrange for an experimental verification of such strong, extra-classical correlations. This is a theoretical issue that can, indeed, be discussed (and is below), but Bell's reasoning does not address it, because he makes an incorrect judgment regarding the difference between conditional and independent probabilities. As a consequence (and because spin- and polarity-based "EPR" experiments do not involve Heisenberg non-commutables), it is not non-classical correlations that are measured by experiments set up to test Bell's relations but rather completely classical ones, which can be accounted for as such (as they are below).

This circumstance is reminiscent of the so-called "Monty Hall Problem," which has become the subject of many psychological studies of cognitive dissonance. The *Monty Hall Problem* refers to a choice involving conditional probabilities, which are readily – and all too commonly – mistaken as being independent. Although it is not isomorphic to the problem at hand (and not even, strictly, to the scenario created by the game show host for whom it is named), the misunderstanding is yet edifying.

The basic idea is that a contestant on a game show is given the choice of one of three doors, behind one of which is a grand prize and behind the others nothing of value. Once a door is chosen it remains closed while the host, with full knowledge of the prize placement, opens one of the two remaining doors, displays a booby prize, and offers the contestant an opportunity to make a new choice. If the contestant has chosen the prize door, both remaining doors have booby prizes behind them, and the host can select one randomly. Of course, if the contestant has not selected the winning door, the host's choice is constrained, because only one of the remaining doors has a worthless prize behind it.

Again, after the host opens a door and reveals a booby prize, he forces the contestant to make a new decision, *viz.*, to either stick with the original selection or choose the remaining, unopened door. The puzzle is: *Should the contestant change doors?* The correct answer is *yes*, the contestant should indeed make the exchange, because this increases the odds of winning from one-in-three to two-in-three. However, most people have trouble understanding this, evidently because they do not assimilate the meaning of the host's constrained circumstances; that he must *always* offer the second choice. Moreover, they often refuse to accept that it is true even after the crux of the issue is explained – that the host's choice with respect to offering an alternative is limited by the player's original selection, even though the offer to change doors appears random.

A similar phenomenon manifests with respect to the Bell Inequalities.

Consider the following parlor trick. A magician produces a 360-degree circular protractor and gives it to a randomly selected volunteer from the audience. He asks the volunteer to turn away from both the audience and the magician, and break the protractor into two randomly sized pieces, which sizes are known only to the volunteer (except that neither piece can be smaller than 90 degrees; per the magician's instruction and thus known to all). The magician then asks the volunteer, still facing away from everybody, to further break the smaller of these pieces in two – randomly, this time with no restrictions on size.

The magician withdraws two opaque envelopes from his breast pocket and hands them to the volunteer, who is instructed to place the largest of the three pieces in one envelope, the smallest in another, pocket the remaining piece and seal the envelopes – all in secrecy (see figure 4-5). The volunteer is then asked to hand the sealed envelopes back to the magician, who produces two intact protractors, two electronic calculators, and two magic markers.

The first volunteer is seated and two more selected from the audience. Each is handed one of the sealed envelopes, an intact protractor, and a calculator for calculating trigonometric functions. The magician instructs each of the volunteers to open their envelope and (1) measure the angular span of the piece within, (2) calculate either the sine or the cosine of that angle, upon their free, secret choice, and finally (3) note the sign (plus or minus) of the result of the trigonometric calculation on their respective envelope. Then, without seeing what the volunteers have done or marked on their envelopes, the magician – in a stunning feat of telepathy – correctly predicts that the two envelopes have been marked with opposite signs – i.e., that one bears a plus sign and the other a minus. The kicker is, the trick is repeated one million times with perfect accuracy.

Figure 4-5 Magician and Protractors

Now, because it is impossible for the magician to predict which of the two envelopes will have a plus or minus sign, the individual probability for a given envelope to have either a plus or a minus sign is fifty percent. And yet the joint probability that one envelope must always bear a plus and the other a minus is one hundred percent. Again, this is because such a conditional probability is not a product of independent probabilities. The fact that each measurement is performed randomly and independently does not affect this fact – in the case of the magic trick, because the relationship between the two pieces, albeit also determined independently (by the first volunteer), remains fixed throughout the measurement procedures. This trick is analogous to the experiments that are conducted to "test" Bell's inequalities, which must be violated for the same reason.

In order to clarify the essentials of Bell's argument and better understand the logical flaw in his reasoning, it is helpful to reduce his scenario to one that is simpler still. In this respect, the so-called "cryptographic approach" introduced by Nick Herbert in 1973 and a somewhat related version introduced by N. David Mermin are helpful. The essential steps of Herbert's argument are here retraced, and Mermin's below. (Although Herbert, like Mermin, invoked his scenario to support the veracity of Bell's theorem, it is actually very well suited for the contrary purpose.)

A source of light emits two, oppositely polarized photons; simultaneously and in opposite directions. On either side of the source and in the paths of the photons – separated by a distance too great for a signal to travel during the time of the measurement processes – two polarizing receivers stand ready. Each is equipped with a filter that can be rotated through 360 degrees, which will thus either pass or stop the incoming photon, depending upon its orientation with respect to the latter's polarization. If a photon gets through, a hit (a positive (+) sign) is recorded, otherwise a miss (a minus (–) sign) is registered. When these polarizers point in opposite directions, the counters on both sides will register the same; both either a (+) or a (–), reflecting the fact that the photons are oppositely polarized.

In Herbert's version of this experiment, it is initially pointed out that when the two polarizers are parallel, the outcome of every run will be (+,+) or (-,-), because the photon's polarizations are correlated. He then notes that if one of the two polarizers is rotated 30 degrees from its original orientation, one out of every four pairs will be anti-correlated, following the quantum expectation that the number of hits will be reduced from four out of four to three out of four – i.e., the cosine, squared, of 30 degrees. Herbert then notes that if the original orientation of this polarizer is re-established and the other is rotated 30 degrees out of kilter, again a result of one out of four misses/anti-correlations will be obtained.

He then asks what the consequence should be of setting both polarizers out of alignment by 30 degrees, so that they are displaced 60 degrees with respect to each other, and notes that in a "classical" (read *uncorrelated*) situation, it should be expected that the result of both sides turning up one out of four misses should yield a joint result of roughly two out of four misses. This corresponds to Bell's proposition that the results of each of two, local/independent polarizer measurements should not depend in any way upon the orientation of the distant polarizer, and thus should be multiplied together to give the joint probability for the correlations. Of course, the cosine squared of 60 degrees gives a probability of three out of four misses, well beyond such a "classical" expectation.

Because of the manner in which it is couched, Herbert's argument is particularly helpful with regard to laying bare Bell's fundamental error. For clarity and completeness, the core of Herbert's argument is quoted directly.

Per one of John Clauser's experiments employing a mercury vapor light source, Herbert imagines each correlated pair of photons to be Blue in one direction and Green in the other, and the polarizers to be calcite crystals, which he labels Blue and Green respectively – with Green stationed on earth and Blue hundreds of light-years away on or about the star Betelgeuse.

"Now we are ready to demonstrate Bell's proof. Watch closely; this proof is so short that it goes by fast. Align the calcites at twelve o'clock. Observe that the messages are identical. Move the Green calcite by α degrees. Note that the messages are no longer the same but contain "errors" – one miss out of every four marks. Move the Green calcite back to twelve and these errors disappear; the messages are the same again. Whenever Green moves his calcite by α degrees in either direction, we see the messages differ by one character out of four. Moving the Green calcite back to twelve noon restores the identity of the two messages.

"The same thing happens on Betelgeuse. With both calcites set at twelve noon, messages are identical. When Blue moves her calcite by α degrees in either direction, we see the messages differ by one part in four. Moving the Blue calcite back to twelve noon restores the identity of the two messages.

"Everything described so far concerns the results of certain correlation experiments which can be verified in the laboratory. Now we make an assumption about what might actually be going on – a supposition which cannot be directly verified: the locality assumption, which is the core of Bell's proof.

"We assume that turning the Blue calcite can change only the Blue message; likewise turning the Green calcite can change only the Green message. This is Bell's famous locality assumption. It is identical to the assumption Einstein made in his EPR paradox: that Blue observer's acts cannot affect Green observer's results. The locality assumption – that Blue's acts don't change Green's code – seems entirely reasonable: how could an action on Betelgeuse change what's happening right now on Earth? However, as we shall see, this "reasonable" assumption leads immediately to an experimental prediction which is contrary to fact. Let's see what this locality assumption forces us to conclude about the outcome of possible experiments.

"With both calcites originally set at twelve noon, turn Blue calcite by α degrees, and at the same time turn Green calcite in the opposite direction by α degrees. Now the calcites are misaligned by 2α degrees. What is the error rate?

"Since turning Blue calcite α degrees puts one miss in the Blue sequence (for every four marks) and turning the Green calcite α degrees puts one miss in the Green sequence, we might naïvely guess that when we turn both calcites we will get exactly two misses per four marks. However, this guess ignores the possibility that a "Blue error" might fall on the same mark as a "Green error" – a coincidence which produces an apparent match and restores character identity. Taking into account the possibility of such "error-correcting overlaps," we revise our error estimate and predict that whenever the calcites are misaligned by 2α degrees, the error rate will be two misses – or less.

"This prediction is an example of a Bell inequality. This Bell inequality says: If the error rate at angle α is ¼, then the error rate at twice this angle cannot be greater than ²⁄₄.[41]

In this scenario, it is relatively easy to catch a hint as to the flaw in the reasoning, and why the anti-locality argument will fail. For if only one of the crystals, say Green, is first rotated by 30 degrees, resulting in a twenty-five percent error rate, and then is rotated an additional 30 degrees in the same direction, the expected quantum mechanical error rate of seventy-five percent will result. This is a manifestly "local" phenomenon, depending only upon the orientation of a single signal with respect to the one polarizer, which can be produced without making changes in the orientation of the second, counterpart polarization measurement. And yet the variation in error rate does not follow from the [misnamed] "classical" prediction that doubling the polarization angle will double the error rate.

Moreover, it is quite clear why it is so – and why it has nothing to do with what is happening on Betelgeuse. Obviously, if a measurement is based upon an angle, and follows a sine or cosine function of that angle, such a result cannot be factored into double that which is obtained at half the angle (see figure 4-6).

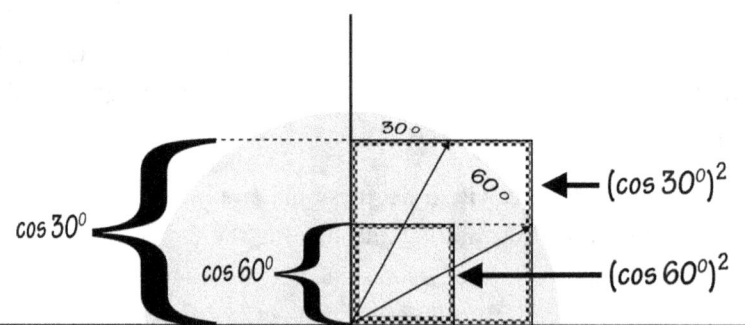

Figure 4-6   The cosine of the angle formed by each of the two pictured radial lines, here measured clockwise from the vertical axis, is the projection of the radial line (taken to be of unit length) on the vertical axis. The square of the cosine is just that – the area of the square bounded by the projection. The square of the cosine of 30 degrees is 3/4; for 60 degrees it is 1/4. If the areas of these squares represent the probabilities of an event occurring – say the likelihood that some object will pass through each of these areas, with a probability of 100% corresponding to the unit square, then it is easy to visualize the variation of the quantum statistics with the angle.

41 Herbert, Nick. Quantum reality : beyond the new physics. Garden City, N.Y.: Anchor Press/Double-day, 1985.

However, an important fact has been neglected. Namely, in practice one does not know the exact polarization of a single photon before it is measured. That is one of the misleading aspects of Herbert's argument. The only thing that can be determined – the only thing that is, in fact, determined via such experiments – is a relation between two measurements. Hence there is no difference between turning one polarizing filter by 2α and turning both by α, half that amount, and the argument that two, independent measurements are being made is very misleading. Only one, joint measurement of two devices is occurring, which can only be described by a conditional probability that accounts for the common, correlated origin of the photon pair.

While it is evident from this example why the quantum expectation value is different from the so-called "classical" value – because it follows from the cosine of the angle between the signal and the measurement vector – it is perhaps not as clear that violation of the Bell Inequalities by "space-like" separated polarizers does not involve non-locality. That is to say, there seems to be some confusion surrounding the expectation that arbitrary changes in the polarizer orientations will yet yield correlated results. The following demonstration, which follows directly from a scenario devised by N. David Mermin, is intended to further clarify this issue. Like the magic trick described above, it does not involve any actual magic.

Mermin's argument is as follows:

"What I have in mind is a simple gedanken demonstration. The apparatus comes in three pieces. Two of them (A and B) function as detectors. They are far apart from each other (in the analogous Aspect experiments over 10 meters apart). Each detector has a switch that can be set to one of three positions; each detector responds to an event by flashing either a red light or a green one. The third piece (C), midway between A and B, functions as a source.

"There are no connections between the pieces, no mechanical connections, no electromagnetic connections, nor any other known kinds of relevant connections... The detectors are thus incapable of signaling to each other or to the source via any known mechanism, and with the exception of the "particles" described below, the source has no way of signaling to the detectors. The demonstration proceeds as follows:

"The switch of each detector is independently and randomly set to one of its three positions, and a button is pushed on the source; a little after that, each detector flashes either red or green. The settings of the switches and the colors that flash are recorded, and then the whole thing is repeated over and over again. The data consist of a pair of numbers and a pair of colors for each run. A run, for example, in which A was set to 3, B was set to 2, A flashed red, and B flashed green, would be recorded as "32RG"...

"Because there are no built-in connections between the source C and the detectors A and B, the link between the pressing of the button and the flashing of the light on a detector can only be provided by the passage of something (which we shall call a "particle", though you can call it anything you like) between the source and that detector. This can easily be tested; for example, by putting a brick between the source and a detector. In subsequent runs, that detector will not flash. When the brick is removed, everything works as before.

"Typical data ... There are just two relevant features:

"I) If one examines only those runs in which the switches have the same setting[42]..., then one finds that the lights always flash the same colors.

"II) If one examines all runs, without any regard to how the switches are set..., then one finds that the pattern of flashing is completely random. In particular, half the time the lights flash the same colors, and half the time different colors.[43]

Now, while the second summary of the data corresponds more or less to the outcomes of actual experiments, the first, due to limits of accuracy – while yet perhaps an accurate expression of the expectations of the

---

42 Author's note: this will occur one third of the time
43 Mermin, N. David. "Is the Moon There When Nobody Looks? Reality and the Quantum Theory." Physics Today 38, no. 4 (1985): 38.

quantum *formalism* – is not isomorphic to the results of actual experiment. Rather, the salient point regarding this issue is that, whereas a purely random distribution of results would be a 50-50 correlation of matching versus non-matching outcomes (summary item II), the outcome of experiments indeed shows a very significant correlation, albeit not one-hundred percent (per summary item I).

For the greatest possible clarity and certainty, the following analysis of Mermin's argument is accompanied by a short algorithm, constructed in *Mathematica* (and listed verbatim in Appendix A, which anyone may run via a free download from the Internet[44]), which simulates these "Mermin-Bell" experiments. This simulation is based on a very simple model, which involves nothing more than the *Malus rule* for computing the occurrence of a hit or miss, depending upon the relative angles formed between the photon polarization and the measurement vector. (The *Malus rule* simply gives the square of the cosine of the polarization angle. For a computer simulation of a more typical type of *Bell-CHSH* experiment – i.e., a simulation of an experiment that has actually been performed in physical laboratories, see *Kracklauer*.[45])

In this "hidden variable" model, polarization is deemed to be real – i.e., to exist before measurement – and to be one hundred percent correlated (or *anti-correlated*) between photon pairs. The measurement process involves determining the orientation of each photon (source signal). The rules are local and independent, inasmuch as the result of each measurement depends only upon information that is carried by each photon and available to each measurement device, separately from, and without regard to, information carried by the counterpart photon, or available to the counterpart measurement device. *No sharing of information occurs between the measurement devices.*

Figure-box 1 shows the *Mathematica* program (text listing in *Appendix A*), followed by the result of a typical test run. The variables reflect a baseball metaphor. The left and right switch settings are given by the variables *abat* & *bbat* (*A* side and *B* side), set randomly in one of three positions, and the orientation of the photons are given by the variable *pitch*, as an

44 free notebook player: http://www.wolfram.com/products/player/
45 Kracklauer, A. F. "Bell's inequalities and EPR-B experiments: are they disjoint?." In FOUNDATIONS OF PROBABILITY AND PHYSICS - 3, edited by Andrei Khrennikov, 750:219-227. AIP, 2005. http://dx.doi.org/10.1063/1.1874573.

```
runs = 10000; (*set number of experimental runs [i.e., iterations of 'Do' loop]*)
AAHit = 0; (*set A hit register*)
ABHit = 0; (*set B hit register*)
HitHit = 0; (*set A-hit & B-hit counter*)
HitMis = 0; (*set A-hit & B-miss counter*)
MisHit = 0; (*set A-miss & B-hit counter*)
MisMis = 0; (*set A-miss & B-miss counter*)
aBat = 0; (*set A 'bat' position [switch-setting]*)
bBat = 0; (*set B 'bat' position [switch-setting]*)
aHit = 0; (*set A hit register*)
bHit = 0; (*set B hit register*)
bats = 0; (*set matching-switch counter*)
aMeasure = 0; (*set A measurement vector*)
bMeasure = 0; (*set B measurement vector*)
(**)
(**)
(*Run simulation [runs] times, then print results*)
Do[testMermin[a], {runs}]; Print["|| TOTAL GG&RR: ", N[((HitHit + MisMis)/runs)], " || Matching Switch Settings: ",
  N[bats/runs], " || GG&RR with same switch settings: ", N[(aHit + bHit)/(bats)]];
(**)
(**)
testMermin[a_] := {AAHit = 0; ABHit = 0; (*re-initialize counters*)
  (*set A and B bats [switch positions: 120 degrees apart]*)
  aBat = RandomReal[]; If[aBat < (1/3), aBat = 120]; If[(1/3) < aBat < (2/3), aBat = 0]; If[(2/3) < aBat < 1, aBat = 240];
  bBat = RandomReal[]; If[bBat < (1/3), bBat = 120]; If[(1/3) < bBat < (2/3), bBat = 0]; If[(2/3) < bBat < 1, bBat = 240];
  (*set signal orientation [pitch] in degrees*)
  pitch = RandomReal[] * 360;
  (*Measure signal relative to A/B measurement vectors [bats]*)
  (*i.e., determine the square of the cosine of the angle between pitch & bat*)
  aMeasure = Cos[(aBat - pitch) Degree]^(2); bMeasure = Cos[(bBat - (pitch - 180)) Degree]^(2);
  (* DETERMINE HIT OR MISS *)
  (* i.e., determine whether aMeasure/bMeasure ≥ .5... signal noise & errors are accounted for by
     introduction of a random real number between 0 & 1, average value of which is .5 *)
  If[aMeasure ≥ RandomReal[], AAHit = 1]; If[bMeasure ≥ RandomReal[], ABHit = 1];
  (*update registers and tabulate results*)
  If[AAHit == 1 && ABHit == 1, HitHit = HitHit + 1]; If[AAHit == 0 && ABHit == 0, MisMis = MisMis + 1];
  If[AAHit == 1 && ABHit == 1 && aBat == bBat, aHit = aHit + 1]; If[aBat == bBat, bats = bats + 1];
  If[AAHit == 0 && ABHit == 0 && aBat == bBat, bHit = bHit + 1]}
```

FIGURE-BOX 1: SEE APPENDIX A FOR TEXT LISTING
TOTAL GG & RR: *0.5* || Matching Switch Settings: *0.333239*
GG & RR with same switch settings: *0.749912*

angle between 0 and 360 degrees. On each experimental run, the program generates a random *pitch* orientation, assumed to point "up" in either the left or right direction, and "down" in the other. So if the right photon's orientation is 270 degrees, the left photon's orientation is 90 degrees, which is to say that left and right are always 180 degrees apart. A hit or a miss is determined as follows. For each measurement, left and right, the square of the cosine of the angular difference between the photon's orientation and that of the appropriate bat is determined. If that value – again, the square of the cosine – is greater than a randomly generated number between 0 and 1, a hit is recorded, otherwise a miss is recorded.

It can be demonstrated by running the program that as the number of runs increases the results tend to duplicate those of Mermin's thought experiment. The number of *green-green* and *red-red* correlations approaches exactly *fifty percent*, and the number of matching switch settings approaches exactly *thirty-three percent*. Contrary to Mermin's expectation, but in accordance with the empirical results of Bell experiments generally, not *every* matching switch setting produces a matching color correlation.

Rather, in this simulation (generating the *random real* as indicated), the number of such correlations approaches exactly *seventy-five percent*.

Moreover, it is precisely here that one finds a key to understanding the differences between orthodox interpretations of the quantum formalism – and especially the expectation of non-locality – and what real world experiments actually have to say. In this connection, the reader is referred to the writings of A. F. Kracklauer, who as briefly noted, has duplicated the results of experiments testing the *Clauser-Holt-Horne-Shimony inequality*, with simulations premised on entirely classical principles. Though perhaps a bit less transparent than the author's *Mathematica* simulation of Mermin's scenario, they are extremely important, because they clarify the distinction between well-established empirical outcomes and quantum expectations.

The mathematician David Hilbert once said that physics is too important to be left to the physicists. On the other hand, von Neumann, also a mathematician of towering status, propagated a misunderstanding not too dissimilar to Bell's. Ultimately, it seems that Einstein's take on the circumstances was the most accurate. In general terms, he felt that the foundational problems of quantum mechanics would not be resolved until physicists could muster sufficient appreciation for the relevant epistemological issues.

§ 4.11.5 Bohm's Confusion regarding Local Causal Action

It was mentioned above that David Bohm considered phase harmony, between the pilot-wave and the singularity whose motion it ostensibly guides, to be a necessary but (dynamically) insufficient condition of that guidance. Unfortunately, this misconception influenced his earliest efforts to elucidate a quantum theory of motion, and it seems to have stayed with him to the end of his life. Because the singularity is associated with a 'real' momentum, and because the operative aspect of the wave field – i.e., the rate of change of the density gradient – does not attenuate over distance, Bohm did not think it possible that such a potential could be a sufficient cause of motion; that is, that the mere form of the field, which, unlike amplitude, does not attenuate with distance, can exert a force strong enough to accelerate a massive particle in the classical sense.

This is reminiscent of another common confusion; namely, that the pilot wave or quantum potential cannot push and pull the particle appropriately, because such a physical cause of motion should necessarily guide the particle away from the anti-nodes and toward the nodes of the field, which is exactly opposite of what actually happens. This misconception is revealed as such in the next chapter, where a framework for a quantum theory of motion [including gravity] is proposed.

Unfortunately, even the most acute minds are subject to confusion of this sort; based on a subtle, tacit assumption – in Bohm's case, that if micro phenomena are to have a local deterministic description, then *mass/inertia, force* and therefore *momentum* necessarily follow Newton's laws for macroscopic objects. In other words, Bohm neglected the salient possibility that momentum, as defined by Newton, may be an emergent property, just as Newtonian force *must be*, per the discussion of section 4.3.

Indeed, there exist relevant precedents for non-Newtonian action principles, and the emergence of mass – for example, in the conceptual foundations of General Relativity and the theory of the *Higgs field*. The *Equivalence Principle* reflects the identity of inertial and gravitational mass, and stipulates that all objects, regardless of mass, fall at the same rate in a gravitational potential. And so geodesics in General Relativity are a function of geometry – which is to say the field gradient. Of course, objects moving with different velocities follow different paths through spacetime, and the rate of spacetime curvature does fall off with distance from gravitationally operative sources. But again, at the quantum scale it should not be assumed that Newtonian mass is an exact, given property, just as $F = MA$ is not. It is *a priori* more likely that mass can be described as a general, emergent characteristic of quantum phenomena.

As is often the case in such circumstances – i.e., when a misconception stands for a long period of time – Bohm held a philosophical position that made him susceptible to prejudice regarding the viability of the quantum potential. As briefly noted in section 4.8, Bohm, in his later years, developed a new interpretation of quantum phenomena, by which he attempted to supersede Cartesian spatial order altogether. His basic idea was that there exists a sub-quantum reality that is non-spatially ordered, but which somehow projects information into spatially extended forms, in the sense that a digital computer projects an image encoded non-spatially onto a graphics display device, which thus manifests a two-dimensional Cartesian order. In Bohm's model, information is likewise not encoded in a form that has one-to-one correspondence with the perceivable components of the projected image, but rather which is related to the image in a manner more like the way that a hologram is related to the three-dimensional image it represents (i.e., the relation between a *distributed pattern of interference fringes* and the *three-dimensional spatial image* that can be recovered from it).

That such an interpretation appealed to Bohm seems attributable to a fundamental confusion regarding the *subject-object dichotomy*. As discussed at length above, it is clear that a physical model of the world cannot be satisfactory from an ontological standpoint, because restricting the qualities of reality to those of "matter" – i.e., "inert substance" lacking any properties other than extension and motion – leaves too much of the *qualia* of experience out of the picture. Moreover, there is a natural tendency to interpret the subject-object dichotomy as a more-or-less objective aspect of reality. And so it seems reasonable that an accurate metaphysical account of reality, which some believe to be the true goal of theoretical physics, must ultimately account for *consciousness*.

Again, this confusion is resolved in the manner discussed at length above. But Bohm felt that his conception of the *Implicate Order*, which is the name that he gave to his notion of the underlying structure of existence, could provide a conceptual basis for the unification of theories of matter and consciousness. As also discussed above, such speculations are not entirely unwarranted; rather, they are for the most part misguided, because they fail to place the subject-object dichotomy in the proper perspective and are empirically very premature.

§ 4.12   Mathematical Simplicity, Beauty, and the Power
and Allure of Symmetry

In addition to the several varieties of confusion and bias addressed above, physicists sometimes describe a another motivation for working without a net – that is, for (ostensibly) giving up physical models in favor of purely mathematical formalisms. This is based on a philosophical prejudice somewhat contrary to that of the "realists" – namely, a belief that the abstract group structures and symmetries that underlie the mathematical relations of physics have a significance of their own. To many researchers, these (*Platonic?*) forms are deemed to be the purest and highest kind of knowledge that physicists can aspire to, and the quest for physical law can be reduced to the search for such forms.

The symmetries connected with the invariance principles of modern physics offer a compelling rationale for theoreticians to take such symmetries to be of fundamental importance. As Eddington once contended, the history of physics can be viewed as the realization of ever more "point-of-view invariant" properties of nature. While it may not be possible for physical constants to be determined solely on the basis of epistemological considerations, as Eddington dreamed, there is yet a substantial foundation for such a contention.

For example, consider the conservation principles associated with the symmetries of space translation invariance (conservation of momentum), space rotation invariance (conservation of angular momentum), and the symmetry of time translation invariance (conservation of energy), as generalized by Emmy Noether. These ideas can be generalized further still, to more abstract operations and multi-dimensional spaces. Thus, the conservation of electric charge can be identified with invariance to rotation of the state vector in a complex space – *gauge invariance* – invariance to a change of phase. Similarly, Lorentz invariance can be identified with invariance to rotation in four-dimensional Minkowski space, as can the mass-energy relation and the constancy of the velocity of light. Moreover, Lorentz invariance extended to changes between arbitrarily accelerating frames gives rise to the inertial geodesic of General Relativity – the gravitational field – in the same fashion that local gauge invariance links the "geodesic" of the electron to the electromagnetic field.

Indeed, the entire panoply of [both quantum theoretical and cosmological] standard model principles and parameters can be shown to follow, in like fashion, from the symmetries associated with the various gauge invariances [of quantum theory and cosmology, respectively]. Symmetry considerations serve a crucial role in the unification of forces – that is, symmetries are not merely suggestive but are an intrinsic aspect of theory, as in the broken symmetries that define the emergence of particles and forces. It is no wonder, then, that symmetries have such an overwhelming significance for the majority of theorists. Consider Roger Penrose on the standard theory:

"... let it be said without reservation that the basic scheme, for all its philosophical difficulties, is an extraordinarily beautiful mathematical structure. The strength of the theory (and here I refer to standard non-relativistic quantum theory, not to quantum field theory) lies not just in the unbelievable range and accuracy of its physical predictions, but also in the mathematical elegance of its formalism. It is here that I have always had difficulties with most hidden-variable theories. To me, it is no help just to improve upon the underlying philosophy of quantum mechanics by the introduction of hidden variables if the price to be paid is the sacrifice of this mathematical elegance."[46]

And Hsu on Symmetry:

"These insights show that [Poincaré's] incomparable understanding of the symmetry properties of the physical world was deeper than anybody else's at the very beginning of the twentieth century. With the help of the enormous impact of special relativity and general relativity, symmetry principles are now generally believed to play a universal and fundamental role in revealing the simplicity of the physical world. In particular, the view that symmetry dictates interactions took root mainly through the works of H. Weyl and of

46 Bohm, David., B. J. Hiley, and F. David Peat. Quantum implications : essays in honour of David Bohm. London; New York: Routledge, 1991, pp 105-106.

C. N. Yang and R.L. Mills in modern quantum field theory and was stressed in particular by Yang. The most spectacular results of the power of symmetry principles shows up in Dirac's prediction of the existence of anti-particles, in the establishment of the unified electroweak theory and quantum chromo-dynamics based on non-Abelian gauge fields discovered by Yang and Mills. Today, one hundred years after Poincaré proposed the symmetry principle of relative motion for all physical laws, symmetry principles in physics have transcended both kinetic and dynamic properties and gone right to the very heart of our understanding of the universe."[47]

And Yang himself:

"The first important symmetry principle discovered in fundamental physics was Lorentz invariance, which was found as a mathematical property of Maxwell's equations, which in turn were based on the experimental law of electromagnetism. In this process the invariance, or symmetry, was a secondary discovery. In his Autobiographical Notes Einstein gave Hermann Minkowski credit for turning this process around. Minkowski started with Lorentz invariance, and required that field equations be covariant with respect to the invariance..."[48]

This attitude is another reason so many thinkers reject the contention that symmetries should be treated as evidence of emergence, and would rather attribute absolute "as-is" significance to select postulates. In general – beyond the key concept of symmetry – there is indeed value to be found in the broader notion of mathematical form. The essential point in this respect is what might be called the issue of *curve fitting*. The limits of experimental and observational accuracy necessitate a certain degree of flexibility with respect to the interpretation of data. Hence, when a graph of relevant data points falls very close to a relatively *simple, functionally defined* curve, it is natural to associate that function with the data. Or

47 Hsu, J. P. Einstein's relativity and beyond : new symmetry approaches. Advanced series on theoretical physical science, v. 7. Singapore; River Edge, NJ: World Scientific, 2000.

48 JingShin Theoretical Physics Symposium in Honor of Professor Ta-You Wu, Ta-you. Wu, J. P. Hsu, and Leonardo. Hsu. "JingShin Theoretical Physics Symposium in Honor of Professor Ta-You Wu." Singapore; River Edge, NJ: World Scientific, 1998., page 61

rather, more specifically, it is natural to associate the data with *the simplest function that can be made to fit* it. And because there must always be some uncertainty in the measurement and consequent distribution of data, the *function* – which, in contradistinction to the data, is precisely defined – tends to have a more meaningful "reality" than the data distribution itself (again, shades of *Plato's Ideas*). Accordingly, when such a function turns up as a prediction of theory, the curve is – quite justifiably – considered to fit the data. From Kepler's musical spheres to Eddington's photographic plates, this is normal scientific methodology... not to mention human psychology.

A corollary to this issue of curve-fitting is that "what's going on outside the window" – i.e., non-methodological observations of uncontrolled, random phenomena; snapshots of everyday experience – is generally irrelevant to science. The majority of experience is too complicated to yield observations of significant general import; "underlying causes" cannot be determined by induction from overly complex data. Therefore, experimentation involves the systematic elimination of as many "extraneous" factors as possible, where extraneous is taken to mean the so-called contingent aspects of a situation – i.e., *precisely those that are not among the pre-conceived factors that are, within the context of operative theory, deemed to be significant*. Thus observation and experiment are conditioned by theory, because theory determines experimental arrangements and the interpretation of outcomes, and curve-fitting is only one particular aspect of the general effort to fit experience to theory. Put another way, curve-fitting is a key part of determining which aspects of a situation are to be deemed "observer invariant." Hence, observer invariance is a subjective construct.

In this sense, the most "abstract" order seems to be the most salient. Accordingly, the mathematical *Theory of Groups* is taken to capture that which Eddington called the "skeletal structure" of experience; considered to underlie the relations of physical events without having the drawback of being tarnished by the idiosyncrasies of the specific system under consideration. In this view, the Kantian "thing-in-itself" makes its presence felt through group properties. If the "actual nature" of whatever it is that human perception of the world is taken to represent is ultimately unknowable – if space and time have no extra-personal significance – then at least the mathematical relations that exist among the phenomena of experience must surely reflect some aspect of extra-personal order, which can thus be captured mathematically (or at least one would think).

But it must be remembered that the notion of abstract order is no less an anthropomorphic objectification than any other ontological hypothesis, and is subject to the same intrinsic representational constraints of cognition as any conceptual construct. In addition, abstract formalisms are not as intuitively tractable as physical models. And in any event, there always remains a need (that is, irrespective of the appropriateness of a given formalism) for empirical evidence, which is grasped via ordinary forms of perception and intuition. This, again, is the primary valid upshot of Bohr's position. It is always necessary to forge an intuitive (and thus irreducible) link between the abstract structure of theory and the more tangible elements of experience. Exclusively mathematical approaches also suffer from the lack of imaginative reach that accounts for a large part of the progress of theoretical research, as typified by the difference between the electrodynamics of Faraday and Maxwell and the theories of Ampere and Weber, *et al.* An important aspect of this value has to do with the manner in which the formalism is connected with experience – i.e., the interpretation of the meaning of theory.

On the other hand, abstract notions of order can help forge relations among concepts that, superficially, seem to lack common ground. For example, the mathematical concept of information can be utilized to link domains that do not even have the property of spatial extension in common. Non-spatial sensate qualities can be correlated, on an hypothesis of information processing, with particular neural events. Of course, without the guidance of physical models all such efforts must seem hopelessly confused – perhaps even pointless. If too much emphasis is placed on considerations of mathematical meaning and/or aesthetic, it is easy to view the idealized objects of theory as more important than the real and therefore true but (necessarily) inaccurate results of experiment.

Thus, symmetry becomes an all-important attribute of theory – its significance not subject to negotiation and its origins not subject to speculation. This attitude toward symmetry characterizes Jeffrey Bub's contention, quoted in section 4.9, that quantum theory is a theory of information rather than physical existents. It is reflected in such frequently expressed opinions as "the symmetries of Minkowski spacetime constitute the meaning and essence of special relativity," and "the symmetries of Hilbert space are the quintessence of quantum theory."

§ 4.13 The Ehrenfest Paradox

Although it is misleading to objectify the symmetries of physics – or any aspect of theory generally – this sort of ideological objectification is very common. For instance, consider such phrases as "the collapse of the wave function" and "the flow of probability." Such expressions are not merely figures of speech. They are symptomatic of the absence of a viable theoretical framework, which is especially troublesome with respect to such fundamental concepts as space and time. Illustratively, when describing time measurements, a physicist might think in terms of "the rate of *Time*" even when what is called for is something more like "the rate of the periodic motion comprising some given phenomenon" (although, of course, *rate* is just a cognitive stand-in for '*Time*,' which is thus not eliminated in any meaningful sense).

Now, whereas the concept of *relative velocity* is readily comprehensible, the notion of the *relativity of time* is just as readily confusing. In fact, it was for this reason one of the principle objects of General Relativity to define spatio-temporal measurements in terms of *events* – where coincidences of observables comprise "a point in spacetime" – rather than abstract notions of space and time. It is therefore a non-trivial concern that confusion regarding the meaning of and relations between such elemental concepts as *space*, *distance*, and *measuring rod* has not been purged from physics. This confusion is at the root of the so-called *Ehrenfest Paradox*, and it stands in need of decisive resolution.

With Ehrenfest, consider a rigid, log-like object – a solid cylinder – rotating about its axis of symmetry. The radius, R, as seen in the stationary frame of reference, is always perpendicular to the cylinder's motion and is therefore always equal to its stationary value, say R_0. But the circumference, $2\pi R$, should appear Lorentz-contracted in the stationary frame. This seems to imply the contradiction that both $R = R_0$ and $R < R_0$ are true. In a real situation such an object, regardless of the material out of which it is constructed, could not maintain its integrity under these circumstances, as the constituent parts (at the atomic and/or molecular level) would have to both (a) condense together in some arrangement of groupings, and (b) stretch apart to permit such groupings, all in accordance with distance from the axis of symmetry. Perhaps more to the point, measuring rods placed along the perimeter of the cylinder to measure its circumference, will necessarily shorten and thus measure a distance that, upon division by 2π, is greater than the expected or stationary radius: i.e., $R > R_0$. It is thus clear that some of the material of the cylinder must stretch apart in order to yield a circumference so measured by the shortened rods. But if the material comprising the circumference of the cylinder is contracted at all points along its tangent, then, just as with the measuring rod, the actual length of the material of the cylinder as measured by the rod should still be the *same* as the stationary value, whereas it is only *the addition of gaps* in the material that yields the larger measurement. And so it is the *sum*: *gaps plus material,* which is taken to comprise the circumference.

The confusion that underlies the Ehrenfest paradox is the identification of "space" or "distance interval" with, alternately, each of the following two related but different concepts: (1) an absolute/abstract quantity of "extension" which, theoretically, can be filled by a suitable material measuring implement[s] – in this case "the circumference" of the cylinder, even if the material of the object does not fully occupy that interval – and (2) such a material measuring instrument *per se*. The meaning of and relations between these terms must be understood in the following sense. If "space itself" were somehow to contract, while the objects in it did not, then a material measuring instrument would appear to be *enlarged* when occupying the "same space" between two objects that it had occupied before the modification, and *vice versa*. However, if the objects that occupy space were to contract and expand along with it, then such spatial distortions *could not be measured* – that is, in the direct manner of laying down a measuring implement.

Therefore, the concept "space itself" is extremely ambiguous – its meaning cannot be accepted as self-evident. For again, if all objects were to enlarge while "space" did not, it would not be possible to distinguish this circumstance from that of its converse; namely, that in which "space shrinks" while the objects within it do not. In fact, any proposition about such a distinction would be characterized as "empty/meaningless metaphysics." In the context of relativity theory, it is understood that a dilatation of "the metric of space" means the same thing as (i.e., is indistinguishable from) a dilatation in any material object occupying that space. And so there is no question of "space itself" suffering a modification that is somehow separate from the modifications of the objects in it – and this is the correct manner in which to approach the Ehrenfest paradox. That is, if "space itself" were somehow to contract, at every point along the perimeter of a cylinder and in the direction tangent to the perimeter at each of those points, then measuring rods placed along the circumference, which must necessarily shrink with the space they occupy, would not appear shortened in relation to the perimeter, and the cylinder would thus maintain its integrity: Its components would not stretch apart, and an observer moving with the perimeter would not detect a defect in the ratio of its length to that of the length of the radius (strictly by measurements made by laying down measuring rods).

Such a circumstance, however, would contradict the findings that led Einstein to General Relativity. For it is precisely the capability of measuring a defect with respect to π in the ratio of the circumference to the radius that characterizes a dynamic metric. And so, again, the *Ehrenfest Paradox* resolves along a similar line. That is to say, the structural integrity of the rotating cylinder must be compromised, and it therefore makes no sense to say that "space itself" is dynamic.

§ 4.14 The Ehrenfest Paradox and the Heuristic Value
of Dynamical Models

Without explicitly referencing the Ehrenfest Paradox, John Bell addressed the underlying problem in a monograph entitled "*How to teach special relativity.*" As Bell describes, this problem baffled the majority of his (physicist) colleagues at CERN.

The problem is illustrated in figure 4-7. There are three parallel space ships – *A, B* and *C* – traveling at the same velocity. A thread of fixed length connects one of the ships (*B*) with the other directly behind it (*C*). Then, at a given moment, the two thread-connected ships begin gently accelerating at exactly the same rate. The question is: "Must the connecting thread ultimately break as a result of a Lorentz contraction?" The correct answer is that it must, but most of the physicists surveyed in Bell's story believed the opposite:

> "This old problem came up for discussion once in the CERN canteen. A distinguished experimental physicist refused to accept that the thread would break, and regarded my assertion, that indeed it would, as a personal misinterpretation of special relativity. We decided to appeal to the CERN Theory Division for arbitration, and made a (not very systematic) canvas of opinion in it. There emerged a clear consensus that the thread would **not** break!" [49, 50]

This is evidently attributable to the same confusion associated with the Ehrenfest paradox. For if "space itself" were to contract, along with everything that it contains, then the thread should not be expected to break. Bell goes on to give his own take on this:

> "Of course many people who give this wrong answer at first get the right answer on further reflection. Usually they feel obliged to work out how things look to observers B or C. They find that B, for example, sees C drifting further and further behind, so that a given piece of thread can no longer span the distance. It is only after working this out, and perhaps only with a residual feeling of unease, that such people finally accept a conclusion which is perfectly trivial in terms of A's account of things, including Fitzgerald contraction. It is my impression that those with a more classical education, knowing something of the reasoning of Larmor, Lorentz and Poincaré, as well as that of Einstein, have stronger and sounder instincts..." [51]

49 Bell, J. S. Speakable and unspeakable in quantum mechanics : collected papers on quantum philosophy. Cambridge [Cambridgeshire]; New York: Cambridge University Press, 1987.
50 Emphasis and exclamation mark in the original.
51 Op. Cit.

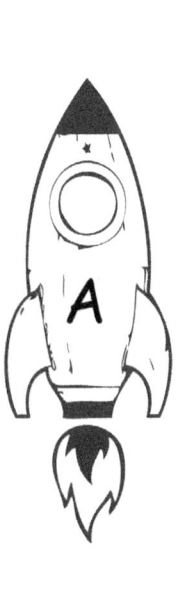
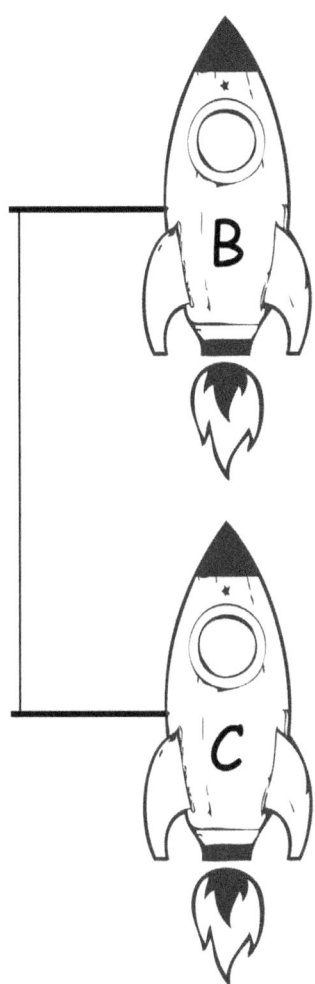

Figure 4-7

In other words, it was Bell's belief that an intuitive understanding of how Lorentz invariance must emerge from whatever underlying dynamics might exist, gives one a better grip of the circumstances than a purely formal grasp of the mathematical relations involved, based solely on the principle of relativity. Consider, in this connection, the following passage from Spinoza's classic essay *On the Improvement of the Understanding*:

"All these kinds of perception I will illustrate by examples. By hearsay I know the day of my birth, my parentage, and other matters about which I have never felt any doubt. By mere experience I know that I shall die, for this I can affirm from having seen that others like myself have died, though all did not live for the same period, or die by the same disease. I know by mere experience that oil has the property of feeding fire, and water of extinguishing it. In the same way I know that a dog is a barking animal, man a rational animal, and in fact nearly all the practical knowledge of life.

"We deduce one thing from another as follows: when we clearly perceive that we feel a certain body and no other, we thence clearly infer that the mind is united to the body, and that their union is the cause of the given sensation; but we cannot thence absolutely understand the nature of the sensation and the union. Or, after I have become acquainted with the nature of vision, and know that it has the property of making one and the same thing appear smaller when far off than when near, I can infer that the sun is larger than it appears, and can draw other conclusions of the same kind.

"Lastly, a thing may be perceived solely through its essence; when, from the fact of knowing something, I know what it is to know that thing ... By the same kind of knowledge we know that two and three make five, or that two lines each parallel to a third, are parallel to one another, & etc...[52]

With respect to the second and third of these modes of knowing, he further illustrates the distinction as follows. Given the numbers 2, 4, 3 and 6, the proportionality between the first and second number is clearly the same as that between the third and fourth. Regarding this knowledge, he writes:

"...Mathematicians, however, know by the proof of the nineteenth proposition of the seventh book of Euclid, what numbers are proportional's, namely, from the nature and property of proportion it

---

52 Spinoza, Benedictus de, Benedictus de Spinoza, Benedictus de Spinoza, Benedictus de Spinoza, and R. H. M. Elwes. Benedict de Spinoza : On the improvement of understanding, the Ethics, Correspondence. New York: Dover, 1955.

follows that the product of the first and fourth will be equal to the product of the second and third: still they do not see the adequate proportionality of the given numbers, or, if they do see it, they see it not by virtue of Euclid's proposition, but intuitively, without going through any process."[53]

Schopenhauer makes the same point more strongly, in a somewhat different, highly interesting way. For him, the very perception of physico-spatial relations constitutes a causal interpretation of the sensations, by which he means that the unconscious mind constructs the phenomenology of experience by means of a causal extrapolation from the stimuli of the senses, so that causality is intrinsically ingrained in the very perception of physical forms, to the same degree as the intuition of space.

"...For this is the sole form and function of the understanding, certainly not the complicated clockwork of the twelve Kantian categories whose invalidity I have shown. All understanding is immediate, and therefore intuitive, apprehension of the causal connexion, although to be fixed it must be reduced at once to abstract concepts. Therefore calculating is not understanding and in itself does not afford a comprehension of things. Only on the path of intuitive perception do we get this through a correct knowledge of causality and the *geometrical* construction of the sequence of events. Euler gave this better than anyone else because he had a thorough understanding of things. Calculation, on the other hand, is concerned with nothing but abstract concepts of quantities whose mutual relations are determined thereby. In this way we never arrive at the slightest comprehension of a physical process. For such a comprehension requires the *intuitive* apprehension of spatial relations by means of which causes operate. Calculation determines how many and how large and is therefore indispensible in *practical* affairs. It can even be said that *where calculating begins, understanding ends*; for whoever is occupied with numbers is, while calculating, a complete stranger to the causal connexion and to geometrical construction

---

53 Op. Cit.

of the physical sequence of events; he is engrossed in purely abstract numerical concepts..."[54,55]

Bell's note on the pedagogical utility of the historical approach to teaching physics clearly echoes Spinoza's clarion call of the *Enlightenment*; it is in the same vein as Schopenhauer's critique of Kant, and Maxwell's earnest response to Newton's missive regarding philosophical absurdities.

Returning to the subject at hand, there exists a time analog of the *Ehrenfest Paradox*, arising from a similar misunderstanding of core concepts. If one employs a Minkowski diagram to graph the course of a faster-than-light signal, one can trace a path that originates at a given point in spacetime and terminates at the same place, but at an earlier time. This is usually taken to mean that faster than light velocities are impossible. However, Kurt Gödel found that General Relativity allows for cosmic configurations with closed time-like paths, thus enabling backward travel in time.[56]

David Bohm – in his book *The Special Theory of Relativity*[57] – employed a Minkowski-diagram, essentially homologous to the schematism of figure 4-8, to show that faster-than-light signaling implies backward-in-time information transfer, and thus nullification of causality. His essential argument, in the context of the depicted scenario, is that if signals can be sent with infinite velocity then observers 1 and 3 (both without velocity in the frame shown) can share information along the line OC – O and C representing the "absolutely simultaneous" spacetime locations of 1 and 3 at the moment they communicate – whereas observers 2 and 4 (moving in parallel directions with identical speeds) can share information along the line OD. Per the diagram, the locations of 1 and 2 coincide at point O while those of 3 and 4 coincide at D. Because observers at the same location at the same time also realize absolute simultaneity, despite the difference in their velocities, it is possible for 3 to communicate a message to 4 when both are at D, and for 4 to send it to 2 at position O, who can then transfer it instantly to 1, who in turn can send it instantly to 3 at point C – where

54 Schopenhauer, Arthur. On the fourfold root of the principle of sufficient reason. La Salle, Ill.: Open Court, 1974.

55 All emphases in the original.

56 Einstein & Schilpp, *Op. Cit.*

57 Bohm, David. The special theory of relativity. Lecture notes and supplements in physics. New York: W.A. Benjamin, 1965.

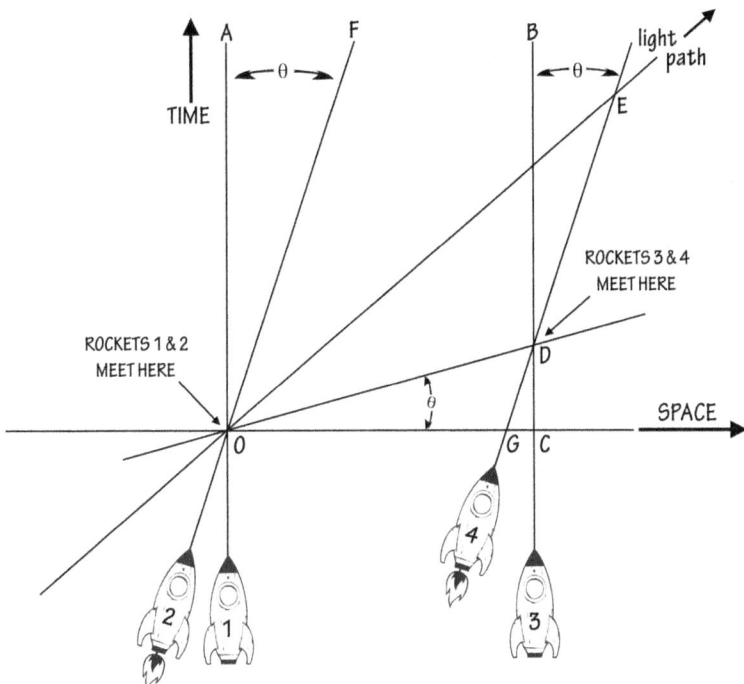

Figure 4-8   Backward in time signalling, courtesy of Minkowski

the information can therefore be received, by its original sender, before first having been passed to 4 at *D*.

Such possibilities do not arise unless it is assumed that light paths are literally *null geodesics* – i.e., that no time elapses (which is to say that "nothing happens") along optical paths. This is a consequence of the Minkowski formalism, which reinterprets the independent concepts of space and time as illusory projections of the unified entity called *spacetime*, which is stitched together by light. As long as one does not accept Minkowski's spacetime construct as something objectively real, it is quite easy to see through such confusion. And while no one would dispute that the primitive, empirically naïve aether theories of the nineteenth century are outmoded, the heuristic value of Lorentz's constructive approach – so nicely illustrated by Bell and Bohm[58] – certainly is not. Not only is the meaning and value of relativity theory much better appreciated in the context of its conceptual antecedents, but so too is its relevance to the future of physics.

58 Op. Cit.

Despite this apparent truism, it is still usually considered taboo, or at least laughable, to suggest that Lorentz invariance might not be absolute. And again, one of the reasons for this constraint on speculation is the tacit tendency to objectify spacetime. It is easily forgotten that the universality of Lorentz invariance is *postulated,* and that any argument that demands its universality is necessarily circular. If one assumes that a Minkowski diagram reflects an objective reality, along the lines of Newton's *Absolute Space* and *Time,* it follows that a signal propagating faster than light can travel "backward in time," so that if the speed of light can be exceeded causality can be circumvented.

Special Relativity emerged from efforts to understand the physical basis for electromagnetism, and at the turn of the last century the state of affairs was such that it could be deemed *pragmatic* – certainly, at least, until a greater body of knowledge might be acquired – to simply accept as "tentative givens" the electromagnetic field as a physical primitive and the velocity of light as a constant, limit value. This is analogous to the manner in which the concepts of *Universal Gravitation* and *Absolute Space* were accepted as givens under Newtonian theory – likewise absent dynamical justification (particularly with respect to the property of *action-at-a-distance*).

However, with hindsight – and especially the knowledge that the classical electromagnetic field is not best interpreted as a physical primitive – this move appears less than optimal. In particular, the deep reliance on treating the symmetries associated with Lorentz invariance as an absolute feature of nature warrants skepticism. This tendency characterizes several other crucial decision points in physics. Einstein was influenced by considerations of symmetry to posit spacetime as a primary element of reality, equating its properties with the gravitational field. In a somewhat ironic turnaround, he soon came to view the generalization of relativity – *intended to resolve the epistemological problems associated with Newtonian Absolute Space and action-at-a distance* – as imbuing space and time with the dynamical qualities of an aether.

This aspect of General Relativity is generally under-appreciated, as is its connection with earlier, aether-based theories of gravity. Its origins are similar, in this respect, to the constructive origins of Special Relativity.

Contrary to the usual storyline, Einstein's dynamic geometrical construct was *not* unanticipated by nineteenth century thinkers. While it is often asserted that the mathematics Einstein adopted for General Relativity had existed well before such an application could be conceived of, this is quite false. Riemann and others, notably Clifford, explicitly identified a dynamic, generally non-Euclidean plenum as a viable model of space, wherein both gravity and matter arise as modifications of geometry. Moreover, Riemann was thinking specifically about gravity as geometry when he formulated the mathematical apparatus that Einstein employed – the key difference, of course, is that Einstein's construction involves the relativistic/Minkowskian merger of space and time. (These ideas circulated in various forms, comprising both purely geometrical as well as purely mechanical and mixed concepts.)

Special Relativity constrains the possible rigidity of matter – because of the finite, limit velocity of light, no object can respond as a whole to a localized force; rather, when one end of an object is moved, motion must propagate through it as an impulse, traveling at a speed less than that of light. Although the theory of quantum electrodynamics was not yet available, primitive ideas about electrons and the field furnished a sufficient, albeit approximate foundation on which the limitation could be understood. However, the special theory of relativity is more restrictive, vis-à-vis the rigidity of matter, than theories of constitutive structure, because strict Lorentz invariance implies that nothing in material bodies can be absolutely rigid – not even the smallest parts. In a sense, it implies that matter is "made of light," which of course would explain the limit velocity, not to mention the matter-energy identity.

As to the equivalence of mass and energy, special relativity suggests the notion that all material objects have an "inner" as well as an "outer" motion. That is, taking the center of gravity of a material object as its frame of reference, the matter-energy relation suggests that within that frame of reference there is continuous activity, a regime of energy that can be transformed into external, observable energy. Moreover, all uniform motion of translation other than that with the velocity of light can be transformed to rest in some inertial frame. Therefore, light has no rest mass or rest energy: Unlike material objects, light cannot be at rest in any frame of reference.

So again, special relativity suggests that matter has rest energy because it is composed of a continuously divisible hierarchy of regimes of scale, all of which can be viewed as containing smaller levels of "things in motion" (up to the limit of the speed of light) – or alternately, again, that matter is made of light.

The symmetries of Lorentz invariance are typical of a broad set of natural laws that are nowadays often termed *emergent*. It is frequently the case with such emergent behavior that it is difficult to ascertain, with certainty, an explanation for the symmetries involved, because they act as a barrier of understanding. In this respect, the self-defeating aspect of the Lorentzian aether that catalyzed Einstein's insight – i.e., that it precludes the determination of absolute rest and distant simultaneity – nicely exemplifies such a barrier. That is, by virtue of the symmetries it embodies, the aether cannot be observed. Nobel physicist Robert Laughlin discusses such "Barriers of Relevance" at some length in his interesting book entitled *A Different Universe.*[59] Of course, to prevent *Theories of Principle* from becoming barriers to understanding it is only necessary to recognize the limits of the regimes to which they apply – or rather, that it makes sense to assume that such limits exist.

Inasmuch as the key ideas of Special Relativity are implicit in the work of Einstein's predecessors, the primary heuristic value of his insight can be characterized as a *recognition of principle per se*; identification of key, broad-stroke features of behavior in a given regime of interest. Thus, in a world where Lorentz invariance holds, certain salient features stand out; such as the variance of inertial mass, time and distance dilatation, the relativity of simultaneity and the equivalence of mass and energy. A "*Principle Theory*" brings such features to the fore and emphasizes their pragmatic consequences. Lorentz did not in this manner recognize the broad-stroke properties that, based solely on the principle of relativity, Einstein's treatment brought to light – even though they were completely implicit in Lorentz's work.

In this connection, Laughlin is very much to the point in his characterization of Einstein's role regarding the *Special Theory of Relativity*:

---

59 Laughlin, Robert B. A different universe : reinventing physics from the bottom down. New York; London: Basic Books ; Perseus Running, distributor], 2006

"The story of Einstein's triumph is so romantic it is easy to forget that relativity was a discovery and not an invention. It was subtly implicit in certain early experimental observations about electricity, and it took bold thinking to synthesize these observations into a coherent whole..."[60]

The symmetries that Einstein's analysis of Maxwell's electrodynamics (and Minkowski's mathematical treatment of Einstein's work) brings out so clearly are in this sense related to the symmetries that "least energy" and "equilibrium conditions" tend to conceal (i.e., as with destructive interference in systems of waves, certain effects of symmetries cancel-out). Symmetries are *effects* – and thus symptomatic – of other, underlying properties; which in turn may be of profound interest and heuristic value. While recognition of the key upshots of Lorentz Invariance was important for its own sake – indeed, it marks a major historical development – it is not logical to take the next, uncalled-for step of stipulating that "hidden properties" – apparently redundant or undiscoverable – are absolutely unnecessary and so should be deemed *non-existent.*

Consider this: Any macroscopic material object is deemed to comprise a collection of exceedingly small, electrically charged entities in electromagnetic equilibrium – an enormous ensemble of exquisitely balanced, ultra-high-tension atomic bomblets literally on the verge of exploding. And so despite the apparent inertness of material objects, physics yet recognizes the existence of a dynamic undercurrent, which the symmetry of positive and negative charge hides from immediate detection. But even if the world were so constructed that the electrical properties of matter were unnoticeable but under the most extreme, artificially contrived circumstances – and if even then it were only marginally recognizable – no physicist would propose, on principled grounds, that this symmetry be declared "absolute," and matter "neutral and inert." Rather, the Newtonian attributes of matter would continue to be viewed, properly, as a collective, emergent property of the symmetry – and the symmetry, in turn, a collective property of the underlying domain.

On the other hand, if it *were* in fact very difficult to detect electrical charge, the property might very well go unrecognized as a consequence

---
60 Op. Cit. page 119

of a dogmatic belief in the "inertness" of matter. It must be remembered that the extent to which theory determines experience is quite significant. Hertz's comments on Newtonian forces are salient in this regard. It was mentioned, in another connection, that he sought to base physics on a new foundation; one which did not employ the concept of force. Here is an excerpt from the introduction to his book on the subject, *The Principles of Mechanics*:

"We see a piece of iron resting upon a table, and we accordingly imagine that no causes of motion – no forces – are there present. Physics, which is based upon the mechanics considered here and necessarily determined by this basis, teaches us otherwise. Through the force of gravitation every atom of the iron is attracted by every other atom in the universe. But every atom of the iron is magnetic, and is thus connected by fresh forces with every other magnetic atom in the universe. Again, bodies in the universe contain electricity in motion, and this latter exerts further complicated forces which attract every atom of the iron. In so far as the parts of the iron themselves contain electricity, we have fresh forces to take into consideration; and in addition to these again various kinds of molecular forces. Some of these forces are not small; if only a part of these forces were effective, this part would suffice to tear the iron to pieces. But, in fact, all the forces are so adjusted amongst each other that the effect of the whole lot is zero; that in spite of a thousand existing causes of motion, no motion takes place; that the iron remains at rest. Now if we place these conceptions before unprejudiced persons, who will believe us? Whom shall we convince that we are speaking of actual things, not images of a riotous imagination?"[61]

---

61 Hertz, Heinrich. The principles of mechanics, presented in a new form. New York: Dover Publications, 1956.

§ 4.15   Summary and Conclusions

1. *Limits of Knowledge, Real and Imaginary.* Because the properties of cognition discussed in chapters one through three are universal features of what was referred to there as the *concept forming process*; because they characterize every representation, and therefore every theoretical construction; and because one of the upshots of these properties is that it makes no sense to seek to identify perceptions and conceptions with any referents – "objective" or otherwise – there can be no absolute standard upon which a dictum to ban "non-observable" or "metaphysical" concepts from science can be founded.

Absent the capability of identifying a concept with an "actually existing referent" to which it refers, there are no criteria upon which the elemental constructs of one or another theory can be deemed to be more or less "real" than those of another; there is no hard and fast basis on which a charge of "unduly metaphysical" can be levied. This, again, is because all concepts can be characterized as *metaphysical* in their operative intent, in the sense that they necessarily carry meanings that cannot be directly/literally "pointed to" in the "things" they are deemed to represent, or in experience generally.

Accordingly, judicious limitation and toleration of the inconsistencies every theory must entail is the salient epistemological issue. And because the inconsistencies inherent to theory grow with theory complexity, those constructs are to be preferred in which the fewest possible logical operations separate the axioms from the propositions deduced on their basis. Note that this stipulation differs from the more commonly recognized maxim to limit the proliferation of premises; which of course is also important. Indeed, it is usually understood that as the number of axioms is diminished – as fundamental concepts are generalized to embrace a wider set of phenomena – the deductive chain from premises to conclusions grows longer. And so, to the extent this is the case, a balance must be struck, a sort of tension resolved, between minimizing the number of assumptions – i.e., what are variously referred to as *undefined concepts/ axioms/postulates/stipulations/premises/etc.* – and the number of inferences that must be drawn from those assumptions in order to describe the phenomena. As will be demonstrated in the next chapter, a more-or-less optimal balance can be realized in the foundations of physics.

Insufficient appreciation of these epistemological relations is at the heart of the many controversies that have marked the development of philosophy and science, particularly since the time of Newton, and has often led to more ready acceptance of, or preference for, such theories as would superficially seem the most *empirical* or *positivistic* – wrongly so in accord with the view here to be developed. Indeed, it may be considered a great irony of science that, in the effort to maintain an apparently phenomenological orientation, or to restrict deviations from empirically grounded concepts as much as possible, the exact opposite result has often been achieved.

And so it is not only when physicists reference such abstractions as "*the geometry of empty spacetime*" that they invoke representations that cannot be directly linked to observation, but also when they refrain from introducing such representations – as, for example, when Newton, unsure of the best explanation for gravity and not wishing to hypothesize broadly, invoked the stand-in mathematical concept of "*a force that acts instantaneously at a distance.*" This is particularly evident with regard to the radical ideas of Copenhagen – such as the notion of observer-mediated resolution of a physical observable's value; an idea intended to link the purport of the concepts *wave* and *particle* in a 'complementary' construct that is irreducible to physical or visualisable description.

Bohr intended his concept of *Complementarity* as a generalization of physical causality; by enabling the ascription of multiple attributes to a set of circumstances in which no underlying "thing" is deemed to exist – in contradistinction to the usual intuitive association of attributes with something substantive. In other words he intended a tentative use of really inappropriate concepts, in a purely operative and thus positivistic approach to describing a set of circumstances, in order to facilitate the calculation of probabilities without invoking any thoughts of an underlying reality; similar to Newton's intended use of the concept of action-at-a-distance. And just as in the Newtonian case, the original *operative/positivistic* meaning morphed into a less tentative and ultimately inconsistent ontological meaning among physicists, because it is unnatural for people (including physicists) to operate without "pictures in their heads," and so to use language without connecting images to words.

A good example of an imaginary problem morphing into a real one in physics (as a result of confusing a positivistic or tentative construction with "something real" in the above sense) is the notion of *electric charge*. The idea of *charge* – manifesting "positive" and "negative" attributes – is quite obviously a convention. It was introduced to enable a schematic description of the phenomena of electricity. But the concept acquired meaning outside its original context, so that ultimately physicists found themselves occupied with ontic questions, e.g., concerning the "stability" of charge – such as: "*How can an electric charge exist as an extended object – shouldn't the 'bits of identical charge' that make up the particle cause it to explode into pieces – and then, in turn, shouldn't the same consideration apply to the pieces?*"

When the conventional, anthropomorphic character of the concept is clear so too is the absurdity of such questions, which are comparable to a Medieval cleric asking: "*How can sexual gender exist? Should not a woman with extremely feminine attributes be highly unstable? Should not a Lesbian implode?*" Such questions make as much sense as those regarding the electron. Again, concepts like electric charge are established by convention; the notions of "attraction" and "repulsion" between charges are *simplifying representations*; as with the conception of "attraction between genders" it is a subjective construct that simplifies the representation of a vast and complex set of phenomena. The concept of charge originated in efforts to describe the interaction of spatially separated objects, which seemed intelligible on the basis of the Newtonian conception of force. It later became convenient to associate the property defined as electric charge with a discrete object (particle) and thus with a position in space.

Note that from an emergentist perspective such confusions do not arise, because ideas such as charge are not understood as fundamental – e.g., as an irreducible cause of attraction or repulsion without further interpretation. For example, a crystalline solid can contain topological imperfections that stand in a mutual relation analogous to that of *anti-particles* – i.e., otherwise identical carriers of opposite charge that annihilate on contact. Likewise, a fluid can contain analogously related vortex filaments, which, like anti-particles, are created and annihilated in pairs; or again, vibrating bubbles that attract and repel in accordance with their phase relations. The topological and wave dynamics of such collective phenomena can mimic the behavior of force-charges generally. Thus, electric charge can be thought of as a *condition that exists within a certain region of space*, and which reacts with certain other regions of space – rather than as a "charged particle" of "no size," whatever such a thing might be. Entities born of positivistic reasoning can easily become monsters if their origins in convention are not kept constantly before the mind.

The upshot of these considerations is that extreme positivism tends to impose arbitrary limits on thought by tacitly introducing a highly restrictive ontology, but under the guise of an "anti-ontological" orientation that ultimately cannot be viable and, accordingly, furnishes only a temporary expedient. In the case of the Copenhagen philosophy, one of the guiding principles would seem to be the ontological confusion that "too small to observe" equals "thing-in-itself" – analogous to confusing the dots on a video display with the reality behind the image, per the image and discussion of section 1.7.1. Moreover, in contravention of the traditional premises of natural philosophy, positivism further complicates things by invoking and/or enlarging – implicitly if not explicitly – the role of the observer and, in particular, the concept of consciousness as a component of physical theory.

Another source of trouble is the oddly common supposition that a "final theory" is in sight – not unrelated to the misconceived identification of an apparent extremum of scale with an ultimate feature of reality. This theme, which evidently emerged with the earliest conception of atoms, seems to gain and lose currency periodically throughout the ages. With the historical alternation of scientific trends – such as the alternating primacy of concepts like corpuscles versus waves, and constructive versus phenomenological description – old ideas are resurrected and reintroduced into theory in a cycle that repeats itself more often than most physicists seem to acknowledge.

The history of science supports a truism quite antithetical to belief in an impending *finale* – namely, that the scale of reality grows, in depth and breadth, with the scope of knowledge. Consider the concept of microbes. It was once considered laughable that tiny living beings, too small to see, could be causative of disease. And to those without a sense of humor, the idea of a "really big show" (i.e., *universe*) was sufficient reason to have Giordano Bruno burned at the stake. Today, physicists reason about the age of the universe and the limits of the Planck scale, particularly the latter, with a dogmatic sensibility not altogether unlike that of some of their forbears. In the light of the epistemological arguments above, such dogmatism – which is inseparable from the orthodox understandings of quantum and relativity theory – should appear in raised relief.

2. *Proper Interpretation and Invocation of the Subject-Object Dichotomy.* At present, misconceptions connected with the so-called *mind-body problem* obscure progress in physics to an even greater degree than in the cognitive sciences. This strange circumstance has grown out of the confusion surrounding the foundations of quantum theory.

The *subject-object dichotomy* is a necessary, overarching heuristic framework – not only for the sciences but for the general conduct of life as well. This is true irrespective of whether epistemological analysis suggests a unification of the phenomenology of experience as neither "observed" nor "observing" – and even if a truly satisfying ontological vision should be available to transcend it. As long as what is called "first-person experience" is such that people face the pragmatic necessity of "orienting themselves in an environment" and "communicating their experiences" accordingly, it is neither feasible nor desirable to altogether relinquish the subject-object dichotomy. (As noted elsewhere, this is analogous to the only salient upshot of Bohr's philosophy; i.e., that it is necessary to describe the arrangement of experiment in the language of "classical physics".) It is therefore imperative to eliminate the confusion associated with it.

This is accomplished by understanding the roots of the confusion in certain general, intrinsic constraints of human cognitive function. *Perception* and *conception* are characterized by the following two conflicting attributes: (1) They are given whole to consciousness, "ready-made," as complete and indivisible aspects of experience. But this *Gestalt* character of cognition is diametrically opposed to a second, equally important and intrinsic attribute, namely (2) Every such construct reveals what is most aptly described as a *figure-ground* duality, inasmuch as a certain aspect of the field-of-attention is brought into focus as *foreground* within a framework, or against a *background*, which furnishes a context that is logically distinct from and yet essential to and inseparable from the meaning of its foreground.

Consequently, every concept harbors an intrinsic *inconsistency* that manifests logical contradictions when analysis of the concept is pushed far enough. Moreover, because perceptual and conceptual representations are necessarily distinct from the experiences they are taken to reference – and carry qualities/meanings that are different albeit related to those experiences – all representations are also *incomplete*.

The *mind-body problem* is an especially poignant example of confusion based on such representational incompleteness and inconsistency, though it also involves another source of difficulty – a peculiar, though almost universally held misconception, *viz*.: the mistaking of a concept of some *physical thing* – in this instance "a brain" – for the ostensive "thing-in-itself" in the Kantian sense (i.e., something deemed to exist objectively; independent of experience). This might be called "sophisticated naïve realism."

Thus the conception of the "brain" takes the role of a sort of second-order representation of the "mind" – *second-order* in the sense that *First*: the concept "brain" can only apply to a particular perceptual experience – for instance, the perception of a particular agglomeration of "soggy-gray-matter." But its *conceptualization* as a "physical object" – i.e., the objectification of the *percept: soggy-gray-mass* as conceived *electrochemical-machine: brain* – is then taken *Second*: to represent something entirely different again; i.e., a "mind," which is to say an "ensemble of experience."

Now the concept *physical object* is just that – a concept. And yet the illusion of naïve realism – that physical objects are real, with qualities just as they are perceived – is extremely stubborn, inasmuch as it is based on immediate perception. So it is easy to mistake the concept of the perceived object – i.e., "the brain" – for "the thing-in-itself that perception of the brain ostensibly represents." But the "thing-in-itself" thus confused with the "object: brain" is also further identified with a *mind* – i.e., something without material characteristics. And so, as with all representations, the concept of *"brain"* necessarily differs from what it purportedly represents.

But because of the second-order nature of the representation – again, *first*: perception of "soggy-gray-matter" as representation of "physical thing that exists out there"; *second*: concept of "thing out there" taken as representative of "a mind" – is in this case so far removed (*qualitatively*) from its object, that the qualitative distinction (and thus confusion) is quite profound.

The intuition that resolves this confusion is that, with regard to the notion of the "*thing-in-itself*: brain" it is clear that "whatever exists 'out there' as brain" must be the same as "whatever exists 'in here' as mind" – i.e., the totality of *first-person experience as such*, sensate qualities and all. The two are identical. It is thus clear that neither naïve realism nor "sophisticated naïve realism" makes for an epistemologically viable ontology. But the concept of reality as "mental" is also a very imperfect, limited construct. While the subject-object dichotomy (employed to overcome naïve realism) is a necessary heuristic device, it is yet incomplete and inconsistent as well. This, again, is a consequence of a general property of cognition. Therefore, in order to usefully employ this dichotomy one must bear in mind these universal powers and [concomitant] constraints of thought, some of the more salient upshots of which are:

a. Representation of experience – either as "physical stuff" or "mental construction" – is again, in either case, *incomplete* and *inconsistent*. However, judiciously employed and taken together, these views provide a powerful paradigm. And while "reality" may not comprise physical objects in space and time, representations along such lines are both necessary and expedient. For example, although a given sensate experience cannot be identified, in a literal sense, with a neurological process, the "mapping" of experience to physical processes in this manner provides for a level of descriptive detail (and an understanding of relations among experiences generally) that could not be obtained otherwise. And so the interpretation of "a set of sensations" as "an apple falling" furnishes much more predictive capability than could be obtained on the basis of a detailed description of "sensations as colors and shapes" – and thus an experience that might be called "*applerama*."

b. On this view, attempts to "derive physical space from pre-space or pre-geometry" do not make sense, because space is not necessarily an attribute of extra-personal reality, but yet definitely exists "subjectively" as a conceptual construction and element of experience. It is therefore unclear why it should be necessary to reinvent the wheel in this fashion. Moreover, even if an argument could be provided for such a move, entirely on grounds of descriptive and mathematical efficacy, the introduction of concepts such as "extra spatial dimensions" and "pre-space" should be unconstrained by *a priori* considerations. For example, it is not necessary that extra dimensions be "small" in order to be invisible, because the *space of physical theory* is not logically constrained to correspond, by one-to-one mapping, to the *phenomenological space of perception*. That is, "real space" need not be isomorphic to the "inner space" of perception. What is taken to be the three-dimensional space of ordinary perception is *invariant* to the detailed nature of models of the external reality, which can be of any configuration whatsoever so long as they are intelligible and can be intuited to represent experience. Therefore, just as only empirical evidence and intelligibility of theory can guide the correlation of a neural process with a given sensation, so only can experience/intelligibility determine the accuracy of a mapping between some theoretical, greater-than-three dimensional space and the phenomenological (i.e., experiential) space of three dimensions.

Nonetheless, if a representation involving more than three physical dimensions of space should be adopted, the question of how topological connections in such a space are to be related to those in three-space is a non-trivial one. In the analogy of television to perception (figure 1-4), the external reality and the camera/receiver/digital-computer/video-software complex that probe and reinterpret that reality are all of three dimensions, while the dot-matrix image is projected on a two-dimensional television screen. Zooming-in the camera on some aspect of the environment might reveal more detail without providing more information about the dimensionality, just as moving closer to the TV screen can reveal more about the dots without necessarily revealing more object features. In like fashion, confusing the Kantian *thing-in-itself* with the physical object seems to be one of the difficulties that impede the application of positivism to physics. The reason it is sensible to employ physical models at all – for example, neurological maps taken to represent cognitive states – is primarily twofold, *viz.*: (i) All experience – every possible observation/experiment – must ultimately be interpreted on the basis of such models (again, this is the only viable upshot of Bohr's philosophy of physics), and (ii) such maps yield the greatest level of detail and intuitive understanding that is scientifically possible, even though they are ultimately incomplete (in the case of the cognitive sciences, this is just as readily apparent as point [i], e.g., the visual perceptions of ordinary experience are extraordinarily difficult to represent, in mathematical language, absent spatial representation – and, ultimately, any such language must be interpreted in a framework of spatial representation). The use of higher-dimensional spaces is of dubious merit in physics. Theories should be limited with respect to the introduction of superfluous mappings; relations that introduce unnecessary complications.

Reality might best be thought of as an infinite-dimensional manifold of sensate and experiential *qualia* organizationally unrelated to the notion of spatial structure, which perhaps has a very limited role as an element or mode of experience associated with humans and certain other sentient beings. But regarding scientific description, given the power and hence heuristic value of spatial intuition it would seem that theoretical constructions should be based on spatial organization as much as possible, even in the cognitive sciences. The more "natural" such constructs are the better, because comprehensibility – even in the limited sense possible via mathematical theory – is the ultimate goal of science. And economy of axioms is only one aspect of comprehensibility. In addition, it is all-important that conceptual constructs not limit the range of experience that physics can embrace (such as the way in which

Copenhagen prevents thinking about regimes deemed irrelevant by virtue of the Uncertainty relations).

c. Because it is not yet viable to attempt to describe experience comprehensively on a non-physical basis, and because the physical paradigm furnishes a framework for detailed and comprehensive constructions [even in the cognitive sciences], physics will continue, for the foreseeable future, to hold its place as the most fundamental of the sciences. And to be logically consistent, physical theories cannot include "sentient souls" as logical primitives. Rather, "observers" should appear in models of physics only insofar as they can be represented as physical constructs. If merely in this sense alone, quantum theory, as usually understood, is incomplete and inconsistent. Physical theory must be observer independent.

d. It makes no sense to stipulate, per Heisenberg, that visualizable physical concepts cannot be applicable at the quantum scale and below, or that theoretical speculations must not reach beyond the limits of scale set by the Heisenberg Uncertainty relations. Agreement with experience and the judicious toleration/limitation of inconsistencies are the primary criteria for theory, and the standard interpretation of quantum mechanics is neither a necessary choice nor a good one.

e. It makes no sense to stipulate that the spacetime continuum of General Relativity represents an irreducible aspect of the world, or contrariwise, that a spacetime continuum must emerge from irreducible quantum properties, which are in turn considered fundamental. For example, if a constitutive construction based on Euclidean foundations can produce as good a map of reality as such based on the stipulation of geometry *per se* as a logical primitive then, from a mathematical viewpoint, there is no reason why conceiving an underlying structure of space embedded/described in an Euclidean manifold may not be considered preferable.

f. It makes no sense to argue that physical theory must determine a unique cosmological model, applicable to the evolution of existence as a whole – i.e., that *boundary conditions must be deducible*. Similarly, it is also senseless to stipulate that physical laws must have an absolute significance and cannot be *emergent* – in a sense, "environmental." Speculating about the ultimate, fundamental building blocks of matter is to equate physics with metaphysics. And the effort to find mathematical forms that constrain the dynamics of the vacuum and its evolution via fixed, ultimate elements, as in the approach of *String Theory*, produces a correspondingly unconstrained "landscape" of possible solutions, the diversity of which is thus inversely related to the degree of complexity and detail of specification regarding the nature and behavior of the building blocks. To link the fundaments of theory with the boundary conditions of the universe is to make such fundaments virtually impossible to find.

The idea of a deterministically evolving universe is not philosophically unsound. However, to the degree that such a construct is to be accurate it must be correspondingly resistant to "compression" in the reductionist sense. This should be understood in the sense of the remark often attributed to Woody Allen: "I'm astounded by people who want to 'know' the universe when it's hard enough to find your way around Chinatown."[62] The apparent limits of the cosmos, large and small – i.e., those that current theory and observation seem to reveal – should not be taken to necessary delimit "everything that exists." Again, no theory should be deemed "final," even if considered as such "only in principle."

g. Despite the philosophic hagglings of the ages, there are indeed aspects of reality that humans can and do have direct knowledge of. In contradistinction to Descartes' illusory "I" (as in, "I think, therefore I am"), everything that is taken to comprise a personal experience by such an "I" is, *ipso facto*, a "thing-in-itself," the existence of which cannot be doubted. Kant's "noumenal" objects of extra-personal reality, deemed to lie behind the veil of percep-

62 Woody Allen. Quotes.net, STANDS4 LLC, 2010. http://www.quotes.net/quote/2283,

tion, must share their existence, whatever the nature thereof, with the feelings, perceptions and thoughts of sentient beings, and thus must have "something" in common with such things. However, science is not yet capable of constructing a model of existence based exclusively on concepts that express the qualities of subjective experience *per se* – i.e., a theoretical model of a sort of "sensorium" or "spiritual universe" absent the characteristic of extrapersonal spatial organization (and it must be accepted as possible that the human intellect is fundamentally incapable of this). And so a science of sentience must be concerned with both the "inner space" of perceptual experience and the drawing of correlations between "physical" and "mental" events generally.

h. The "mind-body problem" resolves in the manner hereinabove indicated; i.e., it is a misconception, not a paradox.

Chapter V

A Framework for Physics

§ 5.1 Introduction

IN the previous chapter, widely accepted arguments against the viability of a locally deterministic theory of quantum physics were examined and invalidated. It was also demonstrated that extreme bias in favor of irreducible conceptual elements – what might be called "Absolute" physical primitives such as fundamental particles and constants – as against the interpretation of natural phenomena and laws as *emergent,* is both empirically and epistemologically inappropriate. In this chapter, positive arguments will be adduced for a unified theoretical framework employing principles of emergent, collective order and local, causal action.

The prime tenet to bear in mind vis-à-vis this enterprise is that the physical constructions invoked are not intended to capture or reflect "the true nature of reality." Physics must not be confused with Metaphysics. Thus, while the ontology of classical dynamics reflects a world consisting of nothing but the mechanical motion of inert colorless objects, which leave "no sound in the wood," human experience is altogether different. It is not only full of color and sound but many other "non-physical" qualities as well: from flavor and fragrance to warmth and cold and feelings generally; not to mention dreams and memories, language and music, humor, etc.

This is especially important to remember in the context of a locally deterministic framework, which may appear crafted to be "realistic" inasmuch as its conceptual elements can readily seem to have ontic significance. However, it must be borne in mind that they do not, and that the invocation of visualizable constructs does not reflect a metaphysical program. Properly viewed, the contention that such an enterprise raises is merely this: that the closer a theory adheres to visualisable constructions the greater is its heuristic value as *map* in the sense of section 1.8.3, with respect not only to intuitive comprehensibility but also logical consistency, efficiency and descriptive completeness as well.

On the other hand, the instinctive appeal of "realistic" physical qualities can mislead speculative thought in a converse manner. The hyper–dimensional constructs of *Superstring* theory sprang from considerations aimed at understanding how a mathematical analogy with elastic tension, a compelling physical intuition, could be made to conform to the requirements of the relativistic and quantum formalisms. The root idea of string theory emerged in this context from an intuitive image of quark confinement as a manifestation of a binding force that increases with distance up to a finite limit, beyond which it rapidly vanishes – like a rubber band stretching and then breaking. And yet it is clear from the epistemological arguments of the previous several chapters that the idea of such a thing cannot be expected to have, literally, an objective referent. Such ideas are anthropomorphisms – conceptualizations of experience appropriate to the objects of everyday perception; reflecting the intuitions that relate tactual and visual sensations as they are integrated via that perception, and colored by qualities that are devoid of meaning outside their perceptual context. While the theory does not treat the many-dimensional construct as an ordinary string or membrane, its spacetime (Minkowskian) formulation nevertheless has a certain intuitive appeal, because it furnishes compact topological imagery for the interpretation of interactions on the basis of Feynman diagrams (e.g., via loops separating and joining).

In 1951, near the end of his life, Einstein expressed his belief that the progress of physics would become stymied for an extended period because, as he put it, "physicists have no understanding of logical and philosophical arguments."[1] In this connection, there seems to be a striking analogy between present-day efforts to construct many-dimensional, super-symmetric "theories of everything" and late-nineteenth century efforts to develop similarly comprehensive aether theories based on intricate elemental mechanisms. With regard to this analogy, the temptation to paraphrase Einstein from his first relativity paper is irresistible – to suggest, with tongue in cheek, that

"The introduction of a 'ten dimensional space' will prove to be superfluous, inasmuch as the view here to be developed will not require a 'string or other surface embedded in such a higher dimensional space' provided with special properties, nor assign a 'compactified manifold' to a point of the empty space in which field processes take place."

It is the author's contention that if the talented practitioners of *String Theory*, *Loop Quantum Gravity* and other currently fashionable approaches to unification and/or quantum gravity consider the ideas presented in this chapter as a viable alternate perspective and starting point for their investigations, they will realize a relatively quick and gratifying reward for their efforts. The mathematical tools and talent at the disposal of today's physical theorists are well suited to the framework. Again, while the dogmatic interpretation of any principle, law, or constant of nature as "Absolute" must be rejected, it is yet a credible conjecture, with good empirical support, that local causal action in three space plus one time dimension is an appropriate context for describing the order that appears in the phenomenology of modern physics.

To avoid repetitiveness the author will not belabor the conventional nature of concepts and premises invoked in the sequel (nor qualify, with quotation marks, every occurrence of an especially suspect concept, e.g., *real, physical, objective,* etc.). Accordingly, the reader is here admonished for the final time to guard against attributing ontological significance to any constructive image or assertion – lest it should appear that, in enthusiastic elucidation of a favorite hypothesis, the author has neglected this imperative himself.

1 Einstein, Albert., Wade. Baskin, and Maurice. Solovine. Letters to Solovine : [1906 - 1955]. A Citadel Press book. New York, NY: Carol Publ. Group, 1993. page 123 (Einstein to Maurice Solovine, Feb. 12, 1951, AEA 21-277)

§ 5.2.a   Early Hints of an Underlying Unity
Part A

As discussed in chapter four, relativity and quantum theory share several interesting commonalities. Both modify the idea of physical reality, attenuating the significance of the concept *physical object* by negating its most essential attributes – the capacity to manifest a definite form, and occupy a distinct position in space and time – while attributing a set of amorphous, wave-related characteristics in their stead. Both theories accord nicely with the intuition that matter comprises stable patterns of energy in a covariant field or energetic substrate – a clean, comprehensible incarnation of the idea that motion and mass are in a sense identical, and that something akin to "pure motion" on a phenomenological level can be understood as detectable patterns in an underlying, otherwise undetectable level of structure. And so the relativistic mass-energy equivalence and negation of rigid structure meshes quite well with the quantum notion of wave-particle duality, the *Uncertainty Principle* and, as Dirac's equation anticipated, all that follows thereon: particle/anti-particle pairs and virtual particles spontaneously appearing and dissolving, transmutating, occurring in resonances, etc. – i.e., the *quantum field* framework.

Both theories originated in the study of profound paradoxes associated with the behavior of light. And, albeit a negative feature – and contrary to the conjecture put forward in the previous paragraph – it is yet a salient similarity that both theories, per their orthodox interpretations, premise that absolute limits of naturalistic inquiry have been reached – final barriers to knowledge due to the existence of physical constraints, which are postulated to be intrinsic properties of matter/energy. Thus, while the evidence is highly suggestive regarding the existence of an underlying structure of space, the symmetries that emerge with and characterize the ponderable phenomena evidently prevent the discovery of any such structure – the existence of which, moreover, is forcefully dismissed in the usual interpretations of both theories.

In the theory of relativity, such an unconditional limit is inherent in the notion that space is featureless, and coordinate systems intrinsically ambiguous. Thus, events in space and time are represented via a non-coordinate-specific, metrically dynamic geometry. Again, as discussed in the previous chapter, the reason for this treatment is the premise that the relativistic attributes of space-time associated with motion and gravity/inertia are taken as axiomatic; they are accepted as given aspects of reality, while space is deemed to possess no constitutive elements or features that can be "measured up" – no fundamental "background" or deeper seat of physical activity that might somehow be determinative of the *observed phenomena* (hereinafter simply *"phenomenology"*). In this view, there can be no experimentally accessible levels of reality that might accord better with the notion of an underlying structure – "Euclidean" or otherwise – deviations from which manifest as the phenomenology. (As hypothesized, for example, in plenum-type aether models wherein fields and even matter can be represented as dynamical features of the material structure underlying and pervading space.)

Similarly, it is an inherent limiting aspect of quantum theory that Heisenberg's Uncertainty Principle is taken to reflect a fundamental feature of reality. Thus, the phenomenology of quantum theory cannot be deemed to arise from a deeper seat of physical activity – i.e., a scale of physical dimensions smaller than that which can be directly probed in accordance with the Heisenberg relations. The epistemological commonality associated with this aspect of quantum and relativity theory is that, in both cases, what appear to be emergent properties of an underlying physical dynamic must nevertheless be treated *phenomenologically*, because the symmetries associated with the phenomena – i.e., the emergent properties – obscure their own causes.

And so, contrary to the taboos regarding the aether, both theories yet imply that the *"Vacuum"* is not without crucially important, physical characteristics. In fact, in their mature incarnations both theories – i.e., *General Relativity* and *Quantum Field Theory* – virtually assert the existence of an all-pervasive energetic substratum of space. As will hereinafter become evident, if these theories had not developed along essentially isolated tracks – each with its own peculiar trends and limitations, informed by a unique set of [largely unchallenged] philosophical prejudices – a unified theory of gravitational and quantum phenomena might have been found in the twentieth century.

§ 5.2.b   Early Hints of an Underlying Unity
Part B

The concept of the quantum-of-action emerged as an indirect consequence of Planck's efforts, at the turn of the last century, to derive an equation for the spectrum of blackbody radiation that was conformable to his
thoughts on entropy and his approach to thermodynamics generally. Although he believed that the ultimate physical reality must be a *continuum*,
Planck addressed the problems that he encountered in his study of the
blackbody spectrum via a discrete counting of energy states; a simplifying
mathematical technique that he introduced to facilitate his calculations.
Thus, the first use of the idea of discrete action was purely as mathematical
artifice and Planck did not, at the time, believe the device to have any significance beyond the traction that it afforded his calculations (somewhat
analogous to Ampere's mathematical treatment of electromagnetic phenomena via the concept of continuous yet countable current elements – as
opposed to Weber's explicit invocation of discrete electric charges). Ultimately, of course, this methodology did lead to the formal concept of the
quantum of action. But it was not Planck who exposed its physical significance, which he certainly did not embrace in 1900.

James Clerk Maxwell's electrodynamics, in conjunction with the statistical-mechanical approach to thermodynamics that he helped pioneer,
seemed to Planck and others to imply a contradiction – namely, that the
equilibrium energy of cavity radiation should be spread over an infinite,
continuous range of frequencies. Although Planck was able to deduce an
experimentally valid, mathematical description of the blackbody spectrum in accordance with the aforementioned statistical treatment, it was
not until five years after this accomplishment that Einstein – more or less
contemporaneously with his work on Special Relativity – wrote a paper
that embraced and promoted the explicit idea of discrete energy packets as a way to explain the photoelectric effect, as well as Planck's earlier
treatment of blackbody equilibrium. Although Einstein's methodology,
like Planck's, lacked an unambiguous physical interpretation, he yet was
able to show that a wide range of phenomena could be described on the

assumption that all energy is somehow carried in discrete forms and, in particular, that light can be treated as an ensemble of such energy quanta. Einstein justified his quantum hypothesis as a heuristic tool for deducing the exchange of energy between radiation and matter. But he believed, as with his work on relativity, that he had uncovered a deep property of nature... once again correctly. While transferences of radiant energy would seem to involve some sort of harmonic resonance between emitter and absorber, under Maxwell's electrodynamics the magnitude of such transferences is deemed to be proportional to the square of the electric field amplitude – whereas both the blackbody spectrum and the photoelectric effect seemed to be explicable only on the assumption that energy is exchanged, not only in discrete quantities, but also strictly in proportion to the frequency of the wave packets.

At this juncture, another deep unity is revealed between relativistic and quantum phenomena, which seems to have gone largely unnoticed until Louis de Broglie recognized it sometime in the early 1920's. However, Einstein was evidently cognizant of it at least as early as 1905. The following quotation is from his first relativity paper:

> "It is remarkable that the energy and the frequency of a light complex vary with the state of motion of the observer in accordance with the same law."[2]

In other words, Einstein realized that energy is tied to frequency via his relativistic treatment of the *Doppler principle*, which is logically independent of his interpretation of the photoelectric effect. Moreover, from an historical perspective this discovery seems quite strange, because it could have been made by anyone much earlier (the relation should, in fact, have been given by Maxwell). The connection between energy and frequency is virtually self-evident, in the same sense that the Doppler relation between frequency and motion is. The equation that Einstein deduced for the energy of light as a function of the Doppler principle – that is, as a function of the relative motion of source and absorber – reduces to:

---

2 Einstein, Albert, H. A. Lorentz, H. Minkowski, and Hermann Weyl. The principle of relativity : a collection of original memoirs on the original and general theory of relativity. New York: Dover Publications, 1952, page 58. English translation of original 1905 German-language paper (published as Zur Elektrodynamik bewegter Korper, in Annalen Der Physik.17:891, 1905)

$$E = E_0 \frac{1 - \dfrac{v}{c}}{\sqrt{1 - \dfrac{v^2}{c^2}}} \tag{5-1}$$

where E and E_0 represent the energy of the light complex as measured in the moving and rest frames, respectively, v is the velocity of the moving frame relative to the light source and c is the velocity of light. For small velocities – in other words, when relativistic effects can be ignored – this equation approximates to:

$$E = E_0 (1 - \frac{v}{c}), \text{ or } \Delta E = -\frac{v}{c}(E_0)$$

But, by virtue of the Doppler effect, the equation for frequency follows the same form:

$$F = F_0(1 - \frac{v}{c}) \text{ or } \Delta f = -\frac{v}{c}$$

where f and f_0 represent the frequency of the light complex as measured in the moving and rest frames, respectively. Therefore:

$$\Delta E = E_0 \Delta f \text{ and } \frac{E}{E_0} = \frac{F}{F_0} \tag{5-2}$$

That is, disregarding relativistic effects, the energy of light is directly proportional to its frequency.

In general, neither relativistic nor quantum phenomena transcend or preclude theoretical description on the basis of readily visualizable concepts. In fact, they can be interpreted quite well in such a manner. Ironically, the direction that quantum theory took may have been largely due to Einstein – not only, as previously discussed, because he did not support with sufficient vigor alternate possibilities that were not particularly appealing to him, but also because he was able to extend the concept of discrete energy quanta abstractly to a wide range of phenomena, entirely on the basis of statistical methods. (Methods which he had perfected in his early research on thermodynamics and Brownian motion. By their application to radiant energy exchanges, he was able to demonstrate that conservation of momentum requires that radiation be emitted in highly localized, directed packages.) Yet he remained among a minority of physicists who were qualified to make a difference while believing in the possibility of a classical or pseudo-classical description of individual quantum events. In this period, neither Planck nor Einstein published any ideas on how to account physically for the quantum hypothesis, although Einstein indicated as early as 1909 that he had been entertaining an image of localized wave packets surrounding and traveling with point particles. Einstein also speculated publicly about what he called *Ghost Waves*, largely bereft of energy, which he conceived to accompany particle motion more or less in the manner of de Broglie's *Pilot Wave*.

§ 5.3 On the Concept of Time in Physics

A key issue in the above discussion of frequency, as well as in all that follows, is the concept of *time* and, in particular, confusion with respect to the notion *"instant of time,"* the intrinsic inconsistency of which is highly problematic in this context. As argued in chapter one on purely epistemological grounds, such idealized objects of conception – like the *dimensionless point* – cannot be taken literally, and can only serve a heuristic purpose. As with all concepts, that of the *instant* harbors contradictory meanings, but inasmuch as it is employed in detailed mathematical calculations it must be handled with extreme diligence.

All physical processes are conceived to occur in a finite time interval. Moreover, during any particular "time-slice" of finite duration (another idealization not to be confused with the instant), a physical process occupies an interval of both time and space. However, this statement is only acceptable without further qualification if the physical process at issue is considered to be "at rest" in the coordinate frame in which it is measured. Adding motion introduces fundamental kinematic and conceptual complications, which are at the root of the relativistic and quantization effects discussed in the following section, and the equivalence of gravity and acceleration with respect to such effects. At the heart of all these relations is a much-overlooked but fundamental connection between space and time as they arise together in cognition, which is obscured and distorted by the notion *instant*.

This connection appears in raised relief in the context of periodic phenomena, and in particular the notion of a *wave-train*, which is inherently bound up with that of *motion* and poses a certain puzzle in this regard. During any finite time period the "space" that a wave-train occupies is changing, and so it is of fundamental importance whether or not its length is measured in a co-moving frame. This is subtly different from, and more fundamental than, the relativistic issues connected with the finite, constant velocity of light (and the quantum mechanical non-commutation of position and momentum). In contradistinction to relativistic effects, the wave-train will be *larger* or *smaller* – measured in a non-co-moving coordinate system – as the velocity of translation is *faster* or *slower*, respectively, in that non-co-moving system. Similarly, the position of the wave-train will be spread over a greater region the greater is its velocity, and thus momentum. This is simply because as the time interval and/or velocity increases, so does the region of space that the wave-train occupies *during a finite time interval*, and some such interval must be stipulated in conjunction with the measurement inasmuch as there are no instants of time.

Implicit in this remark is the characterization of space as something "substantial," because it is only sensible if wave motion is constrained by some physical plenum in which it is embedded; something that "is waving." For if space is deemed to be "truly empty" or "non-physical" these considerations are moot; there is then no basis for the determination that "to and fro" motion internal to some system traces out a greater distance when that system has a translatory motion than when stationary – i.e., the wave motion is relative to nothing but an imaginary and thus arbitrary coordinate system. In order to account for measurable differences between such frames, as occurs for example with respect to relativistic phenomena (e.g., the so-called *"twins paradox"*) there must be some way to distinguish them (contrary to widespread opinion, while this problem is addressed by the generalization of relativity to include arbitrarily moving reference frames it is *not* in fact epistemologically resolved thereby unless it be granted that spacetime is on a physical par with other matter/ energy forms – that is, that the vacuum is in effect "an aether").

And so the following principle must be recognized: Because instants of time do not exist, when specifying the size of a moving physical process or *entity* one must also specify a concomitant velocity relative to the measurement frame as well as a measurement time interval, because – apart from relativistic considerations "proper" (i.e., measurement issues involving the velocity of light) – the size of the space that the entity occupies depends upon both the velocity *and* the time interval associated with the measurement. It is important to note in this context that the notion of the "point particle" is just as devoid of meaning as that of the "moment" or instant of time and, for the same reason, any physical process in space must have an inverse relation between position (volume of space occupied) and velocity (duration of time during which position is assessed).

It should be acknowledged that this fundamental aspect of the relation between time and space measurements is not generally addressed, vis-à-vis the Special Theory of Relativity *or* Quantum Theory. It is subtly different from – albeit logically related and prior to – both the problem of distant simultaneity/clock synchronization that is at the heart of Einstein's original approach to relativity, and Heisenberg's Uncertainty Principle, which is at the heart of quantum mechanics. In Einstein's analysis, the measurement problem is associated with the finite, constant velocity of light, which impacts the measurement process in the well-understood ways. Likewise, in Heisenberg's analysis the measurement problem is associated with the irreducibility of the quantum-of-action. But the problem that arises in connection with "instants of time" has to do with the relation between space and time in a more general sense. It has pervasive consequences, all of which arise from one fact – *viz.*, that motion does not comprise a series or collection of "static" components in "instants" of time.

As long as the significance of this inherent connection between the concepts of space and time is not recognized – which is crucial to a proper understanding of both the relativistic analysis of simultaneity and the quantum mechanical interpretation of the uncertainty relations and furnishes the ultimate logical basis for both the formal unification of spacetime and (as will hereinafter become clear) quantization-of-action – then neither the Minkowskian convention nor the Uncertainty Principle is properly understood, because the underlying epistemological reason is missing. And that reason, again, is the intrinsic unity of the cognitive construct;

the unity of space-time as *ground* in the *Gestalt* construct *motion,* which establishes the moving entity as *figure.*

The concept of motion constitutes a subset of the more general concept *change.* With respect to motion, what is changing is the spatial location of an object. But in order for such a change to make sense – that is, in order for the idea of one object occupying two different locations to be meaningful – the context, or background, needs something against which the difference, and thus the change, can be perceived. Time is this something, which completes the background against which it is possible to speak of "the same thing" at "two different places," rather than "two different things distinguishable by their separate locations." Without awareness of the epistemological unity of space and time with regard to the concept of motion (which, again, is the underlying reason for the viability of their mathematical unification under the Minkowskian formalism), and of the problematic role of the concept of "instants of time," it is not possible to fully appreciate the measurement problem associated with simultaneity and the deep connection with the quantum-of-action.

In summary, with respect to the usual discussion of the relativity of simultaneity, reference is made to the finiteness and constancy of the velocity of light, which makes the correlation of distant events ambiguous. But this is not the most general interpretation of the circumstances; it does not get to the root of the matter. As stressed above, and as will become clear in the sequel, the difficulties connected with the limit velocity of light are symptomatic of a more general problem associated with the notion of instants of time.

§ 5.4   On the Relation between Energy and Frequency

According to Maxwell, light is a continuous periodic fluctuation of the electric and magnetic fields in space and time. Planck, in his treatment of blackbody radiation, utilized the mathematical device of an ideal dipole oscillator embedded in and at equilibrium with a fluctuating electromagnetic field, and imagined the blackbody to be an ensemble of such dipole oscillators and thus equilibria in the cavity radiation.

In this context, the question naturally arises: "How should the interaction between the field and the oscillators be represented?" But in Planck's original work this question was not a consideration. Planck was attempting to establish the second law of thermodynamics on an independent basis; in particular, by employing a continuum construction independent of atoms or other discrete primitives, which he felt were fictions of theory that would ultimately give way to representations based on the principle of continuous structure. His derivation of the correct expression for the blackbody spectrum was based on a probability distribution for the energy across an ensemble of irradiative oscillators, interacting with the corresponding ensemble of field modes that can exist in the cavity, though his calculations relied on a discrete as opposed to continuous energy distribution.

In accordance with the Maxwell-Lorentz theory, energy is proportional to the square of the field strength, and an electron continuously absorbs and radiates energy with its changing velocity relative to the field. Disregarding some of the obvious problems with this notion (such as the infinite energy dissipation associated with closed orbits), and if no thought is given to the mechanism by which the exchange takes place, then, in accordance with the equipartition theorem, an abstract collection of oscillators and an abstract collection of modes of cavity radiation can be considered to be in statistical equilibrium when the field energy is distributed randomly across the allowed *modes* of the field. But because, in principle (as usually assumed in accordance with classical electrodynamics), there are an infinite number of states (i.e., modes) that the field can occupy, the theory has no inherent upper cutoff for the frequency. And so another problem of infinity arises.

It is a misleading aspect of the conventional history that Planck determined to resolve this problem by counting the energy in discrete quanta. It would be more accurate to say that, while he did employ a discrete counting of energy states to make the calculation tractable, and that this treatment indeed helped make the relevant relations more apparent, it was not the discrete counting *per se* that was essential to the solution, but rather the association of radiant energy primarily with frequency, instead of intensity (i.e., the square of the field amplitude). Moreover, if the mechanism of the energy exchange is called into question, subtle considerations come to the fore that apparently escaped the attention of Planck and his peers. Once again, epistemological concerns emerge that are crucial to an adequate understanding of the circumstances – in this case, Zeno's paradox regarding the infinite divisibility of time, which here again, as in much of what follows, is of profound relevance.

Any exchange of energy, between matter and field or otherwise, must take place in finite time. For this reason alone, it is evident that the equipartition theorem cannot be taken to apply to wave interactions in the same manner as to collisions of particles. Because wave phenomena, unlike point particles, are spread in time and space, the only way it can make sense to think that the energy of a wave is associated with the square of its amplitude (e.g., in the case of the electric field, energy is deemed proportional to the square of the magnitude of the electric force vector) is if the wave's entire energy content is conceived to exist in "one crest," so-to-speak, and to ignore all the others, the number of which is proportional, in both space and time, to frequency. Moreover, just as a field within an enclosure has normal modes, so too must a real oscillator. So it is sensible to expect that in order for an exchange of energy to occur, not only must some finite duration of time elapse, but also that some degree of resonance must exist between field and oscillator.

Imagine an ensemble of microscopic vibrating objects, each a miniature model of the *Tacoma Narrows Bridge*, which collapsed in 1940 due to aeroelastic flutter. This is an easily visualizable image that is useful in this context. Imagine further that, whereas the motion of the Tacoma Narrows Bridge was induced by wind, this set of oscillators is driven by water waves, which can be visualized in the manner of waves breaking on a beach. Each oscillator will absorb energy from the wave field via resonance with various field frequencies. That is, depending upon the internal oscillations that each "little bridge" can sustain, certain wave crests will add to a bridge's energy and certain impacts will detract from it – just as with pushing someone on a swing, where "in phase" pushes will increase and "out of phase" pushes decrease the motion. Moreover, this process will occur in time, so that each wave crest that breaks over a bridge can be thought of as conveying one small, discrete bit of the overall energy that will ultimately be exchanged. Naturally, the higher the resonant frequency of the bridge, the greater the number of wave impacts from which it might absorb energy, per unit of time, given that the waves are all moving at the same velocity (for all practical purposes) regardless of their frequency. In principle, a bridge with an infinitely high resonant frequency could absorb the energy from an infinite number of wave impacts, and thus acquire an infinite energy. In reality, such an allocation of energy is impossible.

This imagery is edifying in many respects. Planck's distribution necessarily follows, as it must in any such circumstance, as the higher frequencies take up energy disproportionately to the lower. And so the energy distribution will naturally have an upper frequency cutoff. Thus, while any given crest may be more or less energetic than another, it is yet clear that – in comparison to the energy required to cause a visible flutter of a bridge – the energy of the greatest wave is insignificant (anomalies aside, e.g., a tidal wave). Moreover, the amplitude of oscillation of any real, physical system has constraints. Even if a tidal wave – a quite rare anomaly – were to strike a bridge, the result would not be a smooth, resonant-like absorption of energy. (Indeed, if the bridge could be rigid, and yet sturdy enough to withstand such a singular event, virtually no energy would be absorbed from it). In actuality, the amplitude very seldom reaches that of a tidal wave – rather, statistically, the energy of water waves is distributed over an enormous range of much smaller crests, which again, because of their very large number will, in any more or less stable context, tend to have an average energy that remains fairly constant within that context. Evidently, what is salient is the number of crests and their average energy.

This is all readily comprehensible – extreme concentrations of energy are not the norm in real, physical systems with fixed constraints. Given the enormous range of wavelengths that electromagnetic radiation can exhibit, and the great velocity of light, it hardly seems surprising that the energy carried by radiation – which must be reckoned over a finite time interval – should be proportional to frequency. Again, by means of a straightforward calculation one can – as would seem obvious – reproduce Planck's curve for the blackbody spectrum of energy density as a function of temperature, simply by associating an average energy with each wave "crest" in the sense of the imaginary scenario discussed above.

The simple mathematical relationship between frequency and energy distribution can be easily visualized. The problem of Blackbody radiation is usually discussed in the framework of an isolated cavity, wherein electromagnetic waves interact with the electrons in the walls – which latter coincide with the nodes of the wave field – and form standing wave patterns in the space between the walls. That is, due to the finite size of the enclosure, only certain frequencies, or *modes* – those with wavelengths that are integral divisions of the dimensions of the cavity – are supported. In the so-called "classical" treatment of this problem, which leads to impossible results, the energy is imagined to be equally distributed between each of these possible modes – that is, every frequency, without regard to an upper energy limitation, with a wavelength that fits an integral number of times between the walls of the enclosure.

It is easy to understand why this leads to infinite results, and how to find a clue to the solution without engaging in complex thermodynamic considerations along the lines originally pursued by Planck. Figure 5-1 is an illustration of five plane-waves in an aquarium, the place of maximum amplitude or "crest" of each represented by a plane cross-section of the tank (the number of waves is limited to five in order to keep the analysis tractable and the diagrams visualizable). As previously discussed, if each

Figure 5-1 Five standing waves, depicted here as plane-waves traveling in the left-right direction. The planes represent the places of maximum amplitude. There are six nodes – one at each wall, left and right, and one between each plane..

wave crest in a train of waves carries energy, it is reasonable to consider the energy distributed more or less evenly among them. Inside an enclosure of finite size, each allowable wave mode corresponds to a particular, finite number of crests.

At equilibrium, every possible combination[3] of such modes must be considered a possible distribution, because equilibrium corresponds to the most random condition. This means that the total energy available to be distributed among the waves inside the enclosure corresponds to that which can be carried by five wave crests. Now, the number of different ways in which five objects can be partitioned, that is, the unique ways that five items can be grouped, is represented in figure 5-2. Each playing card represents a particular grouping. So, for example, an ace and a four indicates a set of five items, comprising a subset of four items and a subset of one item.

Figure 5-2 The seven unique ways to partition the number five: one with the number five; one with the number four (four + one); two with the number three (three + two and three + one + one); two with the number two (two + two + one and two + one + one + one); and finally one with nothing but aces. The number one appears 12 times in 5 partitions; two appears 4 times in 3 partitions; three appears twice, in 2 partitions.

3 The appropriate mathematical term is actually *Partition*. *Combination* has a distinct technical meaning.

The next figure (5-3) shows the partitions of the number four. By comparing the possible ways of grouping these small numbers, several general properties of partitions can be gleaned. For example, it is evident that the

Figure 5-3 The five unique ways to partition the number four: one with the number four; one with the number three (three + one); two with the number two (two + two and two + one + one); and again one with nothing but aces. The number one appears 7 times in 3 partitions; two appears twice in 2 partitions.

number of partitions must increase with the number to be partitioned. However, the proportionality is far from linear. For example, there are forty-two (42) partitions of the number ten; two hundred and four thousand, two hundred and twenty six (204,226) partitions of the number fifty; and one hundred and ninety million, five hundred and sixty nine thousand, two hundred and ninety two (190,569,292) partitions of one hundred. It is also evident that the smaller the number the more frequently it will occur, with *one* being the most flexible in this respect. With these ideas in mind, it is easy to understand the blackbody spectra relations – that is to say, the relations between energy density, temperature and frequency.

Analogous to the aquarium in figure 5-1, imagine a blackbody having total energy equal to that which can be carried by a handful of electromagnetic "wave-crests" – say ten. In other words, the blackbody is at a temperature corresponding to such an energy. To calculate the energy distribution curve as a function of frequency, it is first necessary to determine the frequencies that can occur. This means the frequencies with a wavelength that can fit an integral number of times within an imaginary enclosure that represents the blackbody in a manner analogous to the aquarium example above, which fits five wavelengths of the length pictured (the limitation to one direction is here irrelevant). There will be a maximum frequency – a shortest wavelength – corresponding to the total energy of the blackbody as a function of its temperature. The total energy, divided by the average energy of a single wave crest, gives the maximum number of wave lengths that can fit an integral number of times inside the enclosure. Again, in the aquarium example above that number is five. In the case at hand it is ten.

Accordingly, the ensemble of possible frequencies corresponds to the partitions of the number ten, which again represents every unique way that a set of ten wave-crests can occur – i.e., each of those wave lengths that can fit from one to ten times inside the enclosure. It is only necessary to: (a) multiply the number of wave crests in each allowable mode times (b) the number of occurrences of that mode in the ensemble of possibilities (here, again, the number of times the mode occurs among the *partitions of ten* – this gives the total number of wave crests that can occur with each possible mode), and then (c) divide each of these results by the total number of crests that occur in the ensemble of allowable modes – which in the case of the partitions of ten is 420; i.e., the number of partitions of 10, which is 42, multiplied times the total number of wave crests in each partition, which is 10. This gives the probability distribution for the modes in terms of the number of wave crests in each.

Again, it is clear that the smaller the number the more frequent is its occurrence among the partitions. This is basic arithmetic, or rather *number theory*. Figure 5-4 shows a graph of this distribution, which matches the blackbody energy distribution quite nicely – again, on a highly simplified model based on the partitions of only ten objects. Moreover, as the number of partitions increases with the total energy, two important things change. First, the maximum energy density increases and, second, the frequency for which the energy density is a maximum also increases. These relations are clear from figures 5-5 and 5-6, which show the salient

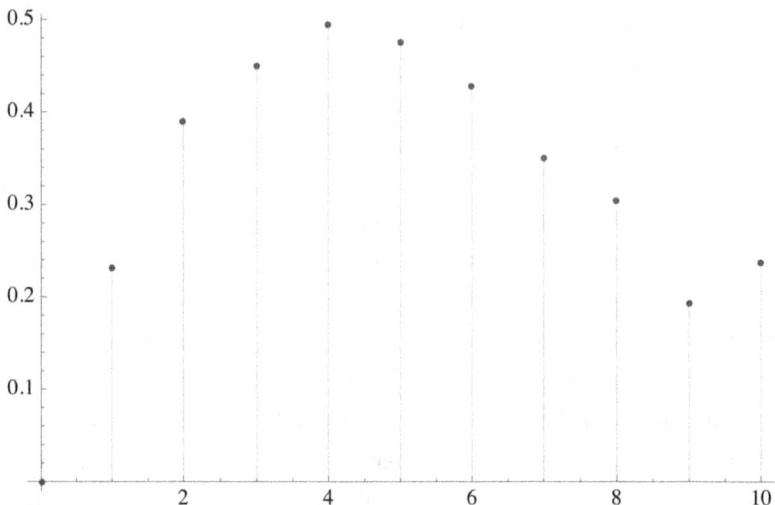

Figure 5-4 Among the partitions of ten, the number one occurs 97 times, two occurs 41 times, three 21 times, four 13 times, five 8 times, six 5 times, seven 3 times, eight 2 times and both 9 and 10 each occur once. The plot above reflects this distribution. The occurrence of each number is multiplied times its energy value – that is, its frequency – in order to obtain the energy/number of crests associated with each mode. These results are then divided by the total energy of the ensemble to yield the values indicated by the vertical axis. Thus one = 97 x 1, two = 41 x 2 = 82, three = 21 x 3, etc. Each of these results, in turn, is divided by 420, which is the number of partitions of ten (42) multiplied times ten (i.e., the total number of "wave-crests" in all the partitions)

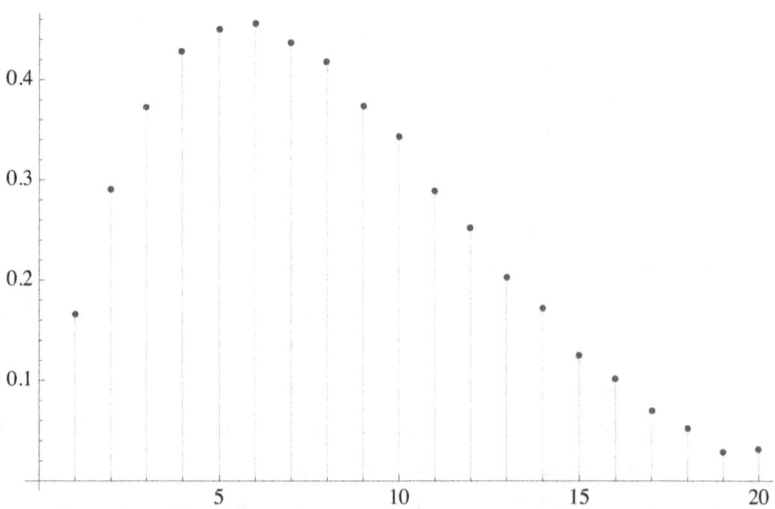

Figure 5-5 A graph of the same relations shown in figure 5-4, here reflecting the partitions of twenty rather than ten.

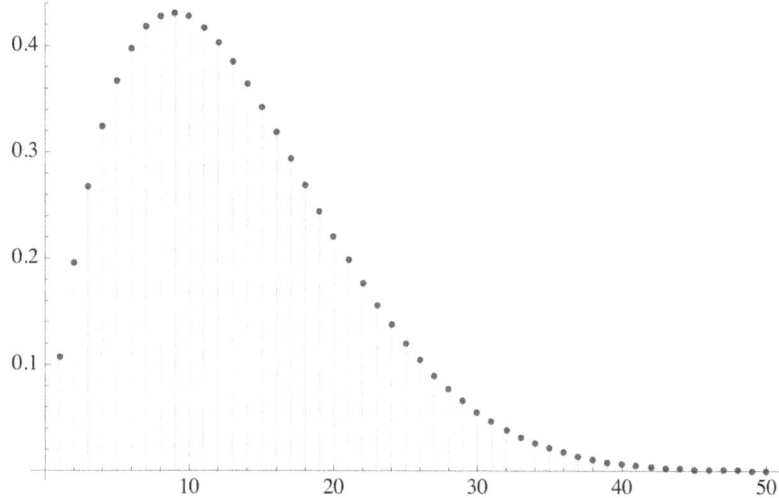

Figure 5-6 Another graph of the same relations, based on the partitions of fifty. These plots illustrate the consequence of treating each individual "wave crest" rather than each wave mode (i.e., a given number of waves that fit an integral number of times within the enclosure) as "equal-opportunity energy carrier." These graphs show how the curve changes with increasing temperature and frequency. Its form can be easily determined on the basis of simple number theory, as described in the text.

values for the partitions of two larger numbers. This matches the pattern of the blackbody spectrum, which shifts in the same manner with increasing temperature.

Einstein's treatment of the photoelectric effect can also be interpreted in this fashion. In the most simplistic kinetic terms, on analogy with the collisions of elastic solids, while it is of course possible to imagine the force carried by the field in a single wave crest to be responsible for knocking an electron out of a metal, a more realistic scenario will involve processes in time. As discussed above, the amplitude of a single wave seems unlikely to be directly related to the energy of an oscillator, which in turn is taken to represent a bound electron(s). Rather, it is reasonable to expect some resonant exchange of energy to occur, as imagined above in connection with the Planck oscillator. So again, frequency would seem to be the relevant factor, on *a priori* grounds. (Although, as mentioned elsewhere, Einstein deduced that there must be an exchange of momentum with these energy exchanges, and it was known by experiment that the absorption of energy by oscillators from the field occurs too fast to come from a smooth, resonant exchange with waves spread widely in space, these factors do not rule out transfers by wave groups such as packets.)

In any real, physical system of waves such as sound, energy is related to frequency. Why, then, does classical electromagnetic theory treat the instantaneous intensity of the electric force as the primary determinate of the energy of the field? In electrostatics, field energy is understandably connected directly with intensity. But how did this notion come to be applied to radiant energy modes, such that a standing wave consisting of ten wave-lengths is deemed to have the same energy as a standing wave of one wave-length, both in the same enclosure? It may well have seemed irrelevant that the electric force in a ray of light undergoes a cyclic change of intensity a billion times per second. Because of the extremely high frequency it might have appeared plausible, within a mechanical context, that the variations of the field vectors are just too rapid to make a difference. Perhaps it seemed as though they should wash out in the manner of white noise – like any process sufficiently random to be characteristic of a stable or equilibrium system. Thermal energy, which consists of random motion, flows only in the direction of increasing randomness, which is thus associated with a decrease of available energy. In 1870, it was perhaps not unreasonable to assume that material systems are not delicate enough to absorb energy from vibrations occurring at a frequency on the order of a billion cycles per second.

In any case, the fact is that Maxwell calculated the energy of electromagnetic radiation on the basis of the average value of the field intensity over the course of a wave cycle. Instead of considering the energy that such oscillations might impart to a system capable of resonating with them, he simply used the mean intensity of the field to calculate the mean energy content of a unit volume of that field, and multiplied the result by the speed of light to obtain the energy flux across a unit area in unit time.[4] Field amplitude, then, came to be considered the primary determinant of radiant energy by a simple misconception, and not by an inherent unsuitability of electromagnetic concepts to micro phenomena. In this connection, the opening remarks of Einstein's 1905 photoelectric paper seem especially poignant, though it is by no means clear that he had come to view things from the perspective given here.

4 J.C. Maxwell, *A Dynamical Theory of the Electromagnetic Field*, Royal Soc.Trans.155.**section108** 1864

"The wave theory of light, which operates with continuous spatial functions, has worked well in the representation of purely optical phenomena and will probably never be replaced by another theory. It should be kept in mind, however, that the optical observations refer to *time averages* rather than *instantaneous* values." [5,6]

As noted parenthetically above, early experiments on quantum phenomena were deemed to have ruled out the requirement of a finite time period for resonant absorption of energy, in favor of more or less "instant" transfers, but these experiments were based on the assumption that wave energy necessarily propagates isotropically and thus always diffuses homogeneously throughout space, even on the micro scale. In other words, not only were ordinary discontinuities in the emission of waves discounted – as well as anisotropic pulses/packets and other universal properties of wave propagation and interference – but so too were any new possibilities regarding wave phenomena generally; linear *or* nonlinear.

In contrast to the usual historical interpretation, the earliest encounters with the quantum-of-action did not necessarily refute visualizable concepts. Rather, they were suggestive of a deeper reality. It merely turned out to be possible to extend the mathematical treatment of the phenomena without a plausible physical model. And so Einstein and others immediately applied statistical methods, based on discrete distributions, to the problem of specific heat. And Bohr extended the concept to Rutherford's solar system model of the atom, without hypothesizing any mechanism to account for its applicability.

But the Bohr-Sommerfeld construction was grossly incomplete. It wasn't until de Broglie's work in the 1920's that any sort of explanation for the quantum of action seemed to emerge. Unfortunately, de Broglie's ideas – though not his equations – were sharply rejected by the physicists in Bohr's camp. Schrödinger's extension of de Broglie's ideas was likewise strictly limited, by same group of physicists, via the probabilistic interpretation that Born gave to Schrödinger's formulation (which, it should be noted, Born resurrected from one of his earlier [failed] efforts to obviate

5 A. Einstein, Concerning an Heuristic Point of View Toward the Emission and Transformation of Light Ann. Phys. 17, 132 1905 Translation into English American Journal of Physics, v. 33, n. 5, May 1965..
6 Emphasis added by the author.

causality via a probabilistic interpretation of physical events... Born displayed a fierce anti-deterministic philosophical predilection well before any apparent empirical justification for such an attitude emerged).

As is now known – thanks to David Bohm's detailed elaboration of a concept first espoused by de Broglie and Madelung[7] – a causal quantum theory could readily have been developed during the nascent years of the orthodox theory's development. It is thus clear that philosophical prejudice, primarily in the persons of Bohr, Born, Heisenberg, Pauli and von Neumann, was largely responsible for the ascendency of the so-called *Copenhagen Interpretation* and the relatively relaxed attitude toward similar formulations (i.e., employing explicitly "anti-realistic" frameworks).

As noted in the previous chapter, Bohr, Born, Heisenberg *et al* claimed to have been inspired by Einstein's approach to Special Relativity, and the belief that he had banished the aether out of strict adherence to a more or less positivistic credo, because the aether is an unobservable element of theory. And von Neumann published a convincing but flawed theorem falsely "proving" that so-called *"hidden variables"* cannot exist – that there cannot be theoretical parameters with knowable values, taken to represent unobservable elements of reality (hence *hidden*) and to have those values in advance of a quantum measurement process, by virtue of which experimental outcomes can be predicted. With the widespread acceptance of this [false] theorem, the probability associated with quantum phenomena came to be established as an irreducible fact of nature, with no possible explanation.

It is edifying to consider that for decades nobody bothered to verify von Neumann's reasoning. Perhaps many were happy that someone else had done the heavy lifting. In any case few suspected, or wanted to suspect, that the theorem might be false. Even as late as the time of Einstein's death, most physicists remained erroneously convinced on this point, despite the fact that by then Bohm and others had exposed the flaw in von Neumann's theorem and had clearly demonstrated, by counterexample, that quantum theory can indeed be interpreted deterministically on the basis of hidden variables.

All of this passed virtually without comment in scholarly circles. Decades after Bohm's decisive work appeared, mainstream peer-reviewed

7 Schrödinger also pursued a physical interpretation, and of course his work was based on de Broglie's.

journals continued to refuse papers on the subject of the interpretation of quantum mechanics, with dictums like 'the debate has been settled against the possibility of hidden variables' and/or 'it is our policy not to accept papers on the subject of hidden variables in quantum theory.'

But with the orthodox restrictions on thought lifted the early developments of relativity and quantum theory appear in a new light. As noted, the wave aspects of matter are highly suggestive, especially in the context of the mass-energy equivalence. And de Broglie's early interpretations of quantum phenomena along deterministic lines – his *Theory of the Double Solution* and its derivative, the *Pilot Wave Theory* – are more than merely suggestive, particularly in light of Einstein's geometric treatment of gravity. Again, these conceptions developed not only independently, more or less in mutual isolation, but also along logically independent lines. It would seem to require willful insensitivity if not ignorance to be unmoved by the intriguing connections between such apparently unrelated theories. And the connections go well beyond relativity and quantum theory – they embrace classical Newtonian physics as well.

§ 5.5.a   The Formal Equivalence between Brownian
Motion and F = MA Dynamics.
Part A.

In 1905, the same year that he wrote his famous relativity and quantum papers, Einstein published important original work on the theory of the Brownian motion. He had at his disposal powerful mathematical methods, similar to those developed by the physicist J. Willard Gibbs, which Einstein had independently re-invented in his early twenties in the course of his researches on the statistical-mechanical theory of heat. While the primary aim of his Brownian motion investigation was evidently to substantiate the existence of atoms, the mathematical concepts he employed have a much wider range of application – particularly his treatment of diffusion.

The law of diffusion, like the closely related law of entropy, can be derived from a very simple mathematical consideration. Imagine a large, rectangular aquarium, containing many small fish that swim about at random in small, irregular motions. If there is no particular pattern to the swimming – i.e., if the fish are for the most part insensible to each other, except on contact, and it is not feeding or breeding time, so there is no reason for the motions to be correlated – each of their paths will approximate a so-called *random walk*. Now, if the number of fish and size of the tank are both sufficiently large then the fish will be fairly evenly distributed throughout the aquarium. Again, this equilibrium condition can be understood on the basis of a simple mathematical consideration.

Consider the aquarium pictured in figure 5-7, divided by three vertical, parallel planes, labeled *A*, *B* and *C*. Again, assuming the motions of the fish to be random, the number of fish that swim past one of these planes – e.g., the central surface *B* – from left to right will, on average, be the same as the number that swim past from right to left. But consider what happens when the fish have congregated to one side of the aquarium, say the right, in order to feed as in the figure. Then, when they resume their regular, random swimming pattern, the number of fish that cross any given plane from left to right – say *C* – will, at first, be zero, while the number of fish that cross from right to left will be large. This is simply because, at first, there are no fish on the left side. Then, as the number of fish between *B* and *C* increases, so too will the number that swim past *B* to the left, and so on. As long as there is any imbalance in the number of fish on one side or the other of one of these plane surfaces, there will continue to be more fish moving across in one direction than the other – until equilibrium is re-established.

The motion of the fish, from immediately post-feeding until equilibrium is reached, is characteristic of a diffusion process, and (again, assuming large numbers) follows the same mathematical law. So if – in the absence of fish – food in the form of very many microscopic granules is introduced in the same manner as above, on the right side of the aquarium between the *C* partition and the wall, it will eventually diffuse throughout the tank. And, at any given time, the net number of granules that moves past a partition, from one side to the other, will be determined by the net difference between the number of granules on the one side versus the other (the granules are assumed to have slightly greater density than the water, and so will have a very small net downward drift velocity).

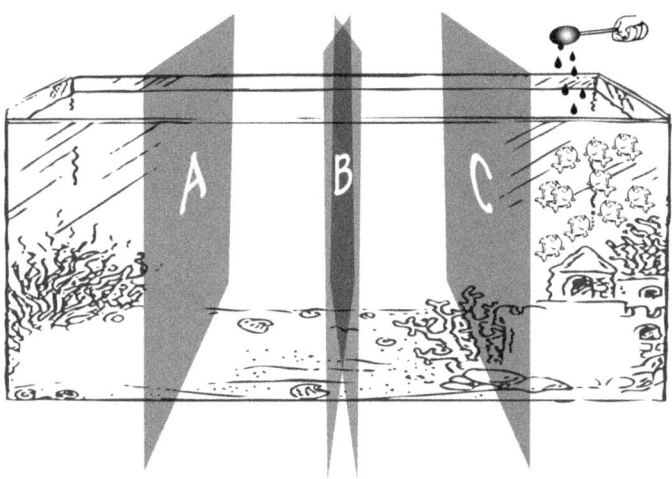

Figure 5-7   Fish have congregated on the right side of the tank to feed. This is a unique, "low entropy" condition. On the average, the fish will be randomly dispersed throughout the aquarium.

In the case of the inanimate food material – ignoring fluid currents – the driving force behind the diffusion is of course Brownian motion, induced by the random heat motion of the water molecules, which impact the granules irregularly and push them about in a manner resembling the "random swim" of the fish. Although the random walk is a well-understood mathematical concept, as is the law of diffusion, it bears extensive elaboration in this context, for reasons that will in short order become apparent.

Imagine that, in the absence of fish, the food is sprinkled evenly across the top. As before, each granule is considered to have just slightly greater density than the water, and so will have a very slight tendency to sink. However, if the difference in density is indeed very small, the greatest part of each granule's motion will be that induced by the random impacts of the water molecules, as opposed to the downward motion induced by gravity – though there will be a net tendency to drift towards the bottom of the tank.

Now consider what conditions must prevail when equilibrium is reached; that is, at the point in time when the distribution of the food across the aquarium becomes virtually unchanging. In the absence of gravity, and all other conditions being equal (i.e., assuming the tank to be large and ignoring surface effects, etc.), the distribution will necessarily be homogeneous – each small region of the aquarium must have approximately the same number of food particles, at any given time, as every other equally small region. But again, in the presence of gravity, there will be a tendency for food to accumulate more densely in the lower portions of the tank.

Figure 5-8   Food particles just slightly denser than water, sprinkled into the top of the tank, will become randomly distributed throughout the aquarium. However, because of gravity, there will be more below the (imaginary) horizontal plane M, which splits the tank into two equal portions, than above it.

Consider the horizontal plane M that partitions the aquarium into two equal parts, top and bottom (figure 5-8). If the initial condition is one in which there are equal numbers of particles above and below this partition then, at any given time, the number of particles that drifts from above to below the plane should, because of gravity, exceed the number that drifts from below to above. But ultimately this circumstance must change, because once there are sufficiently many more particles below the plane to counteract the excess downward drift, simply by random upward motion, then there will again be equilibrium. So it is evident that there must exist a stable, equilibrium configuration of particles with, on average, more in the bottom half of the tank than in the top.

A mathematical analysis of these circumstances is revealing, because one can see quite clearly that motion in a conservative field of force – such as gravity – follows the same law as that of microscopic particles undergoing diffusion in a fluid. That is to say, there is an equivalence between random motion – which follows simple, *a priori* probability considerations – and Newton's law of motion, $F = ma$.[8,9]

In order to establish this equivalence in a manner that is as intuitively obvious as possible, the above scenario is reduced to an even simpler configuration. Consider the aquarium empty of everything except air, and keep in mind the imaginary horizontal plane, M, cutting across the aquarium as above, parallel to the bottom and splitting the tank into two equal parts as pictured. Because of gravity, the density of the air is not constant throughout the tank but is somewhat greater at the bottom than at the top. Therefore, at equilibrium under gravity, there is a balance between the random motion of the air molecules and the force of gravity that causes a more or less constant condition in which, at any given time, there are somewhat more molecules in the lower portion of the tank than in the upper.

Everything about this scenario is simple and intuitively obvious, and yet it is easy to neglect an extremely salient aspect of the situation. At equilibrium, the "excess" flow of air molecules – from below the horizontal cross section to above it, which is necessary to counteract the downward flow induced by gravity – is due solely to the density gradient of the particle distribution. That is, the upward motion, which is microscopically completely random but collectively governed by the distribution gradient, acts precisely in accord with Newton's second law, $F = ma$, which in turn governs the downward motion induced by gravity.

Another way to view this is as follows. Consider the equilibrium distribution of air molecules that is maintained in the tank because of gravity. Now imagine that the gravitational field is suddenly extinguished. In a finite time, the distribution will "relax" and become close to homogeneous. Clearly this behavior exhibits time reversal symmetry. In other words, if a recording of the action is played backwards the behavior will look just like the sudden turning-on of a gravitational field, inasmuch as – in the same

---

8 Quite significantly in this connection, it was demonstrated in the previous chapter that $F = ma$ cannot apply to collisions between rigid objects. Rather, it was determined that $F = ma$ can only be a collective property of such objects.

9 For a relevant formal review that yet does not address this relation *per se*, see: Knapp, Anthony W. "Connection between Brownian Motion and Potential Theory." Journal of Mathematical Analysis and Applications 12 (1965) pages 328-349

finite time – the original equilibrium distribution will be reconstituted. The field and the distribution gradient cause the same behavior, with either the time or space directions reversed. Phrased a little differently, by simple symmetry considerations it is obvious that a distribution of particles, with a density gradient equal to that induced by a conservative field of force, will – absent that field – cause a motion of particles identical to that induced by the field. And so a random walk over an exponential density gradient – which, as will hereinafter be demonstrated, is the form that such distributions in a conservative field must take – is equivalent to a drift velocity induced by a conservative potential, acting in accordance with Newton's second law of motion, $F = ma$.

But again, as demonstrated in chapter four, $F = ma$ cannot apply to individual collisions of rigid Newtonian objects. And so a strange contradiction exists between Newtonian law, as usually interpreted, and the natural phenomena to which the law applies – even though, *en masse*, the phenomena not only conform to the law but apparently must do so, on the basis of *a priori* considerations. Therefore Newton's second law of motion can be understood in a new way.

An Aside on the Concepts of Spatial Density and Subspace

By convention, one thinks of the density of a collection of objects with respect to the manifold in which they are embedded – such as the volume occupied by air bubbles in relation to that taken up by the water in which they float – as the ratio of the former to the latter. This ratio, and the density that it represents, remains the same if the numerator and denominator are changed proportionally; doubling the volume occupied by the bubbles, while doubling the volume of water, leaves the ratio intact. Abstracting a bit, one can liken the volume of water to a volume of empty space. Accordingly, doubling the volume of space within which something is embedded diminishes the density of that thing by half, while doubling the number of objects restores the density to its original ratio.

Of course, space is usually conceived rather differently from a material substance, inasmuch as the density of a material thing can be increased without adding more material to it – that is, by compression – whereas it does not seem meaningful to speak of "compressing space." It would thus also seem meaningless to speak of the "density of space." However, there are contexts in which it is sensible to speak of space as in this manner dynamic, two of which are as follows.

First, space can be thought of as nothing more than an abstraction – a conceptual framework for the representation of relations among ponderable objects. In this purely mathematical sense, space can be thought of as an imaginary continuum subject to the mathematician's whim. Naturally, it is possible in this context to think of space as having an elastic density. Similarly, it is merely a matter of convention if, when speaking of the density of an object in space, one reverses the relation and speaks of the density of space with respect to the object. In this trivial sense, the density of space can change as a function of the objects in it. One can imagine the dimensions of space expanding or contracting, while those of the object remain constant, just as one can imagine the converse – i.e., the dimensions of the object expanding and contracting, while the dimensions of space are unchanged. Empty, mathematical space – without physical characteristics – is as flexible as thought permits.

On the other hand, one can imagine a *plenum* – a space completely filled with something physical, such as a material or field (i.e., a *substrate*) – wherein perceivable objects and processes comprise nothing but ponderable patterns of activity in that substrate. In such a context, the substrate – the substance of which the objects and events are composed – can have the virtues of a "real dynamical space." That is, if a region of the substrate suffers changes in density, so too will the objects embedded in that region. In this sense, any physical thing that is capable of sustaining emergent orders, such as a crystalline solid, can be thought of as a space in which those orders exist. And any such physical space can be dynamic.

Of course, it is possible to combine these views. And so one can mathematically embed a physical substrate – which, in turn, serves as a space for emergent orders – in a continuous coordinate system. In such a circumstance, it is sensible to speak of the physical constituents of the space, and the structure of space that they collectively form, as a "*Subspace*." That is, the dynamics of the substrate ultimately constitute a dynamic of space, and the constituent structure of that space is thus an underlying order – literally, a sub-space. Accordingly, *Subspace* will hereinafter be taken to refer to such a dynamical, constituent structure of space, in which ponderable entities and events subsist as observable patterns in that space (*not* in the sense of a material aether that conforms to mechanical laws, because matter in this context is an *emergent* phenomenon and [at this stage of the development in any case] the concept of *substance* is deemed to be irrelevant and it is only *form, per se,* that is of concern). This model unifies the paradigms of *General Relativity* and *Quantum Theory* on a fundamental level.

Unless otherwise noted, discussions will be in the context of a three-space plus one-time dimensional Euclidean/Cartesian orthogonal coordinate system, with space and time coordinates infinitely continuous, in the sense that no *a priori* limits are assumed regarding maximum or minimum distance scales (the spacetime of GR emerges dynamically from this construct). This is not to say that any assumptions are to be made about the nature of matter/energy, as continuous rather than discrete, or *vice versa*. It is merely being taken as a working assumption that, with respect to any "small" spatial scale – for example, the "Planck length" – it is possible to speak of levels of spatial dimensions indefinitely smaller still. Similarly, with respect to any "large" spatial scale – for example, that of the observable Cosmos – it is herein taken as legitimate to speak of levels indefinitely larger. And of course the same consideration applies to time.

§ 5.5.b The Formal Equivalence between Brownian
Motion and F=MA Dynamics.
Part B.

With the above notions of Subspace and spatial density in mind, consider the equivalence between diffusion and motion in a conservative potential, as described in the example of the air-filled aquarium above. Diffusion is usually treated in terms of a random walk of particles, jostled about by the heat motion of the molecules of the substance in which they are embedded. When a conservative field of force exists, such as gravitation, the motion of the particles exhibits a net drift – under an attractive force such as gravity, in the direction of increasing field gradient (and thus in direction opposite to increasing potential energy). However, instead of speaking of the density gradient of embedded particles, as in the examples of section 5.5.a above, one can employ the converse notion of spatial density and, accordingly, discuss the density gradient of the space in which the particles are embedded. So, for example, under equilibrium conditions and an exponential density gradient, one can think about the "maneuvering room" that the particles have, not as a function of the number of the particles but instead as a function of the "density of the space."

Accordingly, whereas in the conventional way of discussing the distribution under gravity the particle number is described as increasing, per unit volume of space, in the direction of the earth, in the alternate mode of speaking the density of space is increasing instead, and the particles are thought of as homogeneously distributed in that space. Thus, imagining the spatial density to be analogous to the density of air in the atmosphere, as treated in the aquarium example of section 5.5.a, consider the spatial density gradient that would be necessary to account for the actual density gradient of air in the earth's atmosphere – that is, just as in the aquarium example, but under conditions where gravitation does not exist. Idealizing the air molecules as point particles, the form that this gradient must take is obvious – i.e., it must duplicate that of the actual density gradient of the air in that example, and so must follow from the exact same calculation that is required to determine the distribution of air particles, at equilibrium, under the influence of gravitation. This is calculated as follows.

Imagine again the aquarium in figure 5-8 emptied of everything except air. Let the area of the horizontal cross-section of the aquarium, which is cut in the figure by the plane M, to be unity. Consider two heights, one at the level of M, call it h, and one slightly above it, $h + dh$. The pressure must be greater at h because of the weight of the portion of the air in the section of the tank between h and $h + dh$. The weight of that section is nmg, where n is the number of molecules of gas, on average, in the section between h and $h + dh$, m is the average mass of each molecule and g is the gravitational acceleration. The average density, ρ, of molecules in the section between h and $h + dh$ is n/dh. Accordingly, the pressure differential between h and $h + dh$, dp, is given by the following equation:

$dp = -\rho mg\, dh$ where again $\rho\, dh = n$, the number of molecules

That is, the pressure difference is simply the force of gravity on the molecules in the section between h and $h + dh$ (pressure is force divided by area, which latter again is *unity*). Now, the pressure of a gas multiplied by its volume is proportional to the number of molecules of gas times its temperature, where the constant of proportionality, k, is *Boltzmann's constant*:

$$pV = nkT \text{ where } V \text{ is volume and } T \text{ is temperature}$$

therefore,

$$p = kT\frac{n}{V} = \rho kT$$

or

$$\frac{dp}{dh} = \frac{d(\rho kT)}{dh} = -\rho mg = kT\frac{d\rho}{dh}$$

and so

$$\frac{d\rho}{dh} = \rho\frac{-mg}{kT}$$

In other words, the rate of change of the density of the gas with height is proportional to the density, and therefore the density is an exponential function, that is:

$$\rho = e^{-mgh/kT} \tag{5-3}$$

At equilibrium, the downward motion of particles can be described as an acceleration, due to gravity, which is numerically equal to the opposite, upward acceleration of particles from the bottom, attributable, purely statistically, to their larger number. This, then, is the meaning of the formal equivalence between the law of diffusion and $F = ma$, Newton's law of motion. In other words, a statistical drift brought about by a gradient in the number of randomly moving objects is mathematically identical to a drift brought about by the systematic action of a conservative force. And so, with respect to the concept of spatial density, an exponential density gradient has an effect similar to that of gravity.

§ 5.6   F = MA Optics: Classical Wave-Particle Dualism

"*F = MA Optics*" refers to the formal analogy, perhaps first fully appreciated by William Rowan Hamilton, between the transmission of the dynamical action associated with particle trajectories and that associated with the propagation of light rays or wave fronts. One upshot of Hamilton's optical-mechanical analogy is that momentum, in the particle case, serves as the equivalent of the gradient of the index of refraction in the wave case; a relation with profound implications.

Hamilton noticed the similarity between the action principles of Fermat, expressed in terms of the optical path, and of Maupertuis, which relates the paths of optical rays to those of classical objects moving freely under a conservative field of force. Naturally, this can be shown to reduce to a relation between Newton's second law of motion, $F = MA$, and the equation governing the path of light in a medium with a continuously variable index of refraction. This relation with Newton's second law, like that of Brownian motion, can be understood in a simple, visual fashion.

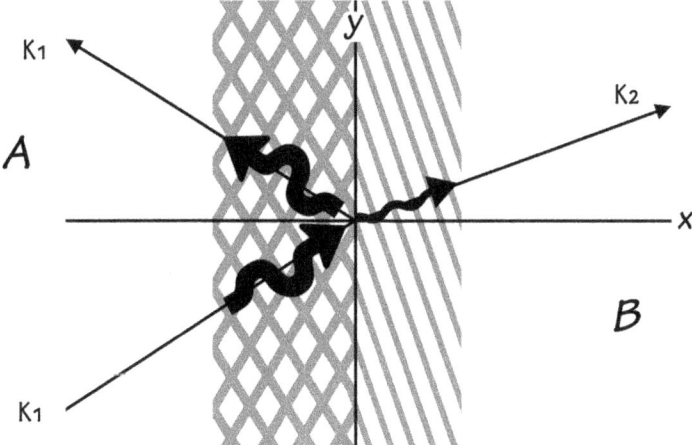

Figure 5-9   That part of wave K_1 that is not reflected slows down at the interface between medium A and medium B, as the latter has a higher index of refraction. Consequently, K_2 not only propagates in a new direction but also has a shorter wavelength than K_1, just as though it were indeed traveling in a space of "higher density."

A ray of light traveling in medium *A* is incident upon the surface of another medium, *B*, with an index of refraction that is greater than that of *A* (see figure 5-9). Imagine that the dimensions of the pictured surface are quite large with respect to the wavelength of the light. Then it will be possible to approximate the construction of an envelope of Huygens wavelets (a concept described in detail in § 5.12 below) with a simple array of lines representing plane waves. Because the index of refraction is inversely proportional to the wave velocity in the medium, this construction reveals the well-known fact that the wave fronts appear to bend at the surface, for the simple reason that the parts of the fronts that strike the surface earliest will be slowed down that much sooner than those that strike the surface later.

Thus, a simple geometric representation provides an intuitive understanding of refraction. Moreover, the detailed actions that underlie the overall "goings-on" is irrelevant to this consideration; as long as certain very general properties hold, it is clear on simple geometric and time considerations – i.e., on kinematic grounds – that the processes must unfold in a specific way.

On the same grounds a further duality can be demonstrated, between the optical-mechanical analogy on the one hand and that existing between Newton's second law and the law of diffusion on the other. Regions of higher refractive index correspond to lower wave velocity, as though such regions comprise a greater distance in the path of the passing wave. And so a region of varying index of refraction can be just as well described, mathematically, as a volume of variable spatial density – a simple transformation, inasmuch as the index of refraction must be directly proportional to the imaginary density of the space.

Notice that such a description of refraction parallels that of Brownian diffusion, as described in section 5.5. That is, in the treatment of Brownian motion above, particle density gradient is equivalent to a reverse gradient in "spatial density," inasmuch as regions that are more densely populated with particles have less maneuvering room, whereas regions that are less densely populated have more. Thus, a surfeit of particles can be viewed, and mathematically described, as a deficit of space, and vice versa. Accordingly, a gradient in the "density of space" has the same effect, with respect to diffusion, as a gradient in the density of particles, but with sign reversed – because particles will tend to move from regions of smaller density to regions of greater density. (See figure 5-10 – which illustrates the concept of a gradient of spatial density versus a gradient of particle density.)

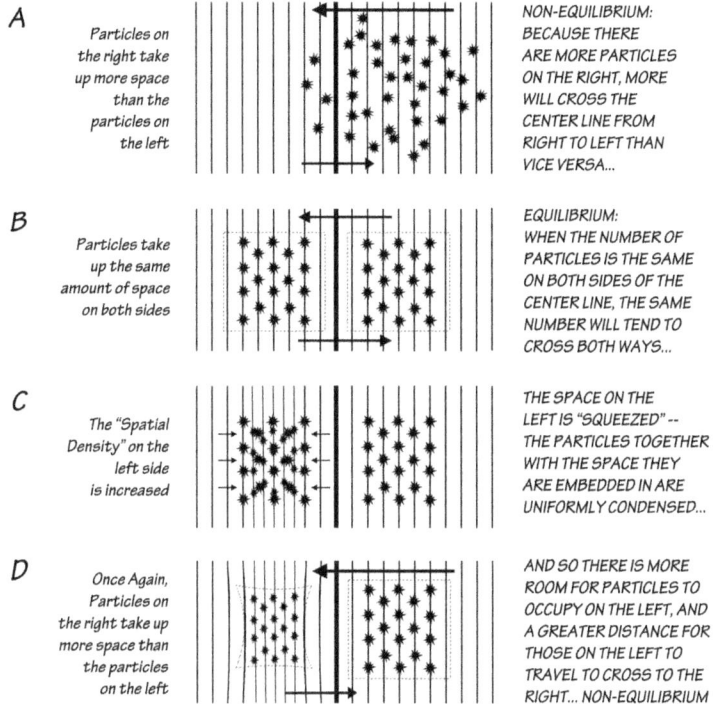

A Particles on the right take up more space than the particles on the left

NON-EQUILIBRIUM: BECAUSE THERE ARE MORE PARTICLES ON THE RIGHT, MORE WILL CROSS THE CENTER LINE FROM RIGHT TO LEFT THAN VICE VERSA...

B Particles take up the same amount of space on both sides

EQUILIBRIUM: WHEN THE NUMBER OF PARTICLES IS THE SAME ON BOTH SIDES OF THE CENTER LINE, THE SAME NUMBER WILL TEND TO CROSS BOTH WAYS...

C The "Spatial Density" on the left side is increased

THE SPACE ON THE LEFT IS "SQUEEZED" -- THE PARTICLES TOGETHER WITH THE SPACE THEY ARE EMBEDDED IN ARE UNIFORMLY CONDENSED...

D Once Again, Particles on the right take up more space than the particles on the left

AND SO THERE IS MORE ROOM FOR PARTICLES TO OCCUPY ON THE LEFT, AND A GREATER DISTANCE FOR THOSE ON THE LEFT TO TRAVEL TO CROSS TO THE RIGHT... NON-EQUILIBRIUM

Figure 5-10 An increase in the density of a region of space means that the entities embedded in it (and the distances between them) become smaller, as measured by an observer outside the region. From the point of view of the occupants of the condensed space, distances – from them to objects outside their [compressed] region – will increase. Therefore, because the occupants must travel a greater distance, relative to their new dimensions, to traverse the bounds of the space they originally occupied, it will appear to an outside observer that velocities within the shrunken space and thus time are slower.

The mathematical equivalence between motion due to force, on the one hand, and diffusion on the other – which latter can be extended to include the concept of a spatial gradient – is thus linked to another remarkable duality, namely, the classical equivalence of wave and particle action. And so the formal equivalence between mechanical and optical action, and the relation of the optical action and the index of refraction to the density of a medium, is analogous to the relation of diffusion to the spatial gradient – i.e., the refractive index is inversely proportional to the velocity of wave motion, as though more or less space were being traversed, depending upon the gradient of the index. Momentum is directly related to the gradient of the index of refraction; i.e., the spatial density gradient.

The significance of this analogy is brought home quite clearly vis-à-vis the relation between the geodesic that is determined by gravity, per General Relativity, and the optical path that is determined by the index of refraction.

§ 5.7   F = MA Optics and General Relativity

Not long after Arthur Eddington confirmed General Relativity's prediction of the bending of light near the sun he published a book dealing with the topic,[10] which he addressed by means of the duality that exists between geodesics of motion in the curved spacetime of relativity theory and light paths in the optical theory of refraction, an analogy that was well understood by Einstein.

In the mechanical-optical analogy, momentum is the counterpart of the gradient of the index of refraction. Hence there is likewise such a relation between momentum and what is here referred to as the spatial density gradient. Now, in General Relativity, the deviation of a geodesic from a rectilinear path in spacetime is associated with a curvature of the spacetime. Therefore, curvature is equivalent to a spatial density gradient; the space-time curvature of general relativity can be treated in terms of spatial density. The curved spacetime of General Relativity can be represented by a physical model of space (i.e., *Subspace*) embedded in a Cartesian coordinate system of three space dimensions plus time, in which the metric of the coordinate system is Euclidean but space, as something physical, has structure – a continuously variable, dynamic density with which the time dilatation associated with curved spacetime corresponds.

---

10 Eddington, A.S. Space time and gravitation : an outline of general relativity theory. Cambridge, U.K.: Cambridge University Press, 1921.

And so there exists a group of dualities between: (1) Newtonian motion in accordance with a continuously acting force or conservative potential – i.e., motion in accordance with Newton's second law; (2) random motion of a set of objects undergoing a Markov process (random walk), which exhibits diffusion in accordance with the density gradient of the object distribution and/or the density gradient of the medium in which the ensemble is embedded; (3) the motion of waves in a material substance, which similarly "diffuse" – i.e., *refract* – in accordance with the density gradient of the substance, and (4) General Relativity's *geodesics* of motion, in accordance with the curvature of spacetime. As will shortly become evident, this group of dualities can be extended to include quantum phenomena as well.

§ 5.8 The Quantum Theory of Motion

Peter Holland dubbed the de Broglie-Bohm interpretation of quantum mechanics *The Quantum Theory of Motion*.[11] This appellation aptly captures the essence of the pilot-wave concept, which is really not much different from Born's interpretation in terms of probability amplitudes. (It should be noted in this connection that the framework here under development is not an extension of any of the proposals put forward by David Bohm. The salient point of contact is simply this: that visualizable processes in three space plus one time dimension can adequately represent the phenomena. On the other hand, while Bohm and most advocates of his position cannot be counted among the supporters of local causal action, it is quite natural/plausible in the context of the developing framework.)

In the Born paradigm, as in de Broglie-Bohm, the field intensity – the density of the probability amplitude at a point in space (the square of the wave amplitude) – corresponds to the probability that a particle will be found at that location. The field is extended throughout space, but the (normalized) integral of the probability distribution is equal to unity, corresponding to the fact that the particle is to be found at a sharply specified position. Moreover, the probability density is locally conserved and "flows"

---

11 Holland, Peter R. The quantum theory of motion : account of the de Broglie-Bohm causal interpretation of quantum mechanics. Cambridge: Cambridge University Press, 1995.

like a current. The tangent to a flow line determines the trajectory of a particle, which is the sense in which the de Broglie *pilot wave* is considered to guide the particle motion.

As mentioned in section 4.11.2, Einstein came across this aspect of the Schrödinger equation in 1927, employing tensor analysis to reveal detailed, fully determined trajectories in the non-Euclidean configuration space of the many-particle system. However, because of the apparent entanglement of trajectories, he dismissed the result as meaningless. But it is clear from Bohm's elaboration of de Broglie's concept that under the mathematical apparatus of quantum mechanics this linking of trajectories is real. Specifically, as discussed at length in the previous chapter, insofar as the Schrödinger wave represents a simultaneous connection among all the points in configuration space, determining all of the particle trajectories, entanglement is an inseparable feature of the formalism.

There are two objections to de Broglie's original guidance interpretation that are often considered decisive. First, as previously noted, Bohm believed that the quantum field could not dynamically determine particle motion because the form alone of the field seems to be operative and its effect is non-diminishing with distance. In Bohm's view, phase correlation is a necessary but insufficient cause of motion. He therefore felt compelled to introduce the concept of *active information*, in accordance with which the particle moves under its own power but follows the form of the field.

The second objection is related to the first, inasmuch as it is also concerned with the dynamics of the guidance condition. Because particle position trends to points of greatest intensity rather than to nodes of the field – as one would expect, for example, with objects embedded in a pressure system (as a buoyant object floats to the surface in water) – many feel that the quantum-potential/pilot-wave interpretation is non-physical, and therefore self-defeating as a deterministic scheme. This objection, although not really a concern from a strictly epistemological perspective, can yet be overcome, as can Bohm's, with the concept of the spatial density gradient. This is because, in the Subspace framework, it is the form of the density gradient that is the operative guidance parameter, and mass and momentum are emergent properties.

The relation between the gradient of the probability density and the gradient of the spatial density is analogous to the relation between the classical potential and Brownian motion discussed in section 5.5. The gradient of the spatial density is taken to guide particle motion after the manner of the probability density. As noted, probability is a conserved quantity in quantum mechanics. And particle motion vis-à-vis the probability amplitude is another member of the set of analogies identified in sections 5.5—5.7, which includes Newton's second law, the law of diffusion, the law of refraction and the law of motion under General Relativity. That is to say, by interpreting the probability density in the sense of the spatial density concept, it is possible to unify the description of every physical mode of action represented under the relativistic, quantum and classical paradigms.

Based upon the simplest assumptions compatible with these ideas, the values and relations between the de Broglie *wave-lengths* of the *electron* and *nucleons*, the *velocity of light*, the *quantum of action*, the *gravitational constant* and the *cosmological constant* emerge in a simple and compelling fashion. Moreover, the link between these constants, the evident acceleration of cosmic expansion (dark energy problem) and the spiral galaxy rotation curves (dark matter problem) becomes particularly perspicuous.

§ 5.9 Lorentz Invariance, Quantization of Action and Gravitation as Emergent Phenomena

In the previous discussions of Special Relativity, the following two points stand out as distinguishing characteristics of the view usually attributed to Einstein as opposed to that of Lorentz. First, the Einstein view establishes the Lorentz transformations as absolute – effective at all scales – whereas the Lorentzian view does not foreclose the possibility that relativistic invariance is emergent, and that signals might propagate at faster-than-light velocities in certain contexts (e.g., "within" the electron, and on non-observable scales generally). Second, and not unrelated to the first point, under the Einstein view the concepts of distant simultaneity and a constitutive structure of space are meaningless, whereas under the Lorentz paradigm the notion of an underlying, fundamental space-time structure is a meaningful prerequisite for the proper understanding of relativistic effects.

History has confirmed at least one aspect of Einstein's view, namely, that Lorentz invariance applies in every so-far measured arena, including those in which electrons and electromagnetic effects generally are not considered primary players; for example, with respect to non-charged relativistic particles. Similarly, history has confirmed Einstein's conclusion that gravitation is linked with relativistic effects.

On the other hand, there are good reasons to believe that the apparent universality of Lorentz invariance is indeed emergent, albeit not in the context-specific manner originally conceived by Lorentz. Moreover, the Equivalence and Uncertainty principles are likewise evidently emergent and "quasi-universal" – i.e., universally valid within wide-ranging regimes of scale.

The considerations behind these contentions are adduced in sections 5.11 through 5.17. Starting from quite basic premises, which are so simple, clear and compelling that it seems reasonable to pursue their consequences, it will be demonstrated that certain crucial aspects of modern physical theory evidently "cannot be otherwise." Rather, as Einstein once conjectured in his figurative mode of expression, it appears as though it may indeed be the case that God had no choices in the creation of the world.

§ 5.10   de Broglie's ideal: a Meta-Law of Nature

Louis de Broglie's 1924 dissertation on Quantum Theory, *Recherches sur la théorie des quanta*, put forward the proposition that the wave-particle duality, which he had theorized to be a property of electrons, should apply to all forms of matter and energy generally, thus reflecting what he called a "*Meta*" law of nature. However, while the inferences he derived from his *Principle of Phase Harmony* were indeed far-reaching, and furnished the foundation for Schrödinger's great leap forward, his analysis did not go quite far enough.

Because it was in a nascent state at the time of the 1927 *Solvay Confer-ence*, de Broglie's interpretation of quantum mechanics was vulnerable to attack by his opponents. It was criticized for being incomplete, inconsis-tent and, accordingly, a superficial imposition on the [narrowly] self-con-sistent but more fully developed Copenhagen formulation. However, if de Broglie had been aware of the deeper principles underlying his discovery, things might have unfolded differently. In particular, Einstein might have been encouraged to take up de Broglie's flag with the same enthusiasm that characterized his efforts to unify electromagnetism and gravity via an extension of the relativistic field.

De Broglie's *Pilot Wave* theory was a somewhat watered-down version of his original vision, usually referred to as the *Double Solution*. In place of the point-particle of the pilot wave model, de Broglie originally conjec-tured that some sort of localized wave phenomenon (evidently nonlinear, and hence in today's nomenclature a *Soliton*) was embedded in the ex-tended pilot field. In his derivative, simplified version, the independently existing pilot wave acts on a massive particle, which in turn is associated with some sort of internal, periodic process. In both cases, the extended wave field guides the trajectory of the singularity by the requirement that singularity and field remain in phase harmony.

One reason de Broglie shifted his hopes to the particle model was be-cause the mathematical problems associated with the nonlinear model seemed well beyond what could readily be addressed at the time. The dif-ficulties were imposing enough in the simplified context of the particle model. On the other hand, this step would seem to indicate that de Broglie was unaware of a profound, truly *Meta* principle underlying his ideas: a group of kinematical properties associated with periodic processes gener-ally.

In section 4.14, it was shown that the matter-energy equivalence sug-gested by special relativity is consistent with the view that all matter has both an internal and an external motion; the internal movement associ-ated with the so-called *rest energy* and the external with the kinetic. If, as seems likely on the basis of both epistemological and empirical evidence, no such thing as a non-composite entity is to be observed in nature, but rather, as seems to be the case on every scale accessible to observation, ponderable bodies/processes are merely persistent or recurrent and thus perceivable patterns of activity, then certain general properties must be associated with such entities or processes.

Among these properties are Lorentz invariance, quantisation of action, the equivalence of acceleration and gravitation vis-à-vis relativistic effects, and laws of motion in accordance with the $F = MA$ format. Again, what appears to be the self-evident nature of these considerations follows from purely kinematic relations, applicable to periodic processes generally.

§ 5.11.a On the Kinematics of Periodic Phenomena: Part A.

In chapter one, the notion of television as a metaphor for visual perception was invoked to illustrate certain aspects of cognition. A similar metaphor is also helpful in this context, to describe certain aspects of periodic phenomena. Figure 5-11 is intended to represent the concept of a periodically expanding and contracting sphere. This concept is developed alongside a schematic, two-dimensional representation of the same idea, imagined to be projected on a television screen (figure 5-12).

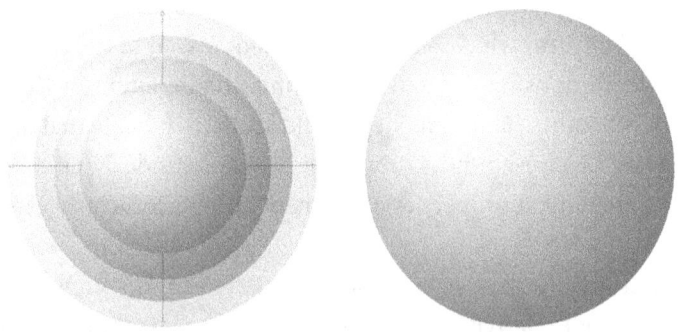

Figure 5-11 This figure represents the concept of a periodically expanding and contracting sphere. Neither dimensions nor proportions of the arbitrarily pictured stages of the process are here specified, as they vary with circumstance as required-by/described-in the text.

Imagine the sphere pictured variously in figure 5-11 to represent a portion of some homogeneous substance comprising a plenum, like the air of the atmosphere or the water of the sea (except compressible), though of course there is no gravity – again, the plenum is homogeneous; pressure and density are everywhere the same. Consider the region of the plenum enclosed by the large sphere on the right, compressed to the size of the small [inner] sphere on the left – the latter taken to have one-half the radius of the former – so that outside the small sphere the density of the plenum remains the same, but inside the density is eight times as great (i.e., volume is proportional to the cube of the radius: $V=\frac{4}{3}\pi r^3$). Now imagine the process reversed, and this cycle of compression and expansion repeated, periodically, at a uniform tempo.

To further simplify this image, the pulsation of the sphere can be represented by the arrows drawn over the spherical shells of the image on the left. Because of the symmetry of the situation, every arrow drawn from the inner to the outer sphere, normal to both, is necessarily the same as every other. And so the process can be thought of as comprising a set of such [animated] arrows – i.e., rays of wave action, like rays of light, reflecting back and forth between the central surface and the outer. Moreover, because of symmetry, one animated arrow is sufficient to represent all the rest.

But now imagine that the pulsating spheroid is given a motion of translation – a uniform motion, say from left to right. The simple symmetry described above is broken, although a new, albeit somewhat less simple symmetry emerges. To see this new symmetry, consider the representations pictured in figure 5-12. In order to further simplify and clarify the salient features, the periodic motion is deemed to occur between the center of the region of the space where the pulsation occurs and the enclosing surface thereof (here reduced to two-dimensions, i.e., a circle). Because the motion of translation is uniform, there is still significant symmetry, so only two aspects of the periodic motion need to be considered – namely, "up-down" and "left-right" (because up and down are identical, only one vertical leg is pictured).

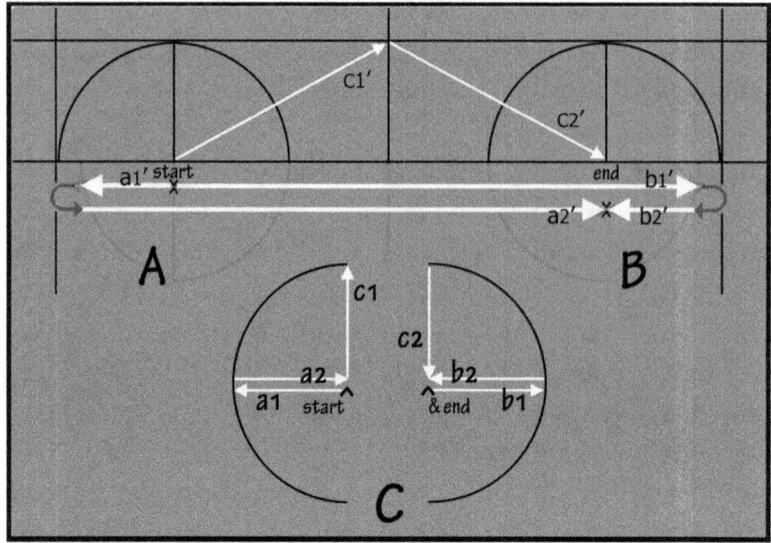

Figure 5-12   Illustration C schematically represents the paths of three "wave-rays," emanating from the center of the circle (which is shown split into semi-circles for clarity), and reflecting from the left, right and top inside surfaces of the circle's perimeter, thus meeting back at the center. The illustration above C, marked A on the left and B on the right, represents the same circle moving from left to right, with speed such that the velocities (paths) of the wave-rays are modified as indicated by the primed letters, corresponding to the unprimed letters in figure C. As described in the text, the proportions are purposefully inaccurate and the center-to-bottom (i.e., 4th) path is not pictured in either illustration – all again for clarity.

It is clear that the paths that the various rays delineate stand in a definite relation to one another. In either case, of motion or rest, each of the sums of the c's must be equal to the sums of the a's and of the b's, respectively $(c_1 + c_2 = a_1 + a_2 = b_1 + b_2$ *and* $c_1' + c_2' = a_1' + a_2' = b_1' + b_2')$.

However, one aspect of the situation is not so unambiguous. A distinction must be made as to whether the two-dimensional illustration represents a projection on a television screen or a physical system. If the paths traced out by the rays are taken to be nothing more than projections on a television screen, then no ambiguity will arise. That is, all mappings between a stationary system and a uniformly moving system, such as the sums of the paths noted above, can be understood in a straightforward manner. However, if the representation is deemed to refer to a physical system, the following question must be considered: "What is the difference between 'rest' and 'motion.'" "How are they distinguished?" In addition, it must be asked: "How much time does it take for a single cycle of the periodic process to occur in the moving system in relation to the time it takes in the stationary system?" In other words, if the ray-paths are merely projections on a screen, they can occur as rapidly as one wishes. But if they represent motions of a physical wave, the time it takes to complete a cycle will be limited by the wave velocity.

This is directly related to the discussion of the concept of time in section 5.3 above – i.e., for any given velocity, the more space there is to traverse the more time it will take to do so. If "space is real" (that is, if motion is not unconstrained as in a simulation via projection, and space is not merely an abstraction) there is "more space" to traverse for a moving than for a motionless system, so the question arises: What conditions are imposed on the mathematical transformations between moving and stationary systems by virtue of the physical characteristics of the space? And then again, how is motion to be distinguished from rest?

§ 5.11.b   On the Kinematics of Periodic Phenomena: Part B.

Putting aside, for the moment, some of the subtler aspects of these matters, two key points connected with the distinction between "motion" and "rest" of an oscillating system must be addressed, one of which essentially recapitulates the argument at the heart of the relativistic interpretation of Lorentz invariance. This, the first, is a direct consequence of the fact that the ray paths crossing the circle no longer travel distances equal to the radius of the circle, but rather a distance that is greater, in proportion to the velocity of the translatory motion. Second, not only must the paths that the rays travel remain equal to one another in length, they must also remain an integral multiple of their common wavelength.

The first of these consequences implies that the shape of the moving circle must be modified, relative to the [fixed] Cartesian coordinate system, from a circle to an ellipse. And, because the overall distance that the rays must travel to complete a cycle necessarily increases with the translatory velocity, time – as measured by these ray cycles – will slow down, in inverse relation to velocity. Moreover, the velocity of the rays relative to the "rest frame" of the wave medium (i.e., the frame in which average velocity of its component parts is zero) will serve as a universal, limiting velocity for the translatory motion of the ray-traced ellipse (in the stationary frame). That is, in order for the time dilation of the moving circle to be real with respect to an identical stationary circle – for the circumstances of the moving and stationary circles to be non-symmetrical – the circle's motion must be distinguishable with respect to the medium in which the rays move (that this must be the case in the physical world, and that general relativity essentially furnishes a framework that provides the basis for distinguishing otherwise relative motions in this respect, will become clear in the sequel).

The second of these consequences implies that the velocity of the ellipse can only change in discrete quantities, determined by the wavelength of the rays. If this were not the case, the reflections from the bounding surface would be out of phase – both with each other and with the source. Another way of viewing this circumstance is as follows. The in-phase rays are the entity in question; just as the regions of constructive interference in pressure waves comprise the acoustic wave packet, and the group velocity is the velocity of the packet, so too are the surfaces of points from which reflected rays meet up at a single point the boundary of the moving circle, and the meeting point its center (i.e., they are *determined* in this manner).

The significance of these constraints is the subject of the next two subsections, but it is already evident that they are nothing other than the conditions of Lorentz invariance and quantisation-of-action. Moreover, it will be shown that these properties must appear in precise mathematical accord with certain *a priori* considerations, and can thus be considered emergent properties of periodic phenomena generally. In other words, the detailed, context-specific arguments of Lorentz, Larmor *et al* deducing Lorentz invariance as a consequence of electromagnetic self induction are *overly specific* – as are all such arguments – they are mooted by general, *purely kinematic* considerations.

Lorentz Invariance follows from the single assumption that ponderable entities comprise periodic phenomena in the space in which they are embedded. And so, as per the scenario described above, observable objects can be thought of as oscillations in an otherwise imponderable underlying

space. While there have been many efforts to show that analogs of relativistic effects can emerge in various physical contexts – including gases,
fluids, solids and condensed matter – the features of Lorentz invariance
are quite general; again, universal aspects of periodic motion.

The same consideration applies to quantisation-of-action. Any context-
specific treatment of these general emergent properties – such as Bohm's
derivation of an emergent, universal constant-of-action on the basis of
collective dynamical considerations[12] – is epistemologically superfluous.
(It turns out that Einstein was, in an *a priori* sense that he apparently did
not recognize, justified in taking Lorentz invariance and quantization-of-
action to be universal physical principles, again, apparently without recognizing their underlying unity.) Most crucially, there must be limits of scale
associated with these properties. That is to say, because these properties
are general and collective, they must be expected to have operative ranges
over some characteristic levels of scale. Moreover, such emergent patterns
can be expected to repeat, like fractals, over various regimes of scale, for
the same reason that fractals do.

§ 5.12 Lorentz Invariance

Consider, again, the example depicted in figure 5-11, a pulsating spheroid (represented in figure 5-12 by a circle in two dimensions). As discussed, the periodic action is schematically represented by directed rays,
emanating from the center in the left-right and up-down directions, and
reflecting from four points on the left, right, top and bottom of the circumference of the circle, respectively, back to the center, where they remain in phase.

Before procceding with this extended analysis of the two dimensional
case, it is relevant to note some general characteristics of periodic phenomena and wave motion. In particular, the requirement that the outgo-
ing and incoming waves remain in phase at the center of the disturbance
needs some clarification, because, while it seems to be an obvious necessity, it is perhaps not entirely clear that this necessity is also sufficient to

12 Bohm, David. "Quantum Theory Radiation and High Energy Physics" Academic Press, Ltd., part
3, 1962 reprinted as chapter 4 of: Bohm, David. Wholeness and the implicate order. London; Boston:
Routledge & Kegan Paul, 1981.

guarantee its fulfillment – i.e., that the condition must be fulfilled in the case of any uniformly moving system, let alone with respect to arbitrary, translational boosts of velocity.

This issue is related to a more general concern, which also arises in connection with Solitons. The marvelous stability of such dynamical systems makes one wonder how it is possible for such a complex of phenomena to form in the first place, let alone remain so tightly organized under all sorts of seemingly adverse circumstances. The simple response to this wonder, which also satisfies the concern regarding linear phase relations, requires some elucidation. On the premise of quite general statistical considerations, it can be shown that the interference effects that underlie these variegated phenomena literally "take care of themselves."

Consider a single, spherical wave disturbance, for example an acoustical wave in a fluid, spreading from a central point as shown in figure 5-13. The figure loosely represents what is known as a *Huygens-Fresnel construction.* Christian Huygens intuited that every point in a medium that is acted upon by a wave can be thought of as the source of a new disturbance. Every point along the progressing front thus becomes the source of a new, spherical wave. Accordingly, the advancing surface of the front can be constructed by regarding each point of that surface as the source of a new spherical wave. The envelope surrounding all of these "wavelets," which is tangent to them along its normals, represents the collective front of the disturbance, which therefore spreads in the directions of those normals.

In an elementary Huygens construction, one must simply ignore the effects of wavelets in all directions other than that in which the envelope progresses. However, Huygens' intuition was ultimately vindicated. As it turns out, if each point is indeed treated as the source of a new wave front spreading in all directions, the collective consequence is yet a single envelope moving in one sense only with respect to the original disturbance, e.g., as pictured above, spreading away from the center. A mathematical analysis of diffraction, as undertaken by Fresnel and later, in greater detail, by Kirchhoff and others, shows that the phase relations are such that the waves moving backwards (toward the source of the disturbance) interfere destructively, while those moving forward interfere constructively, so that the leading surface - the envelope enclosing the wavelets - is in fact an adequate representation of the overall complex of phenomena.

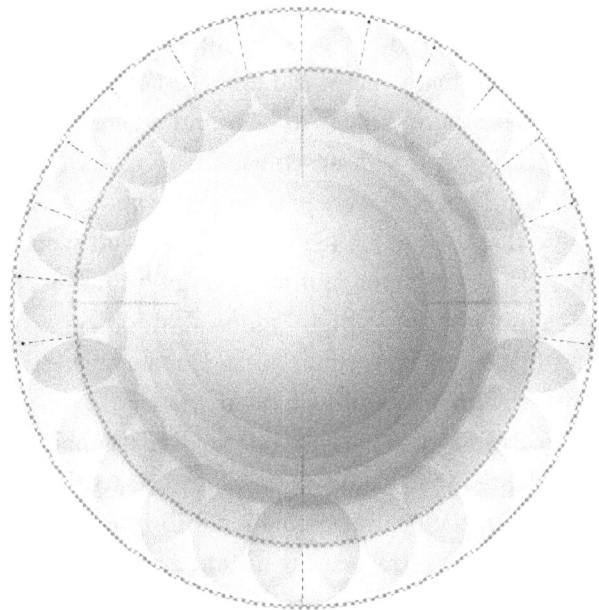

Figure 5-13 A spherical wave front expands outward. Every point on the leading surface is considered to be the source of a new spherical wave. Thus at any given moment the wave front represented by the outermost sphere, which is tangent at every point on its surface to one of the expanding spherical components within, is both the collective consequence of such a group of component waves as well as the source of a new group. Because the component waves expand in all directions, one seeks a reason why the collective result is a wave in only one direction; here, again, outward from the center. It can be shown, in the case of such an expanding wave, that the inward-directed components interfere destructively, whereas the outward-moving components interfere constructively.

In addition, with any group of waves of different frequencies there are places where, for a time, the overlapping waves are close to being in phase with each other, and places where they are not. This point changes as the waves move, so the region of constructive overlap appears as a packet, grouped about the point where the waves are closest to being in phase, which point moves along with what is thus called the *group velocity*. This typical, linear effect occurs partly because waves of different frequencies travel at different speeds. But for the same reason, eventually such linear packets disperse. On the other hand, when nonlinear effects are significant, constructive reinforcement can remain tightly confined and highly stable over a wide range of circumstances.

Just as with linear effects, only those undulations that are in the appropriate phase make a positive contribution to the collective behavior. Moreover, as with ordinary wave packets, this is not at all an unlikely arrangement. Rather, in many real world cases of wave phenomena, at any given time and place there is likely to be found an enormous number of passing/converging waves, with an enormous range of frequencies and phases. Those disturbances that are in appropriate phase contribute to the effect, and those that are not are inconsequential. And so, as a nonlinear wave phenomenon moves along, it draws energy from those undulations with which it is in phase, and simply "ignores" all the others. Again, it is not so difficult in a complex world to find adequate, phase correlated disturbances for the maintenance of such phenomena.

This argument also addresses the above, two dimensional representation of Lorentz invariance. That is, only those reflected rays that return to the center in phase with the central oscillation will interfere constructively. And so only those rays that traverse the distances required by the kinematics of the system need be represented – just as in a Huygens construction, where the backward-moving, negatively reinforcing waves need not be depicted. It is the phase correlated waves that thus constitute the moving, Lorentz invariant entity.

Returning, then, to the matter at hand, when the circle is given a uniform translatory motion, left to right, two crucial consequences follow. These two attributes of the circumstances are:

(1) The four rays no longer travel distances equal to the radius of the circle. Rather, as shown in figure 5-12, each leg of the trip is extended – from the radius of the circle to the length of the hypotenuse of the right triangle that each vertical ray traces out.

(2) In order for all four rays to be in phase when they meet back at the center, the distance that each of them travels must be an integral multiple of their common wavelength.

An examination of the circumstances pictured in figure 5-12 reveals the following relations. Consider first the vertical ray paths – center-top and center-bottom. With respect to the stationary coordinate system, over the course of a half cycle each of these rays traces out the hypotenuse of a right triangle, the base of which joins the initial position at the center of the left circle with the point midway between there and the final position at the center of the right circle. During the course of the second half of the cycle the rays trace out the identical distance, again along the hypotenuses of triangles that are congruent reflections of the first two.

The mathematical treatment is quite simple, as it involves nothing other than the *Pythagorean theorem* of Euclidean geometry. Referring again to figure 5-12, the following relations hold:

$$c_1^2 = (c_1')^2 - (\frac{a_2' - a_1'}{2})^2 \quad \text{and} \quad c_1 = ct_0$$

where c = the wave velocity and t_0 is the time it takes a wave to travel the distance $c_1 = c_2 = a_1 = a_2 = b_1 = b_2$, therefore:

$$(ct_0)^2 = (ct)^2 - (vt)^2$$

where v = the velocity of translation and t is the time it takes a wave to travel the distance $c_1' = c_2'$

Again, this is but a statement of the Pythagorean Theorem: t_0 equals the time it normally takes for a ray to complete a half cycle (i.e., travel a distance equal to the radius of the circle, when that circle is stationary), t equals the time it takes to travel the extended distance, along the hypotenuse of the triangle traced by the ray due to the motion of the circle, c is the wave velocity, which is constant, and v is the translatory velocity of the circle.

This reduces to a relation for time as follows:

$$(ct_0)^2 = (ct)^2 - (vt)^2 \quad \text{or} \quad (ct_0)^2 = t^2(c^2 - v^2)$$

then

$$t = \frac{ct_0}{\sqrt{c^2 - v^2}}$$

and finally

$$t = \frac{t_0}{\sqrt{1 - \dfrac{v^2}{c^2}}} \tag{5-4}$$

Equation 5-4, which expresses the time that it takes for the ray of light to complete a cycle when the circle is moving uniformly (t) in relation to the time that it takes when the circle is stationary (t_0) is, of course, the *Lorentz transformation for time*.

Looking next at the horizontal rays, it is clear that the left-moving ray will reflect off the back of the circle when the latter has traversed a distance equal to:

$$r - a_1' = r \frac{v}{c} \quad \text{and} \quad r = a_1 = a_2 = b_1 = b_2 = c_1 = c_2$$

where v is the velocity of the circle, c is the ray velocity, and r is the radius of the circle. That is, the circle will have advanced the distance ($r - a_1'$) during the time that it takes for the leftward moving ray to intercept it, where a_1' is the distance that the ray covers in the same interval. Similarly, when the ray traveling to the right reaches and reflects from the front of the moving circle, the circle will have moved a distance:

$$ct_1 - r$$

where t_1 is the time that it takes for the ray to reach the front of the circle.

At the time when the forward moving ray reflects from the front of the circle, both the front of the circle and the trailing ray that reflected from the back of the circle will be the distance a_1' from their ultimate meeting point at the center. Moreover, the distance that each of these horizontal rays traverses must equal the distance that each of the vertical rays travels:

$$\frac{2ct_0}{\sqrt{1 - \dfrac{v^2}{c^2}}}$$

where, again, t_0 is the time required, when the circle is stationary, to complete half a cycle – i.e., for a ray to traverse the distance r.

Now, the distance between the point in space where the left or backward moving ray reflects from the back of the circle, and the point in space where the front or forward moving ray reflects from the front of the circle, is related to the diameter of the circle by the following equation:

$$2r = x \cdot \sqrt{1 - \frac{v^2}{c^2}} \quad \text{and} \quad x = a_1' + a_2' = b_1' + b_2'$$

where x is the distance between the two reflection points and again r is the radius of the circle. This is the *Lorentz transformation for distance* measurements in the direction of motion.

It is relevant to recall that, per the discussion of § 5.5 regarding the concept of spatial density, because there is no such thing as an instant of time, the volume of space that an object occupies must be thought of as related to its velocity. In this sense, the Lorentz transformation for distance reflects, literally, the "amount of space" that the circle occupies during the course of an internal time cycle when it is in uniform translatory motion, in relation to the "amount of space" it occupies during the same time cycle when stationary. And so, because the rays represent a real, periodic process occurring in a dynamic physical space, as that periodic process advances through space it both (a) takes more time to complete a cycle, and (b) occupies more space per cycle.

And again, the equations that relate these quantities are the Lorentz transformations for space and time. Moreover, because the internal motion comprises waves of a definite length and number, which in turn comprise the "rest energy" of the system, it is clear that the rest energy[/mass] both (a) varies via the Lorentz transformation for mass, and (b) is quantized by wave number.

§ 5.13 Quantization of Action and the Equivalence of Mass and Energy

As noted above the length of the ray paths, which represent the internal wave motions of the oscillating system, are necessarily integral multiples of their wavelength. Figure 5-14 illustrates the first half of the process (hereinafter 'cycle') by which a periodic phenomenon undergoes a change of position corresponding to a motion of translation, per the analysis of the previous sub-section. The motion is such that, compared to the situation at rest, each wave-ray traverses an additional distance of exactly one wavelength during each half-cycle. In other words, per the illustration, the radius of the outer circle is equal to an integral multiple of the wavelength, that integer being *n*, and the length of the hypotenuse of the triangle is given by *n + 1*.

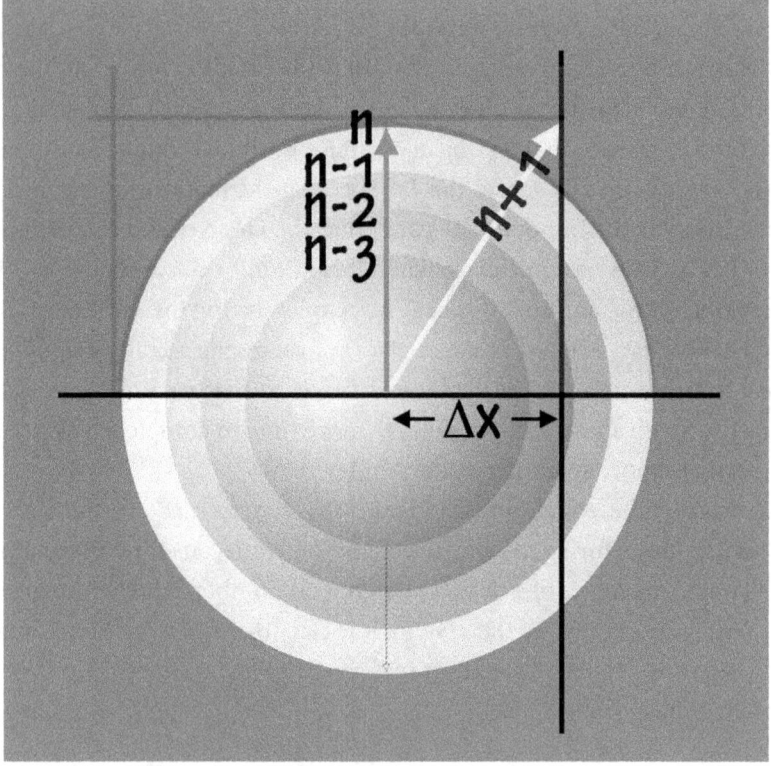

Figure 5-14   The internal wave-number corresponding to the length of the radius of the pulsating spheroid is *n*. In time $\Delta\tau$, the spheroid advances a distance Δx to the right, corresponding to an increase of the internal wave-number by one.

The velocity which the system acquires (corresponding to the change of position illustrated in the figure) is given by:

$$\frac{\Delta x}{\Delta \tau}$$

where Δx is the increment of distance traveled in time $\Delta \tau$. So it follows that:

$$\Delta x^2 = (n + 1)^2 \lambda^2 - n^2 \lambda^2, \text{ where } \lambda \text{ is the wavelength}$$

or

$$\Delta x^2 = \lambda^2((n + 1)^2 - n^2) = \lambda^2(2n + 1) \tag{5-5}$$

Also,

$$\frac{\lambda(n + 1)}{\Delta \tau} = c, \text{ where } c \text{ is the wave velocity}$$

therefore:

$$\Delta \tau = \frac{\lambda(n + 1)}{c}$$

and

$$\Delta x = \lambda \sqrt{2n + 1}$$

Now action is equal to energy multiplied by time. So if the moving periodic process is treated as a Newtonian object (i.e., as having mass), an increment of velocity must involve an increment of action (ΔA) as follows:

$$\Delta A = \Delta KineticEnergy \cdot \Delta \tau = \frac{m}{2}\left(\frac{\Delta x}{\Delta \tau}\right)^2 \Delta \tau$$

where m is mass, or:

$$\Delta A = \frac{m}{2} \frac{\Delta x^2}{\Delta \tau}$$

and, substituting for time:

$$\Delta A = \frac{m}{2} \frac{\lambda^2(2n + 1)}{\lambda(n + 1)/c}$$

or

$$\Delta A = \frac{mc\lambda}{2} \frac{(2n + 1)}{(n + 1)}$$

Therefore, as the wave number exceeds *unity*, the ratio of $(2n + 1)$ to $(n + 1)$ approaches 2. And so, in general, incremental action tends to:

$$\Delta A = m \, \lambda \, c \qquad (5\text{-}6)$$

This, of course, is the value of the *quantum unit of action* – i.e., *Planck's constant* – as a function of the mass and wavelength associated with a physical entity. What it suggests, in this context, is that wave number can be thought of as a measure of the "mass"/"intrinsic energy" of the periodic process being modeled, and that action is quantized accordingly.

This is also in accord with the interpretation of energy as a function of frequency, as discussed in section 5.4, and with the interpretation of the rest energy of matter in terms of an internal motion, as discussed in the previous chapter. Again, this follows from *general physical considerations*. Moreover, by virtue of the increase in wave number with the increase of the translatory velocity of the periodic process, it is evident that mass must increase with velocity. Referring to figure 5-14, and equation 5-4 relating time in two frames, the relation between wave number and velocity is:

$$n = \frac{n_0}{\sqrt{1 - \dfrac{v^2}{c^2}}} \qquad (5\text{-}7)$$

where n is the wave number of the moving process in relation to n_0, the wave number of the stationary process.

Substituting m and m_0 (moving and rest mass) for n and n_0, respectively, gives the *Lorentz transformation for mass*. The equivalence of mass and energy is particularly perspicuous vis-à-vis the above construction for action, where the addition of a single internal wave is considered to be equivalent to the addition of a single mass unit – Δm – such that the effective mass is $\Delta m (n + 1)$, i.e.:

$$\Delta E = \left(\frac{\Delta x}{\Delta \tau}\right)^2 \Delta m \left(\frac{n + 1}{2}\right) = \frac{\Delta m}{\Delta \tau^2} \lambda^2 (2n + 1)\left(\frac{n + 1}{2}\right)$$

where

$$\Delta \tau^2 = (n + 1)^2 \frac{\lambda^2}{c^2}$$

so that

$$\Delta E = \frac{\Delta m(n + 1)(2n + 1)\lambda^2/2}{(n + 1)^2 \lambda^2/c^2}$$

or *approximately*:

$$\Delta E \approx \frac{2(n + 1)\Delta mc^2(n + 1)\lambda^2/2}{(n + 1)^2 \lambda^2} \tag{5-8}$$

In other words, in accordance with the above interpretation of action, when *n* is considerably greater than unity – which is to say, in circumstances where *c* is substantially greater than *v*, and thus internal wave velocity is substantially greater than external, translational velocity – the ratio of *(2n + 1)* to *(n + 1)* will be reasonably close to *two*. Under such circumstances, and in accordance with equation 5-5, relation 5-8 reduces to:

$$\Delta E = \Delta mc^2 = \Delta A \tag{5-9}$$

Accordingly, matter can be thought of as composed of an integral number of internal waves. The amount of matter and thus internal energy – what is called the relativistic rest energy of matter – is given by the number of such waves (and is therefore proportional to the frequency) multiplied by the constant of action.

§ 5.14 The Velocity of Light and Planck's Constant

In substances that support elastic distortions, the velocities with which longitudinal and transverse waves propagate are largely determined by a relatively small number of physical characteristics, the most relevant of which are connected with the classical concepts of pressure, density and the relation between them – that is, the various relations between changes in volumetric dimensions and the density of energy and inertia. In this sub-section it will be shown that, in accordance with the framework under development, a relationship between the velocity of light and Planck's constant can be deduced on the basis of extremely simple considerations, employing generalizations of the above-mentioned classical concepts. Please note that these deductions are not being presented here as a completed theory. The constructions employed are not intended as detailed theoretical models of any real physical system, and the calculations are intended to be interpreted accordingly. (For example, in this subsection no consideration is given to the distinction between longitudinal and transverse waves, let alone any physical characteristics associated with that distinction.) Rather, the purpose at this juncture is mainly to stimulate interest and support the plausibility of the developing framework.

With respect to the behavior of pressure or density waves in a homogeneous system – hereinafter referred to collectively as *acoustic waves* – one of the most important properties is the *Bulk Modulus*, given by the equation:

$$B = -V\frac{dp}{dV} \tag{5-10}$$

That is, the bulk modulus is equal to the *negative* of the quantity: *volume* multiplied times the ratio of the *incremental pressure* to the concomitant *change of volume*. This can be thought of as a measure of an elastic system's resistance to compression, and thus its capacity to store elastic energy. The more pressure that a material develops under elastic compression, in relation to its decrease in volume, the greater its bulk modulus. Wave velocity is proportional to the square root of the bulk modulus, and inversely proportional to the square root of the matter density. And so the greater the matter density, and thus inertia, the slower the wave motion.

A basic equation for acoustic wave velocity, which is surprisingly accurate under a wide range of circumstances, is:

$$c = \sqrt{2B/\rho} \tag{5-11}$$

where B is the bulk modulus (energy density), and ρ is the material density, so that:

$$B = \frac{E}{V} \tag{5-12}$$

and

$$\rho = \frac{m}{V} \qquad (5\text{-}13)$$

and therefore:

$$\frac{B}{\rho} = \frac{E}{m} \qquad (5\text{-}14)$$

or

$$c = \sqrt{2E\big/m} \qquad (5\text{-}15)$$

If ponderable matter/energy is deemed to comprise wave activity in an underlying system, then both the velocity with which waves propagate and the frequencies with which their sources vibrate should be related to the system's elastic properties, and thus to each other. Because the model invoked above (a pulsating spheroid) consists of nothing but a local wave motion – i.e., because, in contradistinction to a mechanical oscillator, such as a mass on a spring, all of the motion occurs within the medium that supports the wave disturbance – the velocity of the periodic movement (internal vibration) is more or less the same as the velocity of waves in the plenum generally. Therefore, in order to be consistent with the general laws governing such systems, the following relations must hold true.

The velocity of wave propagation is given (per equation 5-11) by the square root of the ratio: two times the bulk modulus divided by the matter density of the medium. As noted, the bulk modulus is a measure of the energy density. So the equation for the internal wave velocity is equal to the square root of twice the energy density divided by the matter density. And, as per equations 5-12 and 5-13, because energy and matter density are measured over the same volume – which thus appears in both the numerator and denominator – equation 5-11 reduces to equation 5-15, and therefore:

$$c = \sqrt{\frac{2E}{m}} = \sqrt{\frac{2}{m}\left(\frac{mv^2}{2}\right)} = \sqrt{v^2}$$

or

$$c = v \qquad (5\text{-}15)$$

Where v is the velocity corresponding to the internal motion of the singularity [as it enters into the expression ($\frac{1}{2}\ mv^2$) for the Kinetic Energy], which, as expected, is also the wave velocity of the medium. This means that, if a ponderable entity is in fact a collection of waves – if it consists of motion in an underlying medium – and the rest mass/rest energy of that entity is thus a function of the kinetic energy of the underlying "masswaves," then it follows that *rest energy* will indeed be a function of the *square of the wave velocity*, on general physical considerations.

As with the results of the previous section this is in accordance with the most famous upshot of *Special Relativity*: $E = mc^2$: again, *derived from general principles*. In conjunction with the deliberations of the following few paragraphs, these relations should serve to make the deep connections between relativity and quantum theory more readily comprehensible.

Both the size and mass of what will hereinafter be referred to somewhat loosely as a *nucleon* (i.e., a *neutron or proton*, the masses of which differ by a small amount) are empirically determined values. And while the "radius" of such a particle is not a wholly unambiguous quantity, established in accordance with a unique experimental prescription, there nevertheless exists a small range of values – roughly between *one* and *one and one-half fermi* or femtometres (10^{-15} *meters*) – accepted on the basis of several plausible approaches to a definition of this measure, with the mean value being approximately *1.25 fermi*. In comparison, there is little ambiguity regarding the nucleonic mass, which again is almost the same for both the neutron and proton – approximately *1.67 • 10^{-27} kg*.

If one imagines a singularity of such a size and mass comprising periodic fluctuations in the manner of the pulsating spheroid, one can arrive at a reasonable expression for the frequency and wavelength of those fluctuations. As per § 5.11.a, the wavelength can be taken equal to the radius and, as per the above considerations of this section, the velocity of the fluctuations can be taken equal to the wave velocity of the medium. Therefore, the wavelength of the nucleonic wave can be set at *1.25 • 10^{-15} meters*, from which follows the frequency, here *f*:

$$f = \frac{c}{\lambda} = \frac{3 \cdot 10^8}{1.25 \cdot 10^{-15}}$$

or

$$f = 2.4 \cdot 10^{23}$$

Now, per equation 5-6 of § 5.13, the dynamical action is given by $m\lambda c$. Therefore, in accordance with the determination of § 5.13 above, the smallest increment of action is:

$$\Delta A = m\lambda c = (1.67 \cdot 10^{-27} kg)(1.25 \cdot 10^{-15})(3 \cdot 10^8)$$
$$= 6.236 \cdot 10^{-34} \; joule - sec \qquad (5\text{-}16)$$

or within about five percent of Planck's constant. Moreover, action is equal to the product of energy and time; so if, per § 5.4, frequency is determinative of energy, substituting frequency for wavelength in the above equation (per the relation *wavelength = wave velocity* divided by *frequency*) yields

$$\Delta A = m \frac{c}{f} c = \frac{mc^2}{f}$$

which, because the inverse of frequency is the *period* (i.e., the time of an oscillation), confirms that

$$\Delta A = mc^2 \cdot period$$

or again, the quantum mechanical relation between frequency, energy and the constant of action, in the well known form:

$$\Delta A \cdot f = mc^2 = Energy \qquad (5\text{-}17)$$

As mentioned, there are various ways to determine the radius of the nucleon. While the *Compton wavelength* of the neutron is very close to the mean of the range noted above (1.319 590 898 ± 0.000 000 010 · 10^{-15} meters), this range results from a variety of experimental approaches. And so, just as with Lorentz Invariance, it is possible, on the basis of *a priori* considerations (i.e., that ponderable objects/energy comprise periodic processes in an [otherwise] imponderable energetic substrate) – and in conjunction with empirically determined numbers (i.e., mass and size of a sub-atomic particle) – to deduce a value for the quantum of action that is fairly close to the experimentally determined quantity, while confirming, and adding interpretive meaning to, the relativistic equivalence of mass and energy.

This result, again, is not a deduction of theory *per se*, inasmuch as the model of the pulsating spheroid is only a heuristic device to get a feel for the relations involved. However, it will shortly become evident that the premise of a common origin of Lorentz invariance and quantization-of-action is indeed empirically well-supported.

§ 5.15 Lorentz Invariance and the Principle of Equivalence

Consider the analogy between a continuously variable index of refraction and a continuously variable spatial density, as discussed in § 5.5. Figuratively speaking, from a wave-train's "point-of-view" traversing a region of greater spatial density is like traversing a greater volume of space. It takes longer to cross such a region. And the actual volume of space that the wave-train occupies is *smaller*, as measured from outside the region in a fixed interval of time, the *denser* is the medium.

Therefore, on the basis of the notion of spatial density, one can say that a gradient in that density – e.g., from a region of "normal space" to a region with "increased space per unit volume" – displays the properties of a changing index of refraction. In other words, the velocity of light within such a region, as measured by an observer outside of it, will be slower in proportion to the differential of the spatial density between the regions.

It was also discussed in connection with this concept that, as velocity increases so too does the volume of space that the wave train occupies per unit time. For example, a single pulse (i.e., "wave crest") of light, considered over the duration of one second, occupies during that second a region ~186,000 miles in length. Moreover, all of the relativistic effects associated with a velocity boost can be emulated by an appropriate increase of spatial density.

Of course, velocity only affects spatial measurements in the direction of motion, not isotropically. But a spatial density gradient will also exhibit anisotropic properties, inasmuch as certain effects occur only in the direction parallel to the gradient. Thus, at the interface between two materials with different refractive indices, the direction of the light rays shift, per figure 5-9, because the velocity of the light is only changed on one side of the boundary, and so in the direction parallel to the gradient of the refractive index – i.e., perpendicular to the interface between the materials. In this sense, a beam of light is contracted in the direction parallel to the direction of the gradient, but not "laterally." Similarly, a spatial density gradient will cause a wave-train to contract in the direction of the gradient.

On the other hand, the motion of a clock – i.e., any periodic process – will reflect the average density of the space in which it is immersed, regardless of the direction of the gradient. To understand this, it is only necessary to consider the example of section 5.12. Internal motions must be considered, in general, to be omnidirectional. In order for the wave rays parallel to the direction of motion to remain in phase with those that are perpendicular certain conditions must hold, which are identical to the requirements of Lorentz invariance (and quantization of action). As translatory velocity of the periodic process increases (with respect to the wave velocity that comprises that process), so too does its period.

Moreover, this accounts for the *Principle of Equivalence* as well. Einstein's original reasoning in this connection involved a comparison between two frames of reference – one accelerating uniformly in an "empty Newtonian space," and another "at rest" in such a space but subject to a uniform gravitational field (producing an acceleration equal to that of the first frame). According to Einstein's argument a constant acceleration produces a time dilation, a new rate of time in fixed proportion to that of an equivalent frame at rest. This ratio is equivalent to that which must obtain, under the Special Theory of Relativity, with respect to a uniformly moving frame of reference in relation to another frame that is stationary with respect to it.

Einstein employed the notion of *Absolute Newtonian space* to simplify the considerations, but it actually confuses certain important aspects of the circumstances. For a uniformly accelerating frame of reference cannot produce a constant passage of time – i.e., a rate that remains at a fixed but reduced magnitude in relation to that of a relative rest frame. Rather, because the velocity is constantly increasing, so too must the time dilation. Of course, Einstein's thought experiment is not intended to be ideal in this respect, just as it is not ideal regarding the homogeneity of the gravitational field, which in reality is never exactly uniform over a finite region of space. But certain key relations are nevertheless obscured.

The example that Einstein used in his first popular exposition of relativity gets closer to the salient relations. In that explanation he utilized the image of a spinning disc, and the fact that clocks must run more slowly at the periphery than at the center, because of the time dilation associated with its motion. Here the centrifugal acceleration at the periphery that emulates gravity is a direct result of the motion. So, for example, an astronaut in geostationary orbit 22,000 miles above the equator will feel weightless, whereas someone standing on the equator will not. Although both the spacecraft and the earth rotate with the same angular velocity, the much greater acceleration of the astronaut in comparison to that of the earthbound person is directly correlated with the difference in translatory velocity. Thus the proper role of velocity *per se* as opposed to acceleration is more evident regarding the principle of equivalence.

Moreover, modern laboratory experiments with high-energy particle accelerators confirm that *acceleration is not* the cause of time dilation, but rather *velocity is*, just as called for in Special Relativity. But if the notion of Absolute Space is ultimately untenable, and all motion is relative, in what manner can the magnitude of velocity be measured in order to ensure that the concept of "actual" (i.e., absolute) velocity is the unambiguous cause of time dilation? While the notion of an Absolute rest frame seems untenable, experiments also confirm that, just as with the so-called *Twins Paradox*, when the passage of time for two clocks in relative motion are compared, it is always found that the object that experiences a velocity boost relative to the local inertial frame is the one that exhibits a reduced passage of time. But if acceleration, again, is not the cause of the effect, then how is this to be understood?

Per the examples of sections 5.11 and 5.12, in order for the kinematic arguments to be meaningful, it is clear that the difference between the frame in translatory motion and the frame at rest must be real, distinguishable and measurable. The velocity of the periodic motion must occur at some meaningful (again, measurable) constant rate, so that the translatory motion, which is measured relative to that periodic movement, can also have meaning. This is equivalent to stipulating that the frame in which the periodic motion is measured can serve as a rest frame for motion generally. (Otherwise – that is, if Galilean/Einsteinian relativity were literally true, and there were no differences at all between relativistically linked frames – there could be no basis on which one could say that one of the oscillators [i.e., the "moving" one] traverses a greater distance and thus takes more time to complete a cycle.)

Accordingly, whatever establishes this periodic motion fulfills the function of a substrate of space. Moreover, in order for the notion of spatial density to be meaningful, space need necessarily have properties associated with a substance, or again, perhaps more to the point, a substrate.[13] Accordingly, the notion of spatial density can be realized in terms of the amount of the substrate that these periodic motions occupy, per cycle, in a frame that is at rest relatively to it. And while it may be difficult or even impossible to determine, with indefinite precision, the frame that is at rest [in relation to the average inner motions of the substrate] – e.g., some such motions could be faster than light – the mere fact of that frame's existence will yet be sufficient to account for all of the effects under discussion. And of course, as theory and technology advance, it may well become possible to delve into the inner characteristics of space – i.e., *Subspace.*

One of the virtues of this simple, visualizable model is that it makes the equivalence between gravitation and acceleration easy to understand – or, more accurately, *the equivalence between the relativistic effects associated with motion and those associated with gravitation* (whereas Einstein took the equivalence on essentially "blind faith"). It is evident, as demonstrated in §§ 5.11 and 5.12, that as translatory motion increases, so too does the time required for a periodic cycle to complete. This is understandable in terms of the increased spatial volume that an object in motion experiences as a consequence of that motion, and of the impossibility of "instants" of time – i.e., that, in "real/physical" time, ob-

13 It must be borne in mind that these statements and examples are not to be understood literally; i.e., to mean that particles must be pulsating spheroids or space a material substance. It is only the relations of *form* that are relevant - the underlying *analogies* with such things - the image of a physical plenum supporting simple vibrations is merely *easy to grasp*, and that fully explains its use here.

jects in "real" motion occupy [read: traverse] more space than when at rest – a concept that, again, only makes sense if "real" motion can be defined. Thus, an oscillatory motion that periodically covers a constant spatial interval must, as observed/measured from a region of smaller spatial density, occupy a greater time interval than if it were occurring within the region of smaller density, because in the region of greater spatial density it is covering a larger spatial interval (again, as a result of the increased "amount of space").

The link with gravity is evident. Per the analogy between geodesics of light rays in curved space (under General Relativity's theory of the *geometry of spacetime*) and a continuously variable refractive index (under the theory of *Ray* or *Geometrical Optics*), it is clear that a continuously variable spatial density is equivalent to curvature of space (as per the dualities of $F = MA$ and Brownian motion, and that of Brownian motion and diffusion under a spatial density gradient). And it is just as intuitively clear, in turn, that a variable spatial density is equivalent to a variable velocity with respect to all relativistic effects – from the *redshift* of light to the *advance of the perihelion of Mercury*.[14] Moreover, the association of spatial density with gravity extends the analogy to include a quantum theory of motion, as per the following discussion.

---

14 Detailed calculations confirm that *all* the primary verifications of General Relativity, including the advance of the perihelion of Mercury, follow accurately from this model without *ad hoc* assumptions. Qualitatively, perihelion advance can be understood as follows. In *Subspace* regions of greater spatial density close to the Sun (perihelion being closest and thus spatial density the highest) orbital velocity is slowed relatively to that farther from the Sun (aphelion being farthest and spatial density the smallest). Accordingly, at aphelion the path of the planet turns less sharply than it would if spatial density were greater, and at perihelion it turns more sharply than if spatial density were smaller. Therefore, the overall elliptical path of the planet rotates, in the sense of the orbit.

§ 5.16 Gravitation: Newton's Gravitational Constant,
Planck's Constant of Action and
the Fine Structure Constant

If the relativistic effects of gravitation can be understood in terms of the density of space, and if gravitation *per se* – i.e., the underlying cause of what is deemed the curved geodesics of Einstein's spacetime and Newton's universal attraction – is a function of the spatial density gradient, then how does matter/energy affect this spatial density and thereby these gradients, and what is the connection with a quantum theory of motion?

The answer to this question pulls together the broad conceptual features of the puzzle previously discussed and arranges them into a tight theoretical unity. However, because this next stage in the development of the framework is crucial, and because it depends upon a proper understanding of much of the ground thus far covered, a brief summary and clarification of certain key points is in order.

In particular, it has been noted that in accordance with two premises – one epistemological and the other hypothetical – viz.:

(1) All physical processes – and hence all observations/measurements – occupy finite intervals of time; there is no such thing as "an instant" of time…

and

(2) All ponderable entities/processes may be regarded as comprising dynamic patterns in an underlying, energetic structure of space – i.e., "the vacuum" and "spacetime" are replaced by what is herein referred to as *Subspace*, which occupies indefinitely small and large levels of scale…

It follows [from these two premises] that :

(1) On any given level of scale – i.e., in any particular physical regime – collective properties emerge that are largely driven by the dynamics of the underlying levels of scale.

(2) Lorentz Invariance and Quantization of Action are such collective properties, which emerge together as universal features of periodic phenomena on an appropriate, underlying level of scale. This underlying relationship and unity is herein referred to as *The Common Provenance of Lorentz Invariance and Quantization of Action*.

(3) The relativistic effects of Lorentz Invariance associated with translatory motion are isomorphic to the relativistic effects associated with changes in spatial density; both are consequences of the fact that there are no instants of time and that space has underlying structure, and therefore that the time and space parameters of ponderable entities/events are connected with both (a) relative motion, with respect to the regime of Subspace from which they emerge, and (b) density of the regime of Subspace in which they are thus embedded.

(4) The relations between (a) $F = MA$ motion, the laws of refraction, and the so-called optical-mechanical analogy of classical dynamics that relates them, and (b) the analogy between $F = MA$ motion and the laws of diffusion and Brownian motion, and (c) the analogy between light paths in refractive media and geodesics in relativistically curved space-time, and finally (d) the analogy between quantum motion in accordance with the density of the square of the Schrödinger wave amplitude and all of the above – that is, the interrelations between these several sets of dualities, can be understood as resulting from a common kinematic principle. All are emergent forms of order that can be understood on the basis of the concept of *spatial density* and concomitant *spatial density gradient*.

Again, these several pieces are pulled together neatly by the binding tie of gravity; by the underlying mechanism that causally connects the spatial density gradient with operative and reactive processes/entities generally. This mechanism, which is intrinsic to the model at large, thus accounts for gravity as a *necessary* general feature – as should be expected, inasmuch as gravity can be modeled as a conservative field of force in accordance with the $F=MA$ analogies described above, which includes General Relativity.

Indeed, one of the elegant attributes of the emerging framework is that it is not necessary to invoke additional hypothetical agencies or principles to fulfill this function. Nothing is needed beyond the primary premise vis-à-vis periodic phenomena/waves, and the multitudinous manifestations of such phenomena that can be considered to ineluctably exist (here as physical counterpart of the Schrödinger wave, albeit not in the usual linear, non-relativistic form) – the vast spectra of which permeate Subspace and are an intrinsic feature of the framework.

The Subspace waves that emanate from every periodic process, secular and otherwise, are agencies of coupling – largely indirect – by and between all entities and processes, from the sub-quantum level to the cosmic. Among other things, they mediate changes in the effective density of space and thus time.[15]

Gravity is a collective rather than fundamental property, emergent from essentially three aspects of the underlying dynamic – the *"first cause"* of which is a function of the superposition of waves and of density changes in the carrier of those waves. A continuously variable Subspace density acts as a refractive medium with a continuously variable index of refraction. More abrupt density changes cause more dynamic and multi-varied responses. When a wave encounters a density gradient of any kind, a new wave or group of waves results, the particulars of which depend upon the

---

15 Although only a rudimentary model is developed here, it should be noted that the complex phenomenology of waves is really just beginning to be understood mathematically, and the ever-growing mathematical apparatus associated with this subject furnishes a vast repertoire of behaviors to draw upon – not only for modeling the basic features of the framework, but for far more complex representations. There is a large group of wave properties/behaviors that have only fairly recently become susceptible to mathematical treatment. Although a detailed discussion of these matters is beyond the scope of this treatise, relevant examples will be duly noted in the appropriate context. Waves can manifest behaviors intriguingly similar to a range of phenomena found in crystalline solids and condensed matter, and in connection with phase transitions and esoteric material states generally. Waves and turbulence can simulate everything from Newtonian objects to the dislocations and disclinations that appear in lattice structures. They can manifest patterns on top of patterns, functioning as carrier or substrate for new, emergent orders. But again, with respect to the present discussion the central concern is a simple set of universal, kinematic properties, which for the sake of the clarity of the presentation will be modeled in the most basic appropriate forms.

circumstances and can involve a combination of various degrees of reflection and transmission – i.e., refraction – and thus the creation of a new wave or group of waves.

One of the key considerations with respect to such phenomena can be understood with the help of several closely related, intuitive mathematical concepts developed for their study; the *amplitude reflection coefficient, amplitude transmission coefficient, power reflection coefficient* and *power transmission coefficient*. These quantities, as their names suggest, are related to the relative degree of "splitting" of the wave amplitude and power, respectively, at the boundary of a change of density in the wave medium. They depend, primarily, on a derived property of the medium that has different values on the two sides of such a boundary – a property called the *characteristic impedance* which, for any system that supports periodic motion, i.e., any oscillator, is given by the ratio of the applied force to the velocity of oscillation that results (e.g., for a weight on a spring, the force applied to stretch the spring in relation to the velocity that the weight attains as a result of the applied stretch). For the purposes of this discussion, it is noteworthy that the sum of the reflected and transmitted power at the boundary of a density differential is equal to the power of the incident wave.

In accordance with this general principle – i.e., what might be called *conservation of reflected and transmitted "wave action"* – when a wave crest is incident upon a singularity such as the pulsating spheroid described/ illustrated in the examples above – that is, when a density pulse of a traveling wave is incident upon the "boundary" that separates the ambient/ equilibrium density of a medium from those places, within the singularity, where the density is different (significantly, *either* higher *or* lower) – some of the energy of the wave is reflected and some is transmitted. And, as with any such wave reflection, viewed in Newtonian terms there is an impulse on, or transference of momentum to, the Subspace "substance" at the boundary of the reflection. In other words, any region of a physical system that has a density that differs from that of a nearby region will experience *what amounts to a force* from a wave incident upon it from that nearby region (and in the original direction of that wave). Moreover, the more gradual the change of density the more continuous, smooth and subtle the effect – the difference between *refraction* and *reflection*. (The concept of force is used in this context without explanation or justification in order to simplify the presentation. However, this should not be interpreted to signify that the notion of mechanical force is a necessary physical primitive in the framework being developed, but rather that the *dualities* that exist between the Newtonian concept of force and the collective properties of complex systems allows the use of the term as a sort of intuitive shorthand.)

So it is a property of localized periodic phenomena, such as the pulsating spheroid, that in addition to the incoming wave motions that result from its own oscillation – i.e., those converging waves that comprise what might be called the "compression cycle" of the oscillation – any additional waves that are incident on the singularity, from any direction, will tend to either (a) augment the compression if in phase with the oscillation or (b) suppress the expansion if out of phase with the oscillation. Therefore, on average, the cumulative effect of external waves will be to compress the region of Subspace in which the singularity occurs. Moreover, this applies to any region of Subspace that contains a density gradient; that is, such a region will tend to be compressed by ambient waves, and in proportion to the rate of change of the density.

On account of these properties, the region of space that the singularity occupies is in general denser than surrounding areas. Now consider what must happen if the singularity has any sort of translatory motion. Because the source oscillation comprises a region of increased density, as it moves through space the surrounding medium must "rush in" towards it – and must do so in inverse relation to distance from the oscillating source – in order to thus accommodate the concomitant influx of the medium at those places, successively, that the singularity occupies (see figure 5-15). Therefore, merely by virtue of Brownian motion – i.e., the expected fluctuating movement of all such singularities in Subspace – each must be a sort of local *Subspace sink*.

The density gradient that arises as a result of this behavior can be thought of in the following way. As depicted in figure 5-16 (the proportions are not accurately drawn), the thickness of spherical shells of equal volume varies inversely with the square of the shell's radius. Therefore, as the density of

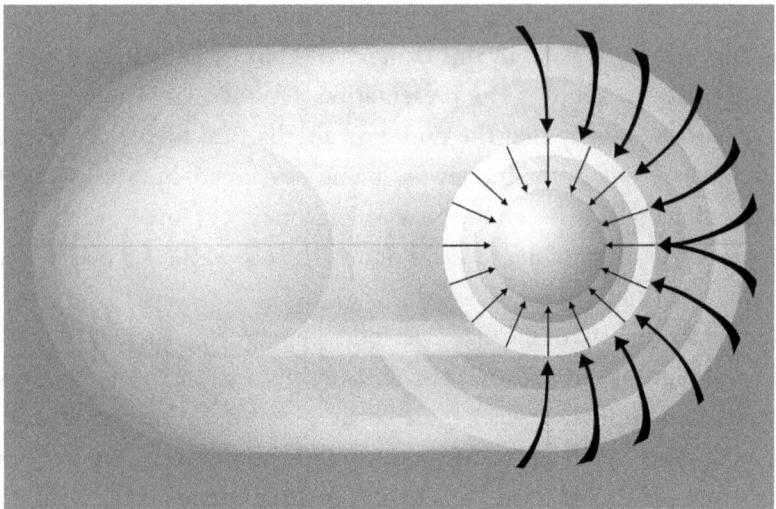

Figure 5-15   The region of space in which the singularity is embedded tends to be denser than the regions surrounding it, because of the compression effect of ambient waves as described in the text. Therefore, as the singularity moves through space, pictured here from left to right, the medium surrounding it and in advance of its path "rushes in" to equalize the density.

$$4\pi R^2 \Delta R = 4\pi r^2 \Delta r$$

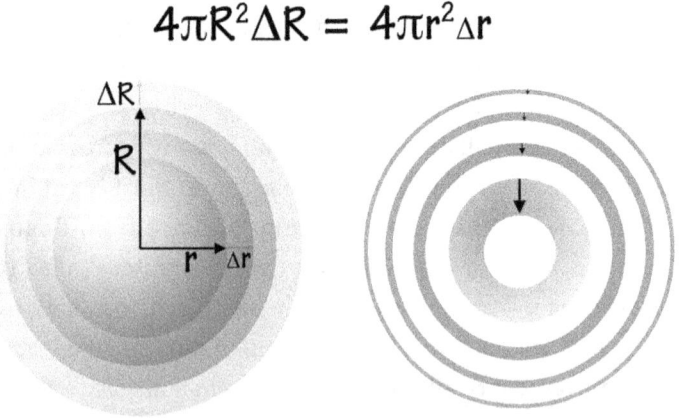

Figure 5-16   As the density of that part of the medium in which the singularity is embedded increases, a density gradient is established in the surrounding space. At any given distance from the singularity, the surface area of a spherical shell increases with the square of the distance from the singularity. Therefore, shells of equal volume have disproportionate thicknesses – again, inverse square with respect to distance from the center of the disturbance.

the medium enclosed within such a spheroid changes, the linear distance between two points along the gradient (i.e., the radius drawn through the points in question) changes in an amount inversely proportional to the distance of those points from the center of the spheroid. Roughly speaking, this corresponds to the dilatation of space associated with the rate of change of the spatial density gradient, as conceived under the developing framework.

On the scale of the singularity this effect is relatively trivial – relative, that is, to the dynamic density gradients throughout Subspace generally (which, as elaborated in what follows, approximate to the infinite collection of "probability amplitude gradients" associated with the usual interpretation of quantum theory, what might be called "the wave equation of the universe" thereunder). But collectively, on the scale of very large numbers of such singularities, the cumulative sink effect (which is inverse-square with distance) can be significant. Moreover, the "first" or "precipitating cause" of this collective behavior is likewise both inverse-square with distance and much more significant on the scale of multiple singularities. In [somewhat crude] analogy with the *Casimir effect*, such sinks tend to be "pushed together," roughly in the manner of Le Sage-type (i.e., kinetic) theories of gravity.

These, then, are the *"three prongs of gravitation,"* viz.:

(1) What might be described as an impulse associated with the overall flux of waves throughout Subspace, which flux exists on account of the multitudinous fluctuations and periodic processes that fill space and which in turn act back upon these processes because of the density gradients that necessarily accompany them. Moreover, on the collective scale, singularities tend to be mutually "shadowed" vis-à-vis this *flux*, and thus appear to be under the influence of a mutual attraction, inverse-square with respect to distance.

(2) A Subspace *"sink"* effect, whereby space tends to "rush in" to the singularity, with concomitant spatial density gradients that are inverse square with respect to distance.

(3) An additional collective action – a cumulative aspect of the behavior that emerges as a *nonlinear, self-gravitation* effect due to the
density gradients in the regions surrounding the singularity, which
are also *acted upon/act like* the singularities and thus, albeit to a
much less significant degree, also "gravitate."

The cumulative effect of these three properties goes by the single name
Gravitation. As Einstein once noted,[16] gravity can effectively be divided
into a "Newtonian component" and a "Relativistic component." That is,
figuratively speaking of the mathematical relations, half the effect of gravity can be attributed to a Newtonian force and the other half to time dila-
tion/relativistic effects generally – in Einstein's formulation, the effects of
curved spacetime; in the present context, the spatial density gradient.

If one considers this idea seriously for a moment – i.e., regarding the
"Newtonian half" of gravity, taken within the context of the Subspace
framework – the model of a localized periodic-process accommodates a
compelling formulation. For by attributing a physical significance to the
de Broglie wavelength of such a singularity, calculation of the "Newtonian"
(i.e., "impulse") aspect of gravitation as hereinabove described yields a
truly remarkable result.

Consider figure 5-17, which shows the cross-section of a sphere with
radius well within the small range of values that have been empirically
determined for the radii of the nucleons – and exactly equal to the de
Broglie wavelength corresponding to the rest energy of one *atomic mass
unit*. Assuming that the wave velocity of the medium is that of light, this
wavelength corresponds to a frequency of $2.25234 \cdot 10^{23}$ cycles per second.
Take this then to be the resonant frequency of a localized periodic process
as modeled above (i.e., with radius/wavelength $\lambda = 1.33102 \cdot 10^{-15}$ *meters*).

If ambient wave momentum and energy are determined in accordance
with Planck's constant in the manner discussed above, then a simple relationship obtains for the interaction between two such periodic processes,
as mediated by the ambient wave field. The most straightforward way to
approach the interaction between singularity and field is to simply multiply the cross-sectional area of the singularity (i.e., its *scattering cross-sec-*

---

16 Einstein, Albert. Relativity: the special and the general theory; 100th Anniversary Edition. New
York: Tess Press, 2005, page 128, Appendix III.

tion) times its resonant frequency. This may perhaps seem too simple, but inasmuch as one of the primary premises of the present model is that each "wave crest" carries energy/momentum, the product of cross-sectional area and frequency gives a salient physical quantity – i.e., *a flux multiplied times an area.*

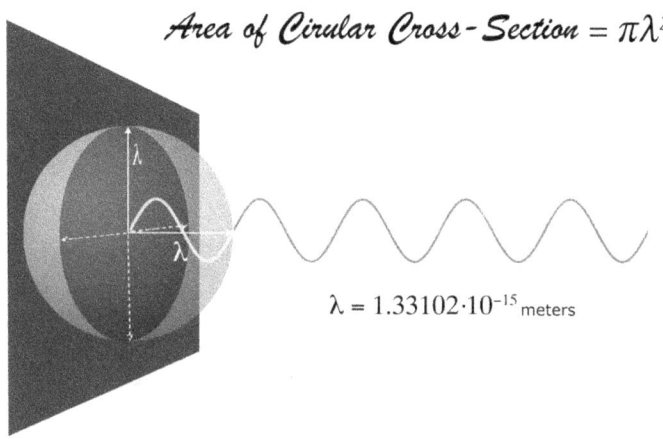

$$\textit{Area of Cirular Cross-Section} = \pi \lambda^2$$

$$\lambda = 1.33102 \cdot 10^{-15} \text{ meters}$$

Figure 5-17   The singularity absorbs momentum from ambient waves at its resonant frequencies – source of the compression of the region of space that the singularity occupies.

In this view, the mutual shielding effect of local periodic processes, as depicted in figures 5-18 and 5-19, should be inverse square with distance. These relations are well-understood (as mentioned in chapter one, mathematical theories of mutual attraction based upon such ideas go back at least to the time of Newton). And so, just as the ambient field is here postulated to compress the region of Subspace in which the singularity is embedded, so too, under appropriate circumstances, it is assumed to draw singularities together.

In addition, one would expect there to be some *coupling parameter*, which represents the probability or efficiency of the *field-singularity interaction.* In electrodynamics, the *fine structure constant* plays the role of such a coupling parameter, inasmuch as it determines the coupling probability or efficiency - essentially the *efficacy* - of *electric charge.* Moreover, if, as many physicists suspect, the fine structure constant is the manifesta-

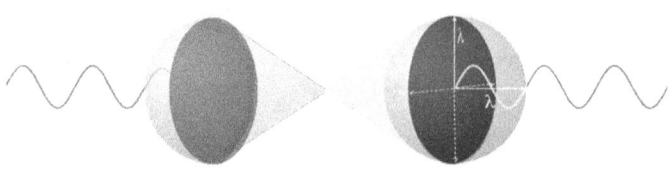

Figure 5-18 Mutually shielding of two singularities.

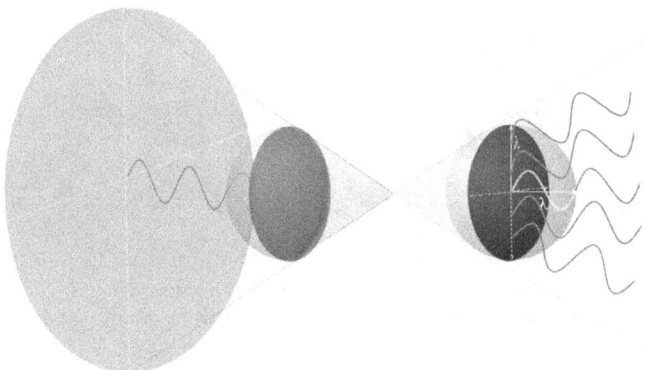

Figure 5-19 The shielding effect is inverse square with distance, on simple geometric considerations.

tion of some truly fundamental aspect of such quantum couplings generally, it should perhaps not be surprising to find that it plays a similar role in gravitation.

As it turns out, this is indeed the case. Based upon the model of gravitation here under development, in which the first-cause among the complex of salient factors is deemed to be an impulse/shielding effect, the fine structure constant determines a simple relationship between Planck's *constant-of-action* and Newton's *constant-of-gravitation* – just as it determines the relationship between Planck's constant and the *electric permittivity* and *magnetic permeability* of the vacuum, and thus the speed of light.

Under this model, the equation for "Newtonian" Gravity is:

$$\pi\lambda^2 f \,\alpha^2 \left(\frac{M_1 M_2}{r^2} \right) = \frac{GM_1 M_2}{r^2} \tag{5-18}$$

Where, again, $\lambda = 1.33102 \cdot 10^{-15}$ meters – here both the *radius* and the *wavelength* of the *singularity* of *nucleonic mass* (*atomic mass unit*, hereinafter *"amu"*). $M_1 \bullet M_2$ represents the product of the *amu* and the *number of such mass units* comprising the two interacting entities, M_1 and M_2, respectively, and thus the product of their masses, while r is the distance between them – that is, between their respective centers of mass. The rest energy of the *amu* is equal to Planck's constant multiplied times the frequency, f, of the *amu's* intrinsic periodic action ($2.25234 \bullet 10^{23}$ cycles per second – again, the frequency of a singularity of such mass, in accordance with the present model). And of course, α is the *fine structure constant*.

Put into words, the *cross-sectional area* of the mass unit, multiplied times the *frequency* of oscillation associated with that mass – thus its resonant frequency with respect to ambient waves – gives what is here interpreted as the *scattering cross-section* multiplied times the *gravitational flux*. In turn, the *efficacy* of this interaction – the *probability* or strength of the flux interaction with the singularity – is given by the *fine structure constant*; which, like mass, enters into the determination of the gravitational effect *twice* (i.e., *first*, it determines the efficiency of the singularity *as shield*; *second*, it determines the response of the singularity to *being shielded*). Therefore it is the *square* of the fine structure constant that appears in the equation. This interaction – again, here taken to be the primary cause of gravity – plays a crucial role in the developing framework, inasmuch as it represents *not only* a primary cause of gravity *but also* the essential form of the interaction between field and singularity generally; and thereby a theory of motion intended to subsume the current interpretation of quantum phenomena vis-à-vis probability amplitudes.

This result alone – the derivation, once again on general physical principles, of the gravitational constant and its connection with the fine structure and [hence] Planck constants – is of course quite interesting. But more follows. Pursuant to this model of gravity, both the acceleration of the observable expansion of the cosmos and the anomalous acceleration phenomena found in spiral galaxies can be viewed as inherently linked. That is, not only are Planck's constant and the gravitational constant closely related, but the gravitational constant and the cosmological constant represent the same physical thing. (In this connection, it is amusing to note that the constant of gravitation is given in units of length cubed divided by the product of mass and the square of time. That is, *Newton's gravitational constant is given in units of acceleration of volumetric expansion in relation to units of matter*). Again, not only can the "*acceleration of the expansion of the cosmos*" be understood in accordance with the above interpretation of gravity – with the empirically determined rate of acceleration anticipated surprisingly accurately – but so too the gravitational anomalies usually associated with "*dark matter.*"

It should here be noted that, just as with the unfounded bias against deterministic interpretations of quantum theory, many physicists have learned to believe that the possibility of any sort of impulse or "shielding" contribution to gravity has been ruled out on the basis of one or another insurmountable objection. But this is not the case. A detailed, mathematical analysis of this matter is beyond the scope of the present work, but suffice it here to say that the various arguments against the possibility of such a component of gravitation have been invalidated. In particular, it is known that as long as only a quite small portion of the energy incident upon an entity is absorbed, the balance of scattered energy can contribute in a disproportionately large manner to a gravitational effect, without causing a correspondingly (non-trivial) heating or mass augmentation of operative bodies, or drag on their translational motion.

§ 5.17 The Gravitational Constant, the Cosmological Constant, the Expansion of the Cosmos, MOND & the Galaxy Rotation Curve.

If gravity is a derivative effect related to the density and density gradients of Subspace – the "first cause" of which is a flux that is shadowed by operative bodies and corresponds, more or less, to what might be called the "Newtonian component" or "Newtonian half" of gravity – then it is reasonable that, on the cosmic scale, such a flux would have a repulsive effect, inasmuch as attraction is a result of internal shadowing. On the basis of this consideration, the equation for the expansion of the cosmos should be given by:

$$A_{cceleration} = -\frac{GM}{R^2} \qquad (5\text{-}19)$$

where G is the gravitational constant, M is the mass of the cosmos – everything inside the expanding boundary/surface – and R is the distance from the center to the surface. In other words, this is the equation for the gravitational force between all of the matter within the boundary of the observable universe and whatever is "on" the boundary, divided by the latter, but reversed in sign. Substituting the relevant numbers, gives:

$$A_{cceleration} = -\frac{(6.674 \cdot 10^{-11})(3 \cdot 10^{52})}{(1.2 \cdot 10^{26})^2}$$

or

$$A_{cceleration} = -1.39 \bullet 10^{-10} \qquad (5\text{-}20)$$

meters per *second* squared, which is quite close to the currently accepted value for the acceleration of the expansion of the universe. That is to say, the current acceleration is given fairly closely by *2G*, where *G*, again, is the gravitational constant. (As noted above, the gravitational constant is given in units of distance cubed divided by the product of mass and the square of time. Or: Newton's constant of gravitation is given in units of *acceleration of volume expansion* per *magnitude of mass* in that volume.) In fact, the overall expansion of the universe can be represented fairly closely by a simple equation such as the following:

$$Gt^2 + .82ct + R_0$$

where R_0 is the radius of the observable universe at [here set arbitrarily at] one billion years after the putative *Big Bang*, t is the time that has elapsed since that [arbitrary] chronological marker and c is the speed of light. Any equation of this form, where G is the quadratic coefficient, gives the quantity *2G* for the acceleration, which as noted is roughly the current observed rate of acceleration of the cosmic expansion.

While it is beyond the scope of this treatise to rigorously demonstrate how this result follows from the model, it can be qualitatively grasped in the following manner.[17] A quick, "down and dirty" image, which gives the desired result but should not be taken literally, is to interpret the flux that is "reflected" by/from all the matter within the "center" of the cosmos to act on the "surface" and thus give the acceleration GM_U/R_U^2, where M_U is the mass of the universe and R_U its radius. Or again, each mass unit can be thought of as both a source and a shield of cosmic flux, so that in nearby regions its effect is primarily as *shield* (because it scatters more than it produces; what it sources is overwhelmed by that from *space* generally, as well as by that from matter at large). Thus at cosmic distances its role is chiefly as *source*; as contributor to the overall flux in the universe. In either sense, expansion of the universe is a corollary to gravity, as it is effectively driven by that quantity of flux which can be associated with ponderable matter/energy; attraction and repulsion have the same root cause, and so are essentially the same thing.

Now, if expansion and gravitation both originate as described, then because it is the cumulative, collective properties that are responsible for the overall effects of gravity, whereas only the direct flux action is a significant factor with respect to expansion, it is to be expected that certain differences should be discernible between the usual predictions of Newtonian/ Einsteinian theory and the observable properties of gravitation (as well as how those properties are interpreted in relation to expansion). Such differences exist, and manifest most dramatically in the regime of small gravitational accelerations.

Both empirically and in accordance with the present framework, the repulsive effects of gravity are deemed to occur between large agglomerations of matter separated by cosmic distances. Again, this is here understood as a consequence of a flux that originates both in "empty" space and in matter dominated regions, which thus only conduces to a significant shadowing effect when the distance between scattering/absorbing bodies is not so large as to wash out the repulsive action of the flux that exists

17 Albeit only partially, neglecting "self-interactions" of the flux and other large scale, dissipative/ modifying factors.

between them. Accordingly, it is with respect to objects situated at the periphery of matter dense regions, and at cosmic distances from other such matter dense regions, where the expansion effect is discernible from gravity, inasmuch as at such places gravitational acceleration is small enough that the expansion-producing component appears as a baseline gravitational effect. That is, when the rate of acceleration due to a large system of matter/energy (such as a galaxy) is small, because of the distance from its gravitational center and, moreover, when that rate is close to the acceleration of the expansion of the universe, the effect of gravity will appear to fall off more gradually than expected on a Newtonian/Einsteinian basis.

The *Mathematica* graph of figure 5-20 displays a simple representation of such relations. Anomalous behavior occurs when $g \sim 2G$: when the acceleration of gravity is near the acceleration of the expansion of the universe. These relations are computed as shown in the graph and described below.

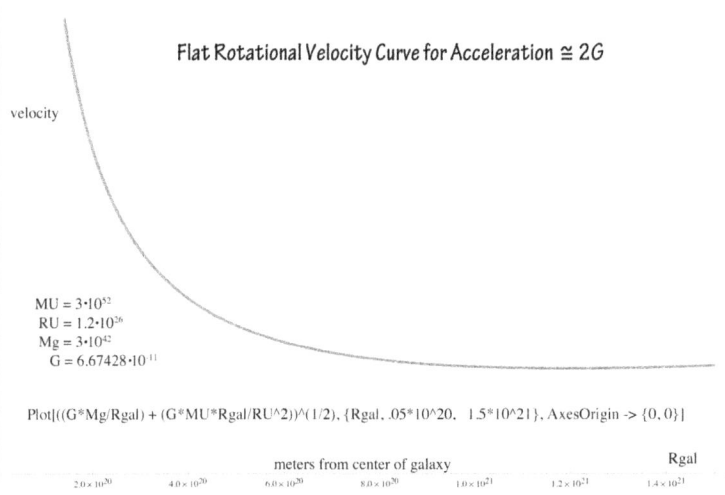

Figure 5-20 *Mathematica* graph showing velocity as a function of distance in a hypothetical galactic orbit. *MU* is the mass in kilograms and *RU* the radius in meters of the observable universe. M_g is the mass of the hypothetical galaxy and *Rgal* the orbital distance from its center. When the orbital acceleration of the object is near $2G$, and its location is cosmic distances from other significant gravitational sources, the expansion effect of the flux becomes a non-trivial component of orbital behavior.

Gravitational (centripetal) acceleration, a, equals the orbital velocity squared, v^2, divided by the radius of the orbit, r. Assuming unit mass for simplicity, acceleration can be equated to net force, which, in the case of anomalous orbital accelerations, is taken to be the sum of (1) the usual Newtonian force attributed to the galactic mass, plus (2) the component of force due to the flux associated with expansion.

This simple calculation yields the results displayed in the graph as follows:

$$A_{cceleration} = \frac{v^2}{R_g} = net\ force = \frac{GM_g}{R_g^2} + \frac{GM_u}{R_u^2}$$

where v is the galactic orbital velocity and G the gravitational constant, M_g is the mass of the galaxy, R_g is the radius of the galactic orbit, M_u is the mass of the observable cosmos and R_u is its radius, therefore:

$$v = \sqrt{\frac{GM_g}{R_g} + \frac{GM_u R_g}{R_u^2}} \qquad (5\text{-}21)$$

And, of course, when the acceleration due to gravity is equal to the acceleration of the expansion of the universe, the expression for gravity is linear in the inverse relationship to distance, and non-linear in the relationship to gravitating mass. That is, for acceleration $g = 2G$:

$$2G = \frac{GM_g}{R_g^2} \quad \therefore \quad M_g = 2R_g^2 \quad \therefore \quad R_g = \sqrt{\frac{M_g}{2}}$$

therefore

$$g = \sqrt{4G^2} = \sqrt{\frac{2G^2 2R_g^2}{R_g^2}} = \sqrt{\frac{2G^2 M_g}{R_g^2}}$$

or

$$g = \frac{G\sqrt{2M_g}}{R_g} \qquad (5\text{-}22)$$

That is, in accordance with equation 5-21, the orbital velocity curve flattens when the gravitational acceleration is close to $2G$; additionally, in this regime equation 5-22 is operative. These relations comprise the central tenet of *MOND*, or *Modified Newtonian Dynamics*, created to describe the "flat" rotation curve of stars in galactic halos. However, whereas Mordehai

Milgrom established *MOND* entirely on the basis of empirical data, equation 5-22 is here derived as a simple relationship between gravity and the acceleration of the expansion of the universe, in accordance with the view that gravity is related to an underlying density gradient of space, caused by a flux, whereby it is also related to the acceleration of the expansion of space – all of which relations obtain because space is treated as a physical system and ponderable events as emergent properties of that system.

In somewhat over-simplified terms, as the primary cause of gravitation is also the primary cause of what appears to be an accelerating expansion of the cosmos, as large agglomerations of ponderable matter are thereby compressed they are by the same mechanism pushed apart – *like filaments of dough between pockets of air in leavening bread.* And so if one looks at a simulation of the evolution of the cosmos, and at an image of its currently conjectured configuration, it appears very much as though visible matter is behaving in just such a fashion, under the influence of expanding "bubbles" of space (see figures 5-21 and 5-22).

And if the primary or precipitating "first cause" of gravitation is just such a flux, it follows that the usual expectations regarding the entropy and anomalous behaviors of gravitationally bound objects must be reconsidered – especially where these two crucial concerns overlap (that is, especially with respect to the entropy of black holes).

For if the primary cause of gravity is essentially *external to gravitating systems*, the problematic issue of gravitational singularities need not *necessarily* arise. Moreover, evidently the entropy of gravitationally bound objects should be deemed *low* rather than high, while a new reason for considering surface area more relevant than volume arises in connection with exceedingly dense matter/energy configurations (i.e., the possibility of a limit regarding the capacity of a system to "scatter/absorb flux," beyond which additional scattering/absorption capacity is superfluous/redundant). This follows from the simple consideration that if the "shell" of an exceedingly dense object is capable of scattering/absorbing a substantial portion of the operative flux, then those constituents of the object at increasing distances from the surface may be irrelevant with respect to this aspect of the gravitational phenom (again, the "first cause" or "Newtonian component").

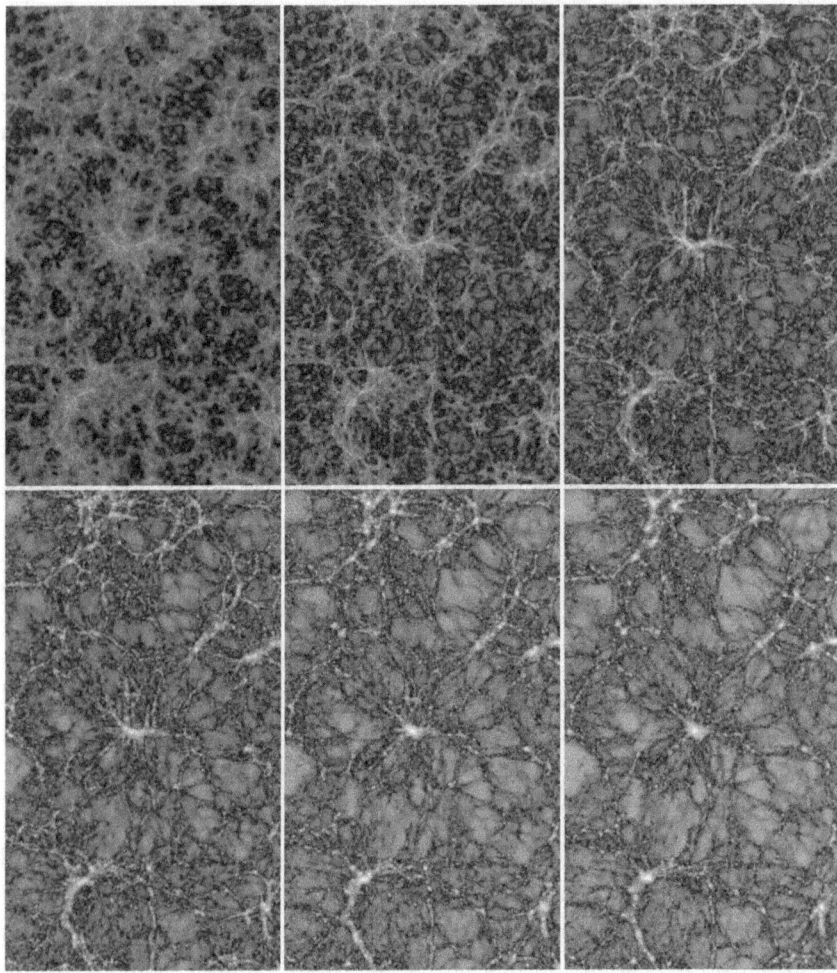

Figure 5-21. Expanding Universe. Filamentary Networks = Matter.
Expansion Proceeds Clockwise from Upper Left
Images based upon animated simulations published by the
Max Planck Institute for Astrophysics, München
www.mpg.de

One more point in this connection. It is believed that the gravitational potential of all the matter in the universe – formally *negative*: the cosmic gravitational energy – may be equal to the *positive* energy of that matter. This is not only readily comprehensible in accordance with the present model but is in fact another compelling reason to take it seriously. For if the rest

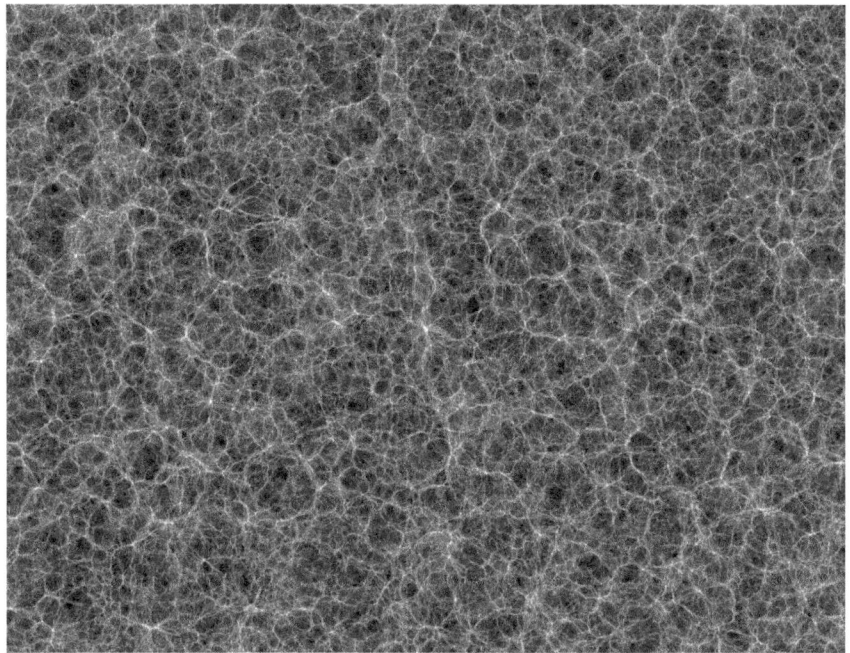

Figure 5-22. *Conjectured Configuration of the Universe "Now"*
Image based upon animated simulations published by the
Max Planck Institute for Astrophysics, München
www.mpg.de

energy of matter is attributed to localized periodic processes, and if gravitational attraction and repulsion are a consequence of an interaction between these localized processes and the flux produced by wave sources generally, then the equality of "negative" and "positive" energy indicates that the energy of gravitation (both attractive and repulsive) and the self-energy of matter are *identical* – i.e., not merely quantitatively equal but *the same thing*. In other words, *the rest energy of matter is equal to the amount of flux that it absorbs from all other matter in the universe*, by *equation 5-18*, which gives the ["Newtonian half" of the] gravitational interaction. In accordance with this model, the "compression effect" of ambient waves emitted from other matter – conjectured to precipitate increased density in regions occupied by localized periodic phenomena – can thus be considered more generally as overall *energy source* of those localized phenomena. This may be deemed the *principle of equivalence of gravitational and inertial energy.* Moreover, by this relation:[18]

$$M_U c^2 = -\frac{GM_U^2}{R_U} \quad \therefore \quad G = -\frac{R_U}{M_U} c^2$$

i.e., if the energy of cosmic matter is equal to the (negative) gravitational potential of that matter, it follows that the gravitational constant G – which, as twice noted above, is given in *units of acceleration of volume expansion* (meters cubed per second squared) per *kilogram of matter* – is equal to the *radius of the cosmos multiplied by the speed of light squared, divided by the mass of the cosmos.* Therefore, if the gravitational and self energy of matter are the same thing, then this simple equation relates the gravitational constant and the velocity of light to these "boundary parameters" of matter/ energy on the cosmic scale and thus, via the fine structure constant and the constant of action, to the corresponding parameters of matter/energy on the subatomic scale (as per equation 5-18, linking the gravitational constant with the mass and radius of a nucleon).

§ 5.18   Discussion I: Towards a New Framework for Physics

In order to flesh out a mathematical theory on the basis of this framework, it is necessary to develop a method for modelling quantum phenomena in detail. As indicated above, this can be accomplished in an intuitively graspable manner by assuming that fundamental particles are localized yet finitely extended processes in Subspace. Toward this end, many helpful insights can be gleaned from earlier efforts to develop a deterministic interpretation of quantum mechanics and [especially] quantum field theory, particularly as pursued by David Bohm and his collaborators. (Bohm always maintained that it was precisely such a prospect – i.e., the possibility of sparking and enhancing alternate lines of research – that was the most important aspect of that early work.) In order to further prepare the ground for this framework,

---

18 Assuming a spherical distribution of matter, a factor of 3/5 arises in the relation for potential energy, but these equations are intended as approximations.

it will be helpful to review some of these key insights, along with those developed in previous sections of this chapter.

To begin, it is known that quantum theory can be extensively re-interpreted along the lines Bohm pursued. Everything from tunneling and spin to the behaviors of fields and anti-particles have been successfully treated in accordance with the deterministic viewpoint. However, two crucial inadequacies of this approach have remained. One is the so-called "non-locality" of the causal mechanism. The other is lack of clarity regarding the nature of the connection between singularity and field – the absence of an adequate interpretation of the "particle" as perhaps, again, a source of and singularity in the field. As a result of these inadequacies, the insight necessary for a unified treatment of gravity and quantum phenomena has also been absent. With the overcoming of these inadequacies, and concomitant forging of a link with gravity, it is possible to build on the underlying unities and phenomenological analogies described in the sections above and establish a unified framework for the treatment of classical, relativistic and quantum phenomena.

It was the great contribution of de Broglie, with respect to his *Theory of the Double Solution*, to have viewed the particle as singularity in the field. Although de Broglie's vision, as Bohm's elaboration of it, is not identical with the model suggested by the present framework, his basic understanding of the phenomenological particle as field singularity, so much like Einstein's vision vis-à-vis the latter's unified field program, is more or less on the same track. Of course, in order to fully establish and understand the connections, especially with General Relativity, it is necessary to develop these ideas mathematically. And today, with the advantage of much-improved mathematical resources, it is possible to explain such a singularity as a *Soliton* – simultaneously a cause and effect of the field – and to understand the field both microscopically, with respect to quantum phenomena, and macroscopically, as generating the collective properties of space-time and gravitation.

Whereas under the usual interpretation of quantum mechanics the square of the amplitude of the Schrödinger wave, normalized to represent probability, is taken to indicate the likelihood of a given observation, such as finding a particle at a particular location, this is equivalent to the view – adopted under the de Broglie-Bohm interpretation – that particles and/or field singularities move in accordance with the gradient of the phase of the Schrödinger wave field. And while this interpretation, in and of itself, does not add anything new to the formalism – and while attributing a potential to the Schrödinger equation certainly does not furnish an explanatory mechanism – the vision that emerges is quite suggestive of how this might be accomplished. For once it is recognized that the [in any case approximate] Schrödinger equation stands in such a relation to observed trajectories, intriguing albeit inexact analogies with other physical models spring to mind, not least of which being that with the geometrodynamical paradigm of General Relativity. But again, it must be recognized that non-relativistic quantum mechanics is intrinsically inexact, and when the relations discussed in sections 5.5 through 5.7 are considered carefully, salient dualities can be seen to exist across multiple domains.

Thus the concept of *Subspace* and the associated notion of *spatial density* – connoting a constitutive structure of space that underlies behaviors that are thus approximately and, properly viewed, tentatively described [on the micro level] via the Schrödinger equation – can account for a wide range of phenomena; roughly speaking, on the basis of "*diffusive*" motion that occurs in accordance with the spatial density gradient. This is analogous to the law of motion under the de Broglie-Bohm pilot wave model, wherein the field represents a potential that guides the particle so that, with respect to an ensemble of particles, the statistical effect of the *probability amplitude* and the *guiding potential* are identical. Accordingly, motion is *away from nodes* and *toward anti-nodes* of the field, and particles are observable processes in the field.

Stochastic formulations developed by Edward Nelson, Bohm, J. V. Vigier and others have employed concepts along similar lines, in the sense that a Brownian-like motion underlies the quantum law. However, in the present framework the quantum field describes the *spatial density gradient* and thus the *effective geometry of space-time*, by virtue of the concept of *Subspace*, which refers to a constitutive structure of space on a given operational level of scale. With respect to the levels of scale accessible to present-day physics, this means the constitutive structure of what is usually referred to as the *Vacuum*, and the phenomenology of physics comprises those ponderable patterns and processes – entities and events – which, given current technological capabilities, are *observable, energetic manifestations* in that underlying structure.

The link with gravity is understood more or less in terms of scattering and absorption processes in Subspace, which have interpretations in terms of the mathematical concepts of *reflection* and *transmission coefficients*, and so in terms of *reflection* and *refraction*. Such actions can also be viewed as *interference effects* as appropriate. Absorption, for example, can be described as a form of destructive interference, which cancels the wave field anisotropically on one side of a localized process.

It has long been known that large collections of dynamically interacting objects exhibit collective patterns of organized behavior – patterns that can be treated as more or less independent, autonomous phenomena – and that one of the most pervasive forms of such behavior is simple (*harmonic*) oscillatory motion. It is also known that in solid (crystalline) structures, periodic disruptions of lattice symmetry cause collective patterns of behavior that emulate particles. These are *phonons*, the so-called "particles of sound" (and of course *dislocations, disclinations* and other more stationary *lattice defects* can emulate particles as well [*phonons* are not such defects, properly viewed, but rather collective attributes of waves]). It is a well-known general result that such singularities must arise when symmetry is spontaneously broken. And while this [*bosonic*] behavior is usually attributed to the underlying quantum mechanics, phonons can be fairly well understood and mathematically treated along essentially classical lines.[19]

In general, it is possible to mathematically describe many complex and beautiful structures as manifestations of waves. Everything from crystalline-like disclinations and dislocations to fractal- and Newtonian-like objects can manifest in the patterns of waves and currents. Moreover, patterns can develop on top of patterns. In particular, the non-linearity of *Solitons* and the mathematical tools available to describe them, along with *Wavelet Analysis* and several other relatively recently developed techniques and technologies, make the elaboration of this vision technically feasible. Although the full details of such a development are beyond the

---

19 See, for example, Spohn, Herbert. "The Phonon Boltzmann Equation, Properties and Link to Weakly Anharmonic Lattice Dynamics." Journal of Statistical Physics 124, no. 2-4 (5, 2006): 1041-1104.

scope of this treatise, only a few simple properties of periodic processes are truly essential to the present framework.

It was discussed above that key aspects of both relativistic and quantum phenomena can be understood as having shared origins. This was referred to as the *common provenance of Lorentz Invariance and Quantization of Action*. This unifying principle extends further, to the phenomena described by Newtonian dynamics and General Relativity, and is here given axiomatic status as *The Principle of Common Provenance*. It is important to keep this principle in mind because, while investigators have found many interesting analogies in the foundations of physics (between for example quantum field theory and condensed matter physics and, more basically, between special relativity and the behavior of defects in crystals), these commonalities should be viewed as special instances of a *meta-principle* that runs much deeper.

If one fails to recognize this deeper relationship, one may be misled to accord the basic forms of quantum and relativistic behavior fundamental status. So it is that unification efforts based on analogies with condensed matter physics simply stipulate the underlying quantum principles as premises, rather than consequences. On the other hand, it is just as misleading to regard Newtonian law as fundamental. The prime advantage, rather, to accepting the underlying meta-principles as primary is the open-minded attitude to which it conduces; with respect, in particular, to the possibility that *all law* is contingent/emergent.

§ 5.19   Discussion II: Conceptual Foundations of Subspace Theory

Again, these ideas are more than merely suggestive. They indicate how a unified theory of classical, quantum and relativistic physics can be established on the basis of a small group of closely related considerations. Moreover, certain fundamental features of *String Theory* and other approaches to quantum gravity are subsumed under these considerations; in non-trivial, thought-provoking ways. Some of the key insights at the center of this group are reviewed below.

Based on purely kinematic relations, it can be demonstrated that the following phenomena arise together, without *ad hoc* introduction of quantum, relativistic or Newtonian properties:

- Lorentz-invariance and the relativistic, geometro-dynamics associated with gravity;

- Quantization-of-action and the matter-energy and particle-wave dualities;

- The apparent non-locality of quantum entanglement;

- The vector potential and the transverse polarization of electro-magnetic radiation;

- The gravitational constant, the cosmological constant, Planck's constant, the fine structure constant and Newton's laws of motion.

And so a small group of seemingly self-evident deliberations suffices to demonstrate the unity of classical, relativistic and quantum physics, while a theory of quantum gravity emerges on the supposition that gravitation is a net, collective effect of the interaction between localized energetic processes/stable formations on the one hand (taken to comprise ponderable matter/energy) with the balance of energy distributed throughout *Subspace* on the other.

Gravity is a cumulative effect of the interaction of the ambient, global energetic field with (a) all of its more or less stable sources, as well as with (b) all more or less stable density gradients generally. That is, wherever a non-trivial spatial density gradient exists outside the bounds of concentrated, ponderable matter/energy, that very gradient (and hence gravity itself) is in turn a gravitational source, in the same sense that spacetime curvature gravitates under General Relativity. And so "Newtonian" gravitation and "Einsteinian" gravitation both arise, in the appropriate regimes, from the same group of considerations – including the relativistic expectation of nonlinear properties, albeit with important distinctions and therefore new, empirically testable possibilities.

General Relativity provides for the description of gravitational effects on the basis of an *action principle*, in accordance with which the *proper time* that an object takes to trace out its geodesic is maximized. This is interpreted as a consequence of the curvature of spacetime under the influence of a gravitational source, which causes time periods to expand and distances to contract. In the corresponding case of Subspace, regions of greater density correspond to places where the time period of oscillatory phenomena is extended, while the spatial volume occupied by such phenomena is condensed. Ponderable bodies correspond to localized, recurrent patterns of activity in Subspace, while the curved spacetime of General Relativity corresponds to density gradients generally. Density gradients and localized processes are thus closely related, though matter is not *univocally* equated with dense regions of Subspace *per se*, as might be the case in a pure field theory of the type that Einstein envisioned. Illustratively, the source of a localized, periodic process may be the conjunction of ambient waves in phase with a local, recurrent phase transition (roughly analogous to the alternating formation and dissipation of a vapor bubble in a small region of a fluid) or any "goings-on" generally, simple or intricate, at levels of scale below that at which the energy source is observable, inasmuch as any change and especially any recurrent change in local Subspace structure will both originate and interact with ambient wave formations.

Quantum Mechanics also allows for a description of phenomena on the basis of an *action principle*, and for a description of matter as a wave-like phenomenon. A particle can be associated with two periodic disturbances, *viz.*: (1) a more-or-less localized periodic process, which denotes particle position and/or momentum, and (2) a distributed field of waves, which can be taken to guide the localized periodic process, inasmuch as places of maximal field amplitudes correspond to places where a local wave-process is most likely to be found. This guidance effect, on the quantum scale, is identical to but different in behavior from General Relativity's [macro] spacetime curvature; its fluctuations are more dynamic and its effects more intense than that of gravity – more or less along the lines of a gravitationally induced "foam" of spacetime on the Planck scale, which has been conjectured to be associated with the *zero-point field*.

In the present model, Subspace waves can interfere with one another in the manner of light, but they can also [partially] scatter from local wave-groups (which latter also tend to mutually scatter, though rather more strongly), and thus take particle-like paths – for example, through slits in material barriers. And so while field waves can tunnel through such barriers, they also diffract upon emergence from [appropriately sized] slits. The conjunctive effect of field and local wave-group interaction is to increase the overall density of any region of space where both are present. Therefore, in places where ponderable matter exists, and fields are [necessarily] interacting with matter – i.e., in places where energy can be *found* – the density of ponderable matter and regions of Subspace corresponding thereto tends to increase.

Increasing Subspace density corresponds with increasing spacetime curvature in General Relativity, and the cumulative effect of energy upon the density of Subspace, operating on scales much larger than that of the quantum of action, comprises the [relatively weak] effect of gravity. Gravitation is thus a relatively large-scale "net" phenomenon, which yet emerges from quantum interactions. And, because the absorption and scattering – the refraction and the reflection of field waves by localized wave-groups – produces "shadows" in the field, the cumulative effect of the field waves upon the density of Subspace will tend to be inverse-square with the distance between operative (absorbing/scattering) wave-groups.[20]

This model also accounts for differences between various transverse wave phenomena, such as those associated with electromagnetic radiation *per se*, and longitudinal wave phenomena, which latter give a physical meaning to the Schrödinger wave equation (or, more accurately, to a non-linear analog thereof). And so the *Vector Potential* has a physical interpretation in this framework, and is connected with a longitudinal component of the complex of phenomena known as *electromagnetic radiation*, just as gravity is associated with longitudinal waves generally.

Localized periodic processes can be distinguished as *Fermionic* or *Bosonic*, based upon whether, as in the case of the former, they are *Solitons*, and their translational velocity is thus always less than the wave velocity of Subspace associated with the given level of scale/set of phenomena un-

20 Thus there is no *fundamental* distinction between gravity and quantum interactions generally, just as with all quantum field phenomena under this framework, but gravity is in a sense the *most general* of such phenomena because it is a *net*, macro scale effect of interactions cumulatively.

der consideration, or, as in the case of the latter, they are collective phe-nomena – analogous to the *phonon* – crucially comprising, among other things (similar in this respect to solitons), transverse as well as longitudi-nal waves, so that their velocity coincides with the transverse wave velocity of those Subspace processes with which they are associated.

The localized, fermionic wave source or *Soliton* occupies a relatively small region of space, wherein the majority of energy associated with the fermion is located. The distributed wave field is phase-correlated with the Soliton. Longitudinal waves can travel at *superluminal velocity*, while transverse effects (at currently observable scales) are limited to the speed of light. Because most observable energy is electromagnetic in nature or crucially linked to electromagnetic processes, the energy of longitudinal waves *per se* do not manifest significantly on the macro scale (except with respect to gravity-related processes).

By virtue of their superluminal velocity, longitudinal waves can inter-act with Solitons in a virtually space-like fashion. While non-locality is *not* a pre-requisite for a consistent quantum theory,[21] this superluminal interaction can account for what might yet appear to be "non-local" ef-fects. For example, with respect to exchanges of radiant electromagnetic energy, longitudinal waves can provide for the establishment of *resonance* between emitting and absorbing particles, which can thus become *phase correlated* before a process of emission and absorption of transverse, elec-tromagnetic waves is initiated. In this manner, the concept of advanced electromagnetic waves as elaborated by Feynman and Wheeler[22] can have a causal explanation exclusively in *"forward time"* (the author does not consider "backward time" to be a meaningful concept, and should be considered in the light of chapters one through three hereof). Charged particles can interact *initially* via longitudinal waves, by virtue of which they become phase correlated (which may be a pre-requisite for a radiant energy exchange to occur).

In this connection, imagine a somewhat over-simplified model[23] of an electric circuit, initially open, comprising a battery and a loop of wire, the

---

21 As discussed at length in section 4-11.
22 Wheeler, John, and Richard Feynman. "Interaction with the Absorber as the Mechanism of Radia-tion." Reviews of Modern Physics 17, no. 2-3 (4, 1945): 157-181.
23 The description that follows is intended to serve a purely illustrative purpose and is not intended as a theoretical description of the properties of electricity.

latter attached to the negative pole of the battery but not the positive (negative here taken to connote a region that has an overabundance of electrons in relation to another, thus positive region). Until the positive pole is connected and the circuit thereby closed no current flows. Now consider this circumstance in terms of *entropy* relations. A current-carrying wire is deemed to contain many mobile electrons moving randomly, in all directions, with a drift velocity superimposed on that random motion in the direction of the current. But when the circuit is open there is no drift, and the motion is entirely undirected. In the present illustration, as soon as the positive pole is connected and the circuit is closed, the conduction electrons in the vicinity of the positive pole begin to drift toward it. Once they do, those a bit farther away can likewise drift, and so on, as the directed motion is thus communicated, like a time-reversed sequence of falling dominoes, back along the wire towards the other pole.

The rate of this communication is very much slower than that at which the change in the electromagnetic field surrounding the wire propagates. This [latter] change is likewise transmitted back along the wire, which acts as an electromagnetic *wave guide*, so that the conduction electrons in the vicinity of the other pole receive the signal to move, via the field pulse thus channeled by the wire, long before the drift velocity would otherwise reach them. So the *current* does not exist until an *"entropy differential"* is established between the two ends of the wire. As long as all of the motion in the wire is random, energy does not flow. But once the circuit is closed, the motion of the electrons near the point of closure is less random than at other places because a net movement in one direction has developed there.

This circumstance is intended to be analogous to the relation between a *radiation emitting* and a *radiation absorbing* electron. The longitudinal, superluminal interaction may not transfer substantial energy, but the phase correlation and thus change in entropy that it establishes may yet be considered necessary to the emission and absorption of electromagnetic energy. Moreover, to the extent that quantum *entanglement* is taken to exist[24] it can be accounted for on the basis of such *phase correlations*, which need not be exclusively confined to uniquely *"unified"* systems, such as pairs of particles that have been created together. Again, it is not necessary for *literally instantaneous* connections to exist in order for apparently non-local

24 In a form, of course, which does not involve literally instantaneous connections.

effects to be observed. Even if the usual, flawed interpretation of the *Bell Inequalities* is accepted, it must be remembered that no experiment can establish that such connections are in fact "instantaneous."

The extended field associated with *fermions* is characterized by a *nonlinear, Schrödinger-like equation*, which is interpreted to have *solitonic* sources and to be frequently changing, both as its various sources move and become correlated and uncorrelated, and as it undergoes stochastic fluctuations due to random Subspace interactions at multiple levels of scale. Subspace solitons can be conceived to arise in multiple ways. They can be shown to be stable, self-organizing forms in accordance with general mathematical considerations regarding such nonlinear phenomena. More particularly, the energy of an individual fermion can be equated with the energy it obtains from the totality of all others, per the relation given in equation *5-18* which determines, by the fine structure constant, the gravitational interaction. Solitons can also be associated, for example, with *local phase transitions* occurring in the underlying Subspace region – again, analogous to the formation of bubbles in a fluid – which can appear and disappear in the manner of virtual particles, as well as become stable/ harmonically-reenforced systems, particularly via interactions with the extended field[25].

Wave forms approximated by the Schrödinger equation (and thus here associated with longitudinal Subspace waves), and those described in accordance with Maxwell's equations (and thus comprising the electromagnetic field) can be tightly unified. It can be shown that they are epiphenomena of the same, deeper processes. Moreover, both the electron and the electromagnetic field, *together with the relations between them*, can be similarly described on such a basis – *i.e.*, as *manifestations of the same field.*[26] The vector potential of electromagnetic theory corresponds directly to the Schrödinger wave of the *photon*. The vector potential is thus here understood as approximately describing a longitudinal component of electromagnetic (i.e., transverse) processes. This is significant, because it is a well known but not well understood feature of quantum mechanics that the vector-potential is "real" inasmuch as it can be shown to be

25 Cf: Akhatov, I., R. Mettin, C. D. Ohl, U. Parlitz, and W. Lauterborn. "Bjerknes force threshold for stable single bubble sonoluminescence." Physical Review E 55, no. 3 (3, 1997): 3747-3750.
26 Bohm, David. Wholeness and the implicate order. London; Boston: Routledge & Kegan Paul, 1981. Ch. 4

operative in regions of space where the electromagnetic field is technically deemed not to exist.[27]

Again, *fermions* are represented in this framework as *solitons* (and are hereinafter alternately referred to as *"localized wave-groups," "solitons,"* and *"soliton or solitonic waves"*), whereas *bosons* are *collective modes* of *distributed Subspace activity*, just as phonons are collective modes of vibrations in solids. Bosons largely comprise transverse wave actions and move at the velocity of light, which is therefore also the limiting, Lorentz-invariant velocity for fermions, inasmuch as the latter involve transverse as well as longitudinal disturbances. (Moreover, because of the *dispersion of Subspace* at operative scales, *the longitudinal velocity approaches the transverse velocity as source velocity approaches that of light and/or frequency approaches a maximal limit.*) Therefore, because (a) the overriding majority of ponderable effects involve, to some non-trivial degree, transverse disturbances and (b) the dispersive properties of Subspace limit the product of frequency and wave velocity – and thus the rate of transmission of *energy* and *information* – the propagation of ponderable Subspace activity is limited to light-speed.

By virtue of the (on average) superluminal propagation of longitudinal waves, the average distance that bosons travel before interacting with fermions tends to be limited, because bosons can become entangled with *virtual particles* (short-lived solitonic wave phenomena) during short time periods in which absorption and re-emission can nevertheless occur. In this way, boson propagation in "open space" tends to involve multiple interactions with virtual particles before a final interaction occurs, the latter corresponding to absorption by or interaction with a stable fermion.

Solitons and bosons also react to stochastic fluctuations attributable to the overall activity of Subspace, which is conceived to be occurring on an infinite range of scales. Planck's constant of action, which is determined in accordance with the considerations of section 5.13, is a collective effect of stochastic and periodic motion, arising from both the quantum level of scale and those below it. The magnitude of the quantum of action is an emergent property characteristic of the boson-mediated level of phe-

---

27 Per the Aharonov-Bohm effect (and macro-scale phenomena associated with superconductivity/Josephson Junctions which so fascinated Richard Feynman [cf Feynman, Richard P., Robert Benjamin. Leighton, and Matthew. Sands. The Feynman lectures on physics : by Richard P. Feynman, Robert B. Leighton, Matthew Sands. Vol.3, , Quantum mechanics. London: Addison-Wesley, 1971. *21-18*])

nomena that comprises the universe as human beings know it – i.e., that aspect/scale of the physical world characterized by the predominance of the phenomena of light (that is, electromagnetic phenomena generally). It is thus determined by the same conditions that establish the velocity of light (and the dispersive properties of Subspace) and the interaction of wave processes generally (which gives the fine structure constant).

For the purpose of illustration imagine, per the discussion of section 5.4, each anti-node of a wave – each "crest" – to correspond to a unit of action. Assume that a finite amount of time is required for interactions in which energy is exchanged. In this sense, the non-commutative relationship between energy and time can be interpreted as a consequence of a finite interaction period – specifically, the time between wave crests, determined by frequency – during which a wave crest or sequence of crests "breaks" upon an object with which it is interacting. Such objects are deemed to have a complex makeup and must respond to the waves in order to absorb energy. Thus, for any given time interval, while the relatively high velocity and low intensity of the longitudinal waves might seem to mitigate their capacity to interact with local wave trains, such interactions are enhanced by phase correlation.

The quantum-of-action – and the non-commutative pairs *position-momentum* and *energy-time* – arise from the model in a straightforward manner. They are properties of periodic phenomena generally, while position-momentum and energy-time constitute essential, what might be called *Canonical* relationships, because they are "*Action Pairs*" so to speak (i.e., the product of each pair is an action, whereas other non-commuting values of quantum theory, as will be seen, are not necessarily connected with intrinsic aspects of wave action, but are rather emergent and derivative with respect to such canonical relationships).

The fundamental connection between quantization-of-action and dispersion was recognized by Schrödinger[28] as early as 1926. The capacity of a wave carrying system to transmit energy is a physical property of that system, and so there must always be a limit on both the velocity of waves and their frequency. These characteristics limit the rate at which information can be relayed, just as they limit the energy that waves can carry.

---

28 E. Schrödinger, Physical Review, **28/6, 1057 (1926)**

On this note, a further set of interesting analogies arise; an additional group of dualities – underlying commonalities – of several distinct domains of physical science. The connection between energy and time, intrinsic to resonant transactions, is also characteristic of certain properties crucial to the operation of heat engines and the early science of heat; from the work of Sadi Carnot to that of Fourier (and which connect the latter's study of heat and its transmission to his mathematical treatment of waves).

Of course, energy and entropy both have conceptual roots in thermodynamics, so it is hardly coincidental that these relationships should be found. More to the point, the flow of heat, like the motion of waves, exhibits a simple but salient feature that Carnot was able to discern in the gross functioning of mechanical heat engines. Although he did not employ the concepts of energy and entropy in his analysis, he yet recognized a fundamental feature of heat engines that exposes, again in a gross manner and on a macro scale, a crucial underpinning of the various dualities hereinabove explored. This key commonality is so simple that it can be expressed in a single image, viz.: that of a saw blade.

The difference between the attributes of a saw blade and that of a sword, when brought to bear on the task of sawing, is intuitively obvious. Put into words, the ripping effect of the saw stems from the space between its teeth. Just as the resonant absorption of energy from a train of waves depends upon the troughs between the crests – and just as the figure-ground of object and space, and wave and trough, constitute the essence of *object* and of *wave*, respectively – so the relation of tooth and gap is of the essence of the saw.

Carnot's fundamental insight regarding heat engines reduces to this: that the efficiency of the conversion of heat to motive power is due to the relation, the difference, between the temperature of the heat source and that of the hcat sink, rather than to any absolute property of the working fluid, such as the amount of heat that it can carry. The essential nature of a heat engine is its *cycle*; the very fact that it cycles, that it establishes a temperature difference between source and sink, utilizes the dissipation of that difference to do work, and then reestablishes the difference; over and over again. Like saw teeth tearing through wood, and crest after crest of a wave adding its energy to a receiver able to cycle in harmony with the relation crest-to-trough, so the energy/entropy differential of the heat engine's source and sink must be established on key. Without the difference between source and sink, tooth and gap, wave and trough, none of these modes of action are sensible. The amplitude of a wave – the height of the crest in relation to the trough – is as the difference in temperature between the source and sink of a heat engine.

Furthermore, the flow of heat across such a differential is governed by the same equation, given by Fourier, that governs the transmission of wave action. And so there is a fractal-like meta-relation between the various levels of operation of the heat engine. That is, the transfer of heat, which is of course a transfer of molecular motion (though not specified as such by Fourier), necessarily follows the same *diffusive form* (which Fourier did specify, based upon the temperature gradient) as the transfer of wave action, per the duality clarified by Hamilton – with further links, via the dualities discussed in sections 5.5 through 5.7 above, to Brownian motion, etc. And then, by the mechanical-optical analogy that exists between the dynamical momentum and the gradient of the index of refraction[/phase gradient of wave motion] inherent to these Hamiltonian dualities, it becomes clear why the Schrödinger equation is formulated in configuration space rather than phase space (that is, because momentum is a function of position).

While this last excursion, into the philosophy of the theory of heat, was perhaps not strictly relevant to the purpose at hand, such general considerations help show how fruitful the underlying concepts of the present framework can be. In comparison, consider how limited the Copenhagen Interpretation and other conventional formulations of quantum theory are, especially in light of how readily/naturally solutions to the many problems that plague such interpretations suggest themselves on the basis of the present model. For example, some physicists have speculated that once the small-scale effects of gravity are understood the so-called "collapse of the wave-function" will find a physical interpretation. This is essentially what the present framework furnishes, along with an adequate conception of micro events generally, wherein the evolution of the wave-equation is clearly determined and the role of gravity completely understood. With mathematical development, the potentialities of the model become even more evident.

In this sense, the Subspace framework furnishes a detailed and intuitively comprehensible interpretation of both the *Spacetime* of General Relativity and the *Vacuum* of Quantum Theory. But again, the concept of *Subspace* – comprising unlimited scales on which variegated behaviors and phase relations can emerge – suggests possibilities that diverge from those of the usual field concept. The degrees of freedom include the *Order Parameters*[29] of all emergent phenomena; the *Collective Variables*[30] that define stable/recurrent patterns. And there is no essential distinction to be drawn between "emergent" and "fundamental" elements – and hence no necessity to postulate the existence of fundamental elements, or to otherwise define the "substance" of Subspace (e.g., in such a way that might allow for "tears" or "holes" at a fundamental level, as per some interpretations of the continuum under general relativity). Moreover, it is not necessary to treat Subspace as an ideal continuum in that (as with any concept) continuous structure can be regarded as a limit property, in this case a mathematical limit to which a composite structure approximates by virtue of its elements becoming increasingly densely distributed on diminishing scales of size (inasmuch as, again, such elements can be patterns in another level of structure). And conversely, discrete structure can be described on the basis of a continuum. One only need reflect on the infinite dualities of mathematical description and the irreducible qualia of personal experience to understand that the notion of a single substance of reality is no more 'substantial' a concept than any other limited, anthropomorphic tool in the cognitive toolset, and that the search for "ultimate elements" – stringy, membranous or otherwise – is not necessarily a viable scientific enterprise.

In this connection – regarding the relation between emergent and fundamental properties and the concept of physical elements (again, along the lines of strings/membranes) – it is important to remember the epistemological conclusions of earlier chapters. It is evident, on purely logical grounds, that the concept *"fundamental element"* is an artifact of thought

---

29 Haken, H. "Synergetics." Naturwissenschaften 67, no. 3 (3, 1980): 121-128.
30 Bohm, David, and David Pines. "A Collective Description of Electron Interactions. I. Magnetic Interactions." Physical Review 82, no. 5 (6, 1951): 625-634.

and cannot be attributed ontological significance (and that contradictions *must* arise on the basis of attempting to rigorously define such things). But just as this simple tenet is easy to forget in its application to the most basic aspects of theory, it is even easier to lapse into naïve realism regarding more sophisticated theoretical conceptions; for example, such as *entropy* and *reversibility of dynamical law*.

If the reader will pardon another brief excursion, the relation between these two particular ideas poses a strangely sticky pseudo-problem, the resolution of which can be rather edifying, much like the mind-body confusion. The concepts of *entropy* and *dynamical law*, not to mention *time*, are no less subject to the inherent incompleteness and inconsistency of conceptual constructions than are any other artifacts of cognition. Therefore, that they imply logical contradictions should not surprise. But the idea of *reversible dynamical law* is especially problematic, inasmuch as it is based on the peculiarly suspect Newtonian conceptions deconstructed in chapter four. As therein demonstrated, the Newtonian notions of *force* and *action-reaction* are inconsistent with the dynamical laws that are ostensibly premised on them. And yet the time reversible character of those laws follows directly from the symmetry of *action* and *reaction*. On the other hand, if $F = MA$ is given an emergent interpretation on the basis of the diffusion-related concepts suggested above, reversibility is no longer an implication. As is generally the case, it is only by idealizing the mathematical equations of Newtonian physics as *Absolutes* – as "*Laws of Nature*" rather than regarding them as "*Cognitive Constructs*" – that the confusion arises.

The same problem is evident in connection with the concepts of *Absolute Space* and *Time*. Per the analyses of chapters one through three, every object of perception and conception is embedded in a figure-ground context, and objectifying the figure as something independent is contradictory. *Space* can be thought of as *ground*, with respect to which the "object in it" acquires meaning as *figure*. Or, reversing the relation, space can be thought of as foreground, *but only in relation to some object[s]*. This is the source of Kant's problematic, which he called an *Antinomy of Pure Thought* and attributed to an inappropriate conjunction with the concept of *infinity* (that is, by giving space an extra-personal meaning and attributing the possibility of infinite versus finite magnitude). But more to the point, the specific root of the problem is separating space from its more appropriate binary, figure-ground context; as happens with Newton's *Absolute Space*. In other words, *space* is as yin to *object* as yang, or vice versa, in the figure-ground construct: *object/not-object*. But the construct: *space/nothing* (i.e., no space, no object, *no nothin'*) is insensible. Again, in Kant's view this removing of the subjective construct of space from its proper, "finite" context is key. But infinity is not the root problem here; it is rather the misconceived abandoning of the fundamental *figure-ground context* in which the intuition of space makes sense.

And of course the same problem arises with respect to time. Time is that attribute of thought necessary, as counterpart of the concept of space, to the furnishing of a context for the conception of the "unchanging object" – that aspect of experience that "remains the same" when situated in different spatial contexts. One of the binary, figure-ground attributes of the concept *time* can be thought of as the relation: *instant* versus *interval*. *Object* has meaning literally as instantiation of "instant" – something unchanging and thus outside time (moreover, it is only in terms of the *static*-ness of "object" that the notions of "instant" *and* "object" are sensible). This complex of concepts – space/time/object – embodies contextual meanings that cannot be separated, and objectifying any aspect of the complex as some-*thing* independent is necessarily contradictory.

It is relatively easy to see, in contradistinction to the concept of space, how that of time is necessarily illusory. For without the notion of an *instant*, which is self-evidently a contradiction, the notion of a temporal interval has no meaning. That is to say, every perception of duration involves an arbitrary division of experience, a separation of some part of experience from some other part, i.e., as "*an event.*" But precisely because the division of events is arbitrary so too is the concept of time. Put somewhat differently, time is not composed of instants, because instants do not have duration. Time, rather, must be thought of as composed of *events*.

But that which constitutes "an event" is subject to nothing but discretion: it can be an experience – such as looking at a picture or watching a movie – a lifetime; or perhaps an expanding cosmos. In the case of the latter, it is just as arbitrary to say "the Big Bang lasted a tiny fraction of a nanosecond" as to say "an event is not 'watching a movie' but rather 'looking at a picture'." Because the *instant* is the only non-arbitrary division of this sort – yet it does not exist. Assuming the standard cosmological theory correct, it is clearly only unambiguous to say: "the *Big Bang* is *still Banging*" and the presently observed "*accelerated expansion of the universe*" and the "*Bang*" are the *same thing*: one "event."

Time, it is clear to see, is no less arbitrary a concept than any other, such as the labeling of a row of bushes "a fence" should the labeling and the bushes serve that discretionary, subjective purpose. To divide experience up time-wise is arbitrary; everything is clearly "in the present," the "eternal now." In other words, it is only the smallness of human perception that makes the partitioning seem necessary.

And for this reason, just as with the concept *physical world,* the notion of time serves a very valuable heuristic purpose. While any argument about the "direction" or "arrow" or "reversibility" of time is clearly meaningless, if the object of discussion is the *meaning of these concepts* as such: *their implications,* there is sense in inquiring after the relation between dynamical law and the law of entropy, or how physical law embodies the notion of the arrow of time. Questions in this vein can sharpen the understanding.

In this respect, the sense of the directed-ness of time is particularly well reflected by the present framework. This is manifest in two distinct ways. First, dynamical, $F = MA$ law is superseded by a model steeped in the notion of directed processes related to probabilistic concepts. Such concepts, in turn, involve the most extreme *a priori* aspect of mathematical thought; the awareness of self-evident truth based on the intuitive sense of symmetry. Second, the mathematical treatment of wave action, as exemplified by the Huygens construction, embodies and thus establishes a temporally unidirectional attribute of physical phenomena at a fundamental level.[31]

For example, as discussed in section 5.12, modelling wave action on the basis of the Huygens-Fresnel construction is possible because the backward [concentrically inward] directed wavelets associated with the expanding spherical envelope of the disturbance interfere destructively, whereas the outward-directed parts of the wavelets interfere constructively. This characteristic of wave action is at the heart of the Feynman sum-

---

31 Of course, these two aspects of the framework are very closely related, as noted in the discussion regarding the early theory of heat.

over-histories approach to quantum field theory, which treats the waves merely as representations of probability amplitudes. However, the underlying dualities of the description, also related to the classical wave-particle duality (and the stationary integrals associated with the transmission of dynamical action) are fundamental properties of wave action.

And so without any special assumptions, the framework has built into it the sort of temporal directed-ness that characterizes the phenomenology of experience, but which seems controversial on the basis of purely mechanical concepts. In addition, the usual interpretation of the entropy of gravitationally bound systems needs to be revisited, because the primary cause of gravity is *not intrinsic* to such systems but rather *extrinsic* to them. Accordingly, the entropy of an extremely dense system, such as a black hole (the nature of which must in turn also be reevaluated), should be considered low rather than high. Moreover, the overall relevance of the concept of entropy stands in need of reconsideration.[32]

This is already clear on the basis of Steven Hawking's analysis of black holes and the consequent possibility of Hawking Radiation, which indicates that entropy can flow cyclically, as it were, across multiple levels of scale. Thus the macro action of gravitation can drive systems into closely bound, low entropy states, whereas on a micro scale virtual particle-pair creation can redirect that trend. If this were not the case, and if black holes did indeed represent maximal entropy states, the regular dissipation of that entropy would be paradoxical (because the "fluctuations" that drive the dissipation are recurrent and directed, like the action of a reverse direction 'entropy rectifier', e.g., *Maxwell's Demon*[33] – this is an especially important consideration in the present context).

In this connection, the association of black hole entropy with the surface area of the event horizon must also be reevaluated. Yet, whereas the role (if not concept) of the event horizon might need to be reexamined, a new possibility arises in connection with the relation of surface area to volume vis-à-vis very dense configurations of matter/energy. Because the

<hr>

32 It should also be noted in this connection that the interpretation of gravitation as extrinsic rather than intrinsic rules out the possibility of another absurdity; namely, the notion that matter/energy can be created "for free" at the gravitational center of the cosmos insofar as gravitational energy is taken to exactly balance non-gravitational energy, as "negative energy" versus "positive."

33 Maxwell, James Clerk. Theory of heat J. Clerk Maxwell. Text-books of science. New York: D. Appleton and co., pp. 338-339, 1872.

primary action of gravity is associated with the capacity of matter/energy to couple with a flux – which has been seen, in accordance with equation 5-18 (§ 5.16), to be given by a definite coupling parameter (namely, the fine structure constant) – it follows that there may be a limit to the gravitational effectiveness of concentrated matter/energy, because within dense configurations the effect might be expected to fall off rapidly with distance from the surface.

Finally, this relation – between gravitation and the fine structure constant – must be seriously considered. Not only does it relate Planck's constant and the properties of light to gravitation, and explain equality of gravitational and inertial energy, it makes the introduction of the gravitational constant (and, if the framework is to be taken seriously, the cosmological constant), heretofore so crucial to physics, unnecessary and therefore superfluous. Together with the *Principle of Common Provenance*, this is one of the most salient of the above derived conclusions for the future/foundations of physics. (Moreover, *this relation characterizes the other known forces* – for example, the coupling parameter in quantum chromodynamics, associated with the behavior of virtual quark-antiquark pairs vis-à-vis vacuum fluctuations, can likewise be dynamically related with the mass/energy of the relevant entities and with Planck's constant of action.[34])

The overriding concern with respect to all of these considerations is that they neither be attributed ontological significance nor interpreted as signaling a return to mechanical concepts. Quite the contrary, the nature of the proposal must be recognized as something very different. For on the one hand it is suggested that local causal action and directly visualizable, three-space/one-time constructions have a place among the foundational conceptions of theory, and on the other that it is not such constructions *per se* that are primary but rather: (a) certain *a priori* mathematical considerations based upon them, and (b) the emergent *order parameters* that characterize the collective behaviors such considerations reveal. Collective behaviors are in an important sense more fundamental than the hypothesized causes of them, because they can be viewed as *a priori* consequences of an ensemble of such hypothetical causes, all equally viable (hence the existence of dualities). As alluded to above, the essential aspect of these *a*

34 Batchelor, David, A Semiclassical Derivation of the QCD Coupling, arXiv:0905.0013v1, 4/30/09

priori mathematical relations is the core intuition that ultimately informs all considerations of *probability* and *symmetry*.

In fact, the concept of probability is inseparable from that of symmetry. That which is self-evident in calculations of probabilities derives from intuitions of symmetry. Just as the probability of "heads versus tails" is "50-50," so the tendency to maximal entropy is the tendency toward maximal symmetry. Indeed, one might go so far as to say that the answer to Einstein's wonder that "the most incomprehensible thing about the world is that it is comprehensible," and Wigner's "unreasonable effectiveness of mathematics in the natural sciences"[35] is that human thought is guided by symmetry relations, and when analysis reveals dualities among experiences and mathematical relationships, it is by the observation of symmetries that such dualities are discerned. Put another way, mathematico-logical problems are characterized by an absence of insight regarding relevant symmetry relations, and solutions to problems follow upon the discovery and exploitation of such relations.

Permit one final excursion, and consider the following puzzle. One morning at sunrise, a monk leaves his mountaintop abode and descends a long, winding trail to a village in the foothills, which he reaches before the end of the same day. The next morning, again at sunrise, the monk leaves the village and begins his ascent back to the mountaintop, along the same path, and again reaches his destination the same day. Prove that there exists a point along the trail that the monk crosses at exactly the same time of day on both his ascent and his descent.

The solution to this puzzle, as with all mathematical problems, requires the recognition of a symmetry – one that uniquely characterizes the relations in such a manner that the solution appears self-evident (here given in *Appendix B* in order not to deprive those who have not previously encountered this puzzle the pleasure of solving it). Because of the salient role of symmetry in the life of the mind – in aesthetic, intellectual and ethical matters (i.e., in the recognition of *beauty*, *truth* and *justice*, respectively) – and because there has been no other significant guiding intuition to follow as a consequence of the relinquishment of physical modelization, physicists have

---

35 Wigner, Eugene P., The Unreasonable Effectiveness of Mathematics in the Natural Sciences, *Communications in Pure and Applied Mathematics*, 13, 1-14, 1960

come to put inordinate emphasis on the importance of abstract symmetries.

While this is understandable, the role of symmetry must be viewed in the appropriate context. Symmetries are ubiquitous aspects of experience, and in any given problem-solving situation it is crucial that the appropriate symmetries be found; those that are uniquely relevant to the particular purpose and circumstance. Consider another example. Figure 5-23 depicts a geometrical problem used by Wolfgang Köhler to demonstrate a somewhat different but closely related idea. Here is the problem and solution in Köhler's own words.

"...There is a circle with the radius *r*, and in this circle I construct a rectangle (Figure [5-23]). The problem is the following. If I now draw the line *l* within the rectangle, what is the length of this line? Not everybody will be able to give the right answer immediately. And yet, the answer is extraordinarily simple. Just as in many other cases, we find the solution by adding something to the given material – in our particular case, by adding just one line. The given line *l* is a diagonal of the given rectangle. But rectangles have two diagonals. If we add the second, what do we find? The second diagonal extends from the center of the circle to its circumference and is, therefore, the radius. Now, since the two diagonals of a rectangle are, for simple reasons of symmetry, always equally long, our line *l* must necessarily also have the length *r*, the length of the radius, whatever particular shape the rectangle may have. Once more, the solution of the problem is a matter of discovering a relation – to be sure, in this problem a relation discernible only after the second diagonal has been added to the given material. Once this has happened, we recognize, of course, why the addition of the second diagonal and the corresponding new relation bring about the solution. In other words, once the material has been properly changed, we understand perfectly why the addition of the second diagonal gives us the required answer. This is what we call *insight* in thinking.[36]

---

36 Köhler, Wolfgang. The task of Gestalt psychology. Princeton, N.J.: Princeton University Press, 146, 1969.

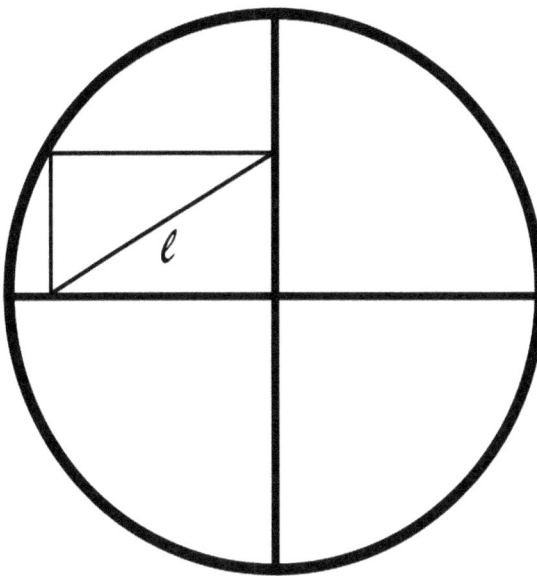

Figure 5-23 An image similar to that from the Köhler text

Note that, just as with the discussion of chapter two regarding Cantor's *diagonal slash* procedure, insight refers literally to *seeing a new relation*, one that of course may be considered implicit in the given circumstances, but which is really added thereto; that is to say, it involves a creative act. And so with the discovery of symmetries. It is only certain uniquely relevant symmetries that ever need be found, and thus it is not symmetry *per se* that is at issue. What is of concern, rather, is the capacity to find those specific symmetrical relationships that furnish leverage in connection with the overcoming of a given cognitive obstacle.

Accordingly, symmetries are neither independently existing features of the world nor the proper object of theoretical physics *per se*, but rather aspects of experience as creatively envisioned by the theoretician; and thus necessarily, by simple logic, emergent properties. It is incumbent on the researcher to seek those unique symmetries that are of especial value with respect to *intelligibility* – in the broadest sense; which is to say, inclusive of concerns regarding *explanatory satisfaction*. Such concerns must neither be diminished nor dismissed, and certainly not characterized as "meta-physical."

A final word on historical trends regarding reductionist and emergentist ideas. The following lines are taken from an extraordinary piece of work that fell into the author's hands just as the present treatise was going to press. Discussing the history of the mathematics of nonlinear processes, and in particular the opposition that its earliest practitioners faced, Alwyn Scott describes John Russell's decade-long study of the so-called *solitary wave*, beginning in 1834, and his ultimate vindication, in 1895, by Kortweg and de Vries:

> "By this time, however, Russell was resting in his grave and interest in his solitary wave had waned, as we see from three facts. Based on the views of Stokes, first, the importance of the solitary wave was discounted in the *Encyclopedia Britannica* of 1886 [306]. Second, Horace Lamb's opus on hydrodynamics allots a mere 3 of 730 pages to the solitary wave [522]. Finally, there were only about two dozen citations of reference [833] from its publication in 1845 to the beginning of the nonlinear science explosion in 1970 [306]."[37]

The final sentence of this excerpt is stunning, and so Scott's reference [306][38] is included below. Of course, it should not be surprising that the "explosion" of interest in nonlinear phenomena occurred when it did, inasmuch as this explosive growth has closely paralleled that of the digital computer. It is difficult to overestimate the significance of this latter phenomenon, especially as it too is still in a nascent stage. But this provides a good segue to the next chapter, as the role of numerically controlled technologies will prove to be of greater significance still in the emerging science of sentience.

---

37 Scott, Alwyn. The nonlinear universe : chaos, emergence, life. The frontiers collection. Berlin; New York: Springer, 46,2007.
38 Filippov, A.T.: The Versatile Soliton (Birkh"auser, Boston) 2000

Chapter VI

A Framework for the
Science of Sentience

§ 6.1 Introduction

IN this chapter, a framework is presented for a unified though somewhat limited approach to the science of sentience; limited, that is to say, in terms of the aspect of the subject that is addressed. One must respect that the variegated, wide-ranging fields of the cognitive sciences – from the neurobiology of the visual system to linguistics – are at once vast and highly specialized, and that researchers must be free to employ such theoretical conceptions as are occasioned by their respective empirical domains, as individual discretion advises. As Einstein once remarked in a related context, the researcher knows best where the shoe pinches. In this sense, philosophers and other generalists should never seek to limit the concepts that scientists find useful.

It is only with respect to the links that experiment reveals between mental processes and organic systems that the present framework is aimed at facilitating unification of the various cognitive fields. Toward this end, it seeks to establish a method for describing, quantitatively and in extensive detail, the phenomenology of first person experience, including the *qualia* of that experience in all its subtleties – from tactile, visual and aural perception of the physical world to inner feelings and emotions – even such complex tactual/muscular perceptions, for example, as accompany learning a physical skill, such as riding a bicycle. Again, the ultimate goal of this approach is to make plausible the possibility of a wide-ranging unification of physiology, and thus physics, with the science of sentience; which is to say, to make this goal seem both feasible and meaningful to cognitive scientist and philosopher alike.

The epistemological analyses of chapters one through three revealed that the fundamental aspect of cognition there referred to as *concept formation* (or the *concept forming process*) is intrinsically limited with respect to logical consistency and completeness. It was demonstrated that this limitation makes an epistemologically irreducible contribution to the character of human knowledge and thought, insofar as any given percept or concept is purported to carry a representational meaning; or, put somewhat differently, insofar as *image* or *idea* is deemed to reference some "thing" or "aspect of experience," and to the extent logical inferences are connected with such referential purport. All knowledge and thought is limited accordingly; intrinsically and hence universally.

It was also shown that, despite these limitations, which manifest in axiomatic constructions as the [corresponding] defects of *incompleteness* and *inconsistency*, carefully crafted representations can yet carry enormous heuristic and pragmatic value, and are the necessary means by which experience is simplified and codified, and thus comprehended, often with remarkable predictive power. It was further shown that conceptual models involving the notion of "the physical" are among the most valuable in this sense; where by *physical* is meant ideas that ineluctably involve the intuition of *extra-personal spatial extension generally* – without, that is, any specification as to the particular nature or forms that "matter/energy" might or might not take; i.e., without regard to "what it is" that "occupies" or is extended "in" space.

Of course, from the point of view of the cognitive scientist these aspects of cognition are just that – *aspects of sentient experience*, the very subject of inquiry. Together with all other aspects of experience that can be identified and described, such cognitive processes comprise "*the phenomenology of mind*" – again, the object of study. So the limitations of cognition appear as both artifact and attribute – in the former sense, as appurtenance of theory; in the latter, as intrinsic aspect of the natural world. Accordingly, while the phenomenology of first person experience seems qualitatively incommensurable with models based on physical ideas (again, involving "external spatial extension"), theoretical conceptions of *subjective experience as such* have the virtue of being intimately connected with what might be called "a particular subset of reality" or "set of things-in-themselves," and are thence capable of furnishing a bedrock upon which both a theory of cognition and, ultimately, a unified theory of what is now usually thought of, dualistically, as "objective" and "subjective" reality might be constructed. From this standpoint, a theory of experience may be characterized as a model of the universe as *"Sensorium"*— i.e., a totality of all experience, or a self-contained cosmic stream-of-consciousness, as it were, absent any need or implication of an external cosmic observer/mind. (Although it might be more appropriate to think in terms of an "infinite-dimensional manifold" as opposed to serial streams.)

Such a theory might be viewed, with some justification, as "scientific Metaphysics," insofar as experience, *sui generis* – which is known immediately and thus with certainty to exist – is the acknowledged object of study, making the enterprise a science of *ultimate existents*: hence, *Metaphysics*. This is a turnaround of Kant's original idea regarding the world of "actual being," unknowable but somehow abstractly reflected in the relations of mathematico-physical theory. Ironically, the "physical world" of naïve realism is the quintessence of first-person experience – quite literally, "the stuff that dreams are made of" – whereas whatever exists outside personal experience must yet be in the same category as this "stuff," in the sense that it too exists, and therefore is not *a priori* necessarily different, in any "substantial" sense, from subjective experience considered as such.

In addition, it is clear that what might be called "physical models proper" – that is, in the sense of the theories of physics, which hang on the intuition of extra-personal spatial extension – have unique heuristic and pragmatic value, and furnish a means for acquiring predictive knowledge on a level of detail that seems unthinkable by any other method. And so some combination of concepts is called for, derived from both "physical" and "mental" modes of representation. This must be accomplished with a judicious toleration of the limitations of theory, bearing in mind the powers and weaknesses of conceptual constructions generally, as well as the particular limitations of the subject-object dichotomy. While there are inherent defects in this paradigm, it is quite well suited to both everyday life and the current state of scientific knowledge.

Therefore, until science reaches a stage where the correlation of "physical"-based and "mental"-based models yields a solid indication of how a transcendence of the subject-object dichotomy might be effected, both types of construction must continue to be employed. At least for the near-term future, some form of qualified dualism would seem to be an inescapable aspect of scientific theory, at least in the sense that the correlation of the phenomena of subjective experience with "physical" models – such as the mapping of the perception of a musical tone to neural and/or other processes in/about the brain – remains a fundamental goal of science; with the always operative proviso that such dualistic theoretical constructions are not intended to convey anything of the "intrinsic nature of things," but rather are intended to indicate relational orders, or what might be called *patterns of analogy*, and always with some degree of incompleteness and inconsistency.

Beyond this caveat, and the practical problems of application, there is another, albeit somewhat superficial difficulty facing researchers who would pursue such lines, insofar as the scientific goal of correlating mental and physical events tends to raise controversial issues among many philosophers, for reasons discussed in the first three chapters hereof (regarding in particular the misconceived mind-body problem). And while professional researchers in the neurosciences tend to ignore such controversies (as they must), there is yet no solid philosophical paradigm to guide and interpret their work.

Happily, such a philosophy is ready to hand, and grows naturally out of the following consideration. While naïve realism is an unsuitable basis for theoretical physics, those very qualities that make the viewpoint difficult to abandon from a psychological perspective – i.e., its instinctive nature and common-sense appeal – also make it an ideal point of departure for the study of subjective experience. This is because the world of naïve realism is, to a significant extent, the world of personal experience, especially with respect to the structure of that experience. That is to say, the order of personal experience is, in a salient sense, the order of "things as they are" (i.e., as they are perceived).

Moreover, even those aspects of experience that do not seem "contained within" the physical world – such as emotion and thought – are yet intricately related, in descriptively relevant ways, with physical perceptions. For example, sounds and sensations are "placed," perceptually, within the framework of subjective space in accordance with the kinesthetic sense. Neurological studies suggest that this sensibility is pervasive, highly organized, and clearly reflected in the brain, as illustrated by the existence of so-called *grid cells*, which evidently reflect inner "Cartesian coordinate systems."

Indeed, a little introspection reveals the sense of space to be so pervasive that virtually every sensation, and even thought, can be associated with a location. For example, when "thinking in words" the "inner voice" might seem situated more or less in the center of the head, near the "location" from which sounds heard through stereo headphones also seem to emanate. Or when one visualizes an image, it may appear overlaid on the immediate visual field/environment, which one thus has a sense of "looking through." And it is known that the direction in which one turns one eyes when thinking about how to answer a question, left or right, is related to which brain hemisphere is primarily engaged.

In ways that will hereinafter become readily comprehensible, physical concepts can be applied to many crucial features of subjective experience, *qua* experience. For example, the space-time relations among perceptions of "things outside the skin" follow the laws of classical mechanics and optics, adjusted for perspective. Accordingly, the "inner world" comprising those perceptions can be embedded in a sort of many-dimensional mathematical space – hereinafter *Subjective Space* – the first three dimensions of which are the "Cartesian" directions of *inner space*, the fourth of which is *personal time*, while all the rest reflect the multitudinous "degrees of freedom" of sentient experience generally – the colors, sounds, feelings, etc., that comprise the life of the mind.

On the other hand, whereas the phenomenology of experience is a directly known subset of "things as they are" (again, *experience exists*, and is thus a "thing-in-itself" in the Kantian sense), and whereas spatial concepts are applicable to experience, it would seem to be the case that: (1) whatever it is that is conceived to exist "outside experience" (Kant's original meaning of *thing-in-itself*) is "of the same essence" as experience (i.e., both are in some sense "made of the same stuff," to whatever extent, if any, this mode of thought is appropriate – perhaps more appropriately, both are in the same category, i.e., the class that reflects the *property of existence*); (2) the order of both "experience" and "extra-personal existence" is, to some limited degree of accuracy, reflected by the known "natural laws" of the physical sciences; and (3) it remains to be seen, as an empirical matter, to what extent if any these two distinct spaces – i.e., (a) the mathematical *Subjective Space* of personal experience, and (b) the mathematico-physical space of extra-personal existence – can be embedded in the same manifold (that is, from the point of view of *efficacy* and *intelligibility* of theory; the speculation is not intended in a metaphysical sense). Not until this third point can be answered sensibly can one likewise discuss the question as to how unification, of the psychical with the physical, should best proceed (i.e., whether a unified theory of nature should even hinge on spatial concepts).

At any rate, until unification becomes imminently viable, it is appropriate to pursue an empirically grounded science of sentience as such – irrespective of how, ultimately, spatial and spatio-physical concepts must fare. Just how this can be done, along the lines suggested above, is the subject of what follows.

§ 6.2 The Phenomenology of Mind

Introspection furnishes direct knowledge of experience. And yet, inasmuch as introspective knowledge, like any other, involves representational, conceptual processes, such knowledge must be limited. This is the incontrovertible upshot of chapters one through three. And so the notion that personal experience comprises an "inner/subjective world" is based on the same subject-object dichotomy (i.e., anthropomorphic perspective) that informs the complimentary side of the paradigm regarding the existence of an "external/physical world." Indeed, it might appear that an objective characterization of personal experience is, in principle – i.e., merely by definition – impossible. However that might be, anthropocentric conceptions are yet a part, an essential aspect of experience, and as such are no less worthy of investigation than whatever may constitute the "elemental aspects" or "fundamental principles" of the psyche, if such things can be identified (or rather, *prove sensible as elements of theory*).

It should be noted in this regard that, as again with all concepts, the notion of fundamental or irreducible elements is an abstraction that does not connote meaning or bear information content *per se*. Moreover, as discussed at length, the object of perception or conception always requires a definite context in relation to which some meaning or information content can be assessed, and perhaps even quantified.

As meta-example of this circumstance, consider the abstraction called the *mathematical concept of information*. The relevant context involves the additional, quantitative notion of probability. In this context, the notion of uncertainty is crucial, and refers to a quantifiable lack of knowledge regarding some source of information; information that can be thought of, in turn, as codified or represented in a quantifiable set of data. A measure of just how much information is represented in such an information source or data set – or a quantifiable subset thereof (e.g., "one-half of the information") – is, accordingly, a measure of how much uncertainty regarding the content of that source exists prior to one's knowledge of it. As noted, however, the *meaning* that the data set ostensibly represents is not "*in*" the data – rather, the data merely enables some interpreter, an agent capable of framing it in the appropriate context, to "extract" the information/meaning that it is deemed to codify (or again, *represent)*.

Under the paradigm of information theory *meaning* as such is not quantified. Just as with the figure and ground of perception, there must exist a context for the data in order for it to be meaningful, or represent information. This context can be thought of as an algorithm for the interpretation of the otherwise "raw" data. Without the algorithm, or context for "reading" the data, the latter is meaningless.

In a similar sense, the all-pervasive tinge of anthropomorphism, which characterizes the comprehension of experience generally, is a necessary precondition (context) for the realization of meaning. And though such meaning is necessarily limited by the constraints inherent to human cognition, this does not negate the positive information measure of the knowledge that such interpretation yields. Whether something is thought of as a "physical entity" or a "mental construct," the extent to which information content is associated with that something, and can be quantified as such, is the ultimate concern in this regard.

What is essential, therefore, is that (1) a method be found not only for accurately describing and categorizing but also for *quantifying the meaning* or semantic-content of consciousness, and (2) *relations* between experiences be discovered, *which can* likewise *be described mathematically*. And, inasmuch as the format in which the overall content of first-person experience is embedded is the framework of Subjective Space – set on the foundation, as it were, of the first four dimensions of "inner" space and time – and because that format is communicable; i.e., because it sets the context for perception of the shared or "external" world as well – it provides an accessible, common scaffolding for the piecing together of the skeletal structure and ultimately flesh of a descriptive framework for subjective experience, as that structure is uncovered via an appropriate archeology of the mind. (These general, admittedly somewhat vague remarks will become clearer in the paragraphs that follow.)

§ 6.2.1 Describing Experience I:
Definitions of Information

At first thought, it does not seem possible to describe and quantify subjective experience in the same detailed sense that, for example, a physicist might so represent the position and velocity of a classical physical object. When one thinks of what it means to make a measurement, one always has in mind a circumstance in which a given physical entity is compared to another in a precise, controllable manor. However, deeper reflection reveals that this concern is unwarranted, inasmuch as every such "physical measurement" is, at the same time, a subjective experience. Calling something "physical" simply means to focus attention on one particular, conceptual aspect of a larger, overall set of experiences (in the sense of Mach's

"ABC's")[1]. Thus every physical measurement *is* a subjective judgment – a detailed comparison between two or more aspects of an experience, such as the perception of the coincidence in space of the edge of one object with a marking on another (e.g., a line on a ruler).

And so *the laws of classical physics describe the structure of personal experience*. Accordingly, these laws, including especially the laws of ray optics, are encoded in the software algorithms that generate the interactive, first person experiences delivered by immersive 3-D computer games. And so, again, it is in this sense that Mach alluded to the duality of physical and psychological perspectives as nothing other than two different ways of viewing the relations in and among experiences. The *content* of experience is essentially the same in both cases, whereas the interpretive *context* is different. Therefore, the laws of classical physics are applicable to this content.

Moreover, the distinction associated with the adjudication of experience as "physical" or "mental" does not affect the relevance of the general mathematical concept of information, which can obviously be applied to "subjective entities" as readily as it can be applied to "physical things." Thus, if a graphic designer creates an image with a computer, effectively encoding a representation of that image as a digital file, the image need not be instantiated as a directly perceivable attribute of some physical substrate, such as an arrangement of colored ink on a piece of paper, or emanations of various frequencies of light from a graphical display device, in order for a set of data sufficient to uniquely characterize that image to exist, and to be quantified. In other words, the content of the digital file, apart from its manifestation as an image on a screen, quite obviously has a quantifiable measure of information content – i.e., its *file size*.

But because meaning is determined by context, the amount of data contained in a digital graphics file is not in itself a quantity representing some sort of "absolute information measure" that can be directly connected with a corresponding image. Information can be thought of in at least two distinct ways. First, again, is the meaning *represented* by the data – the semantic sense of information, which Claude Shannon characterized as "irrelevant to the engineering problem."[2] (This aspect of the concept is of

---

1 As described in the extended quotation from his *Analysis of the Sensations* in § 3.7
2 Shannon, C E. "A mathematical theory of communication. " The Bell System Technical Journal, Vol. 27, pp. 379–423, 623–656, July, October, 1948: see Introduction

course salient, and can be treated in theory as something akin to the idea of *knowledge*.) Second, there is the sense in which it is possible to speak of and measure the data bandwidth of a communications channel. This is the attribute of information that is closest to the idea of an "absolute" measure of "raw data" – a sort of conserved substance, akin to energy (and of course it requires energy to carry information) – corresponding to the engineering aspect of the circumstances under Shannon's paradigm. It is essentially the same thing as the *degrees-of-freedom* required of the substrate or carrier of the information.

The data set encoded in a computer graphics file, via software that an artist employs to create an image, is a store of information, but not information that represents an image *per se*, inasmuch as the information is not directly interpretable as a visual perception but rather is a unique group of instructions specific to a certain class of computer apparatus, in turn running a particular set of software packages – instructions that tell the hardware/software combination how to represent the image as light on an electronic display or as ink on a piece of paper. In this respect, a computer file is no different from a dictionary entry for a word. The ink on the paper, even the words viewed as such, in no sense contain the semantic content associated with the words; the *concept*, which is a non-denumerably infinite set of potential instantiations of some meaning.[3] Like a dictionary entry, a computer file requires interpretation by an intelligent agent.

Because the computer, like the mind that interprets a dictionary entry, has a certain amount of "intelligence" built into it – that is, because it is an entity capable of interpreting data (in particular, strings of binary bits in a digital computer file representing sequences of ones and zeros) – the data set contained in the graphics file is not as large as it would need be if the computer were "dumb"; that is, if it did not already have the means to *decode the file* – translate it into display/print instructions – and if display and print devices, in turn, did not have the respective means to transcribe those instructions.

Therefore, in order to quantify (meaningful) information, the context under which the information is interpreted is a crucial part of its speci-

---

3 More or less the sense in which Roger Penrose associates the threshold between coherent and collapsed Schrödinger wave with what he deems the non-computable aspects of human thought, as discussed in chapter four (if the author has understood him properly) *cf* page 105, page 50 and reference regarding the latter.

fication. Different quantities of data are needed to represent information in different interpretive contexts. In practice, there must always be some such context, which is why there is no such thing as an absolute quantity associated with a given source of information. Moreover, in practice, what is measured is the uncertainty associated with some data set (considered independently of context; once again, "the engineering aspect of the problem"). Disregarding specific technical conventions, this might be viewed as the relation between: (1) what *is* encoded in a given data set and (2) what *could* be encoded in the given data set. Or it might be thought of as a measure of the relation between (1) the configuration or state of the interpreter, given knowledge of the data set, and (2) the measure of all possible configurations or states of the interpreter (i.e., all those states [a] given the knowledge represented by the data set, plus all those states [b] absent that knowledge). So if the interpreter is seeking the answer to a yes or no question, the answer will reduce the uncertainty about the underlying information from fifty percent to certainty.

The concept of information can play a key role in the quantitative description of experience. However, before this function can be explored, a short detour is required in order to develop a few additional ideas, similarly crucial to the endeavor.

§ 6.2.2 Describing Experience II:
The Concept of Subjective Space

A key theme running throughout this treatise is the heuristic value of visual representation, which rests on the power of spatial intuition. While great stress has been laid on the suitability of this form of representation for theoretical physics, the root of this suitability extends to another domain. For it is evidently related to the fact that experience is framed in a spatial-temporal context. In this regard, the notion of spatial intuition (and hence the term *visualizable*) must be understood in a very general, *a priori* sense, per the fact that persons born blind nevertheless have an intuitive sense of space, which is therefore to a significant degree independent of the gamut of sensate properties associated with visual perception. And while *time* is

often referred to as the "*inner*" sense and *space* the "*outer*" (Kant's terminology), space and time are yet inseparable, intuitively given attributes of
experience, and so all perceptual content can be embedded within a four
dimensional manifold.

If the reader sits quietly still for a moment and, during that interval, takes
an inner inventory of the thoughts, images and sounds – the feelings and
sensations of every kind that transpire – it will be found that it is possible
to place these experiences within a three dimensional space, in and about
the "self," the sense of which seems to be located more or less at the center
of that space (many [especially western] people "feel" this location to be
immediately behind the eyes and between the ears). So if one is sitting on
a chair in front of a desk and computer, typing on a keyboard, one hears
the sounds of the clicks at those points where the fingers feel the pressure
of the keystrokes and the eyes in turn see the fingers striking the keyboard.
If there is a sensation of something in the stomach, say hunger, that feeling
is located closer to the center of the self – closer, that is, than the wrists
and keyboard. Likewise, feelings in the feet are sensed to be lower in space,
where the feet can be felt and seen to touch the floor, etc. Indeed, if the experiment is pushed far enough, one tends to find that all elements of experience can be localized in this fashion – including such subtleties as verbal
thoughts; somehow "audible" yet internal verbiage, which one imagines
"hearing in one's head" – as noted, more or less in the same location where
one hears sound from a stereo headset. This is somewhat similar to the
virtual sound of one's own heartbeat, which comes from the tactile sensation of pressing one's hand to one's chest. Or again, if a thought has visual
components, the imagery seems superimposed on a portion of the space
immediately in front of the eyes. This portion of the external visual field
then seems "out of focus" or "washed out" while the fantasy imagery pre-
vails. (Hence the feeling of being "looked through" when in the presence
of someone thus preoccupied.)

The upshot of these rather rudimentary considerations would seem to be
that, while not every feature of experience may be of an inherently spatial
nature, most if not all may yet be "located" within the context of the inner
sense of space – either directly or in connection with other, more explicitly
spatial aspects of experience. The very fact that such a spatial discrimina-

tion among experiences is possible implies that they carry some sort of signature – i.e., *information* – that can be interpreted as indicating spatial location. Therefore, spatial information would seem to be encoded in or overlaid on sensations and subjective experience generally, so that everything can be linked to a location. And a similar set of correlations can be seen to exist with respect to time. Anything that can be remembered – as anything that can be recorded generally – can be placed within a temporal framework. (In this connection, one need only contemplate the stereotypical representation of temporal order via a "time-line" to recognize how pervasive and crucial the sense of space is to human cognition. Although time is qualitatively quite distinct from space, one really only truly feels that one "understands" temporal relations when envisioned in such a fashion.)[4]

This last point, regarding time and memory, links up with another key fact about the cognitive role of spatial intuition. It is known that the process of long-term memory formation is intimately connected with core brain processes similarly crucial to the sense of spatial orientation. As noted in the next section, special neurons called *grid cells* impose rigid Euclidean grids on the perceptual environment.[5,6] Prior to this discovery, it was known that neurons called *place cells* in the hippocampus encode spatial locations in the environment, but it was not known how the information was derived for the encoding of these cognitive maps. It is also known that the hippocampus is involved in the formation of long term memories, and because *grid cells* are proximate to and feed directly into the hippocampus, the close tie between these primary cognitive functions – spatial intuition and memory – is quite evident, indicating a strong connection

---

4This relation must be considered salient vis-à-vis a *Theory of Meaning*, which as previously noted is outside the scope of this work. Suffice it here to say that the sense of understanding with respect to time relations, pursuant to their representation along a line, is a trivial yet powerful illustration of what it means "to understand." In this context, it is clear that the relationship between temporal order and the parts of a two-dimensional manifold (*line*) is perforce one of *analogy*. Spatial order is "concrete" only in the sense that it is *visual*, whereas time is "abstract" only in the sense that it is *not*. To say that a time-line aids the understanding of time relations clearly means that it provides a *representation*, one moreover that is about as qualitatively distinct from its object, in the sense of chapters one through three, as a representation can be. And this of course, again, emphasizes the overriding importance of spatial intuition in the phenomenology of mind.
5 Hafting, Torkel, Marianne Fyhn, Sturla Molden, May-Britt Moser, and Edvard I. Moser. "Microstructure of a spatial map in the entorhinal cortex." Nature 436, no. 7052 (6, 2005): 801-806.
6 of mice and (*presumably*) men

between memory and'spatial perception, and therefore a strong connection between spatial perception and experience generally.

§ 6.3 The Dimensions of Subjective Space

The pervasiveness of the sense of spatial orientation suggests that people can explicitly connect subjective "space coordinates" with descriptions of subjective experience generally. But accuracy of measurement, as in physical science, is of the utmost importance, so it might seem that a scientific description of inner spatial relations is a hopeless goal. However, it turns out that this is not the case at all – actually quite the contrary.

As noted in the previous section, the mind is instinctively keen to maintain its sense of spatial orientation – its kinesthetic awareness of the body generally. Moreover, this kinesthetic sense is fluid; as the body's external orientation changes, through translations and rotations, and as its internal relations change, through twists and bends, the overall sense of orientation changes appropriately, so that the mind is generally able to interpret all of the body's relative positions with respect to the environment (with exceptions of course – one can experience tactual and kinesthetic analogs of optical illusions). So the existence of accurate information, *per se*, is not necessarily of concern. Rather, the problematic issue is how to impose detailed coordinate values on inner space. In other words, the problem is one of *measurement*.

Now, subjective or perceptual space can be represented as contiguous with external space; with respect to certain types of experience, it is even appropriate to speak of a mixture of the two realms. So again, when calling to mind a visual memory with eyes open, it is not without sense to speak of a region of the external space into which the memory image is "overlaid" so to speak; a region that is temporarily "translucent" (as, again, when someone who is lost in thought appears to be looking through the people and things in their line of sight).

And so it would seem that the realm of inner space may be described in much the same way as that of physical space. Actually this should not be much of a surprise, inasmuch as a description of one's "physical environment" *is* a description of first-person experience. The only difference between external and internal spatial dimensions would appear to be the interpretation of the metric. In other words, whereas the space that is sensed to be "under the skin" may be *topologically* well ordered – and topological information is vital – the geometry may not necessarily be rigidly Euclidean. (Think of those old-fashioned, phrenological-style maps of the "homunculus" depicting disproportionately sized regions of the brain associated with the perception and motor control of various parts of the body, per figure 6-1. In a related sense, some parts of *Subjective Space* can be considered more "dense" than others.)

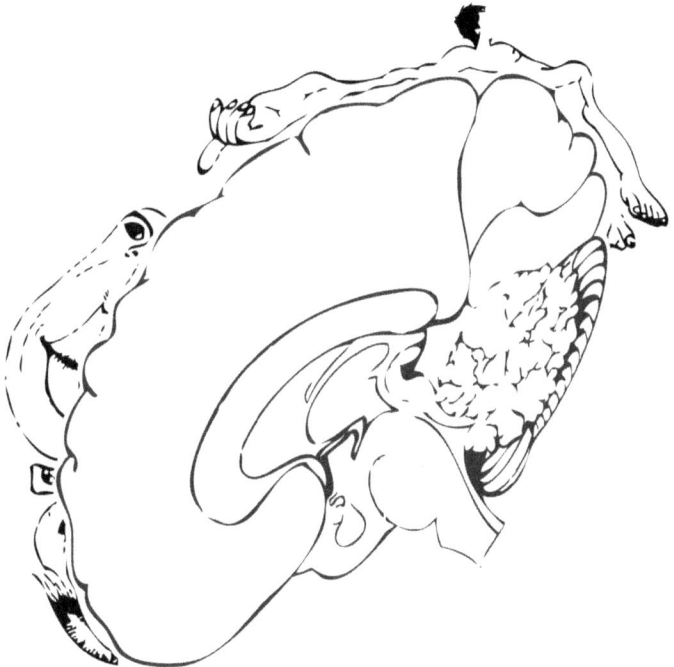

Figure 6-1 A crude illustration of the mapping of sensory-motor processes in the brain. The relative sizes of the brain regions associated with the body parts do not reflect the relative sizes of the respective body parts. Similarly, the representation of body space may be "denser" than the representation of external space.

On the other hand, everything that is perceived to be outside the skin is indeed rigidly Cartesian, and the sense of depth/parallax compensates for the distorted optical perception of eyesight. Thus, a round table that is skewed by the angle of sight to appear elliptical is nevertheless *perceived* as *"round."* (It may be helpful to think of this in terms of a loose analogy with gravity, inasmuch as the space closest to the center of the body is metrically distorted, in the manner that gravity distorts the space around concentrations of matter/energy.)

In this sense, the function of *grid cells* seems to correlate quite well with the concept of a Cartesian map of external space. These cells, found to be active in the brains of mice and rats when running mazes, fire at regular points in the animal's external environment. These points correspond to a regular geometric (Euclidean) pattern, in the absence of any corresponding markers in the environment at those points – as though a coordinate system were literally being superimposed on visual, tactual and aural perceptions by the brain.

Given that a pervasive sense of space exits, the only real problem, as indicated above, is one of measurement. That is to say, how can one obtain "the measure" of subjective experience, so that it can be represented/embedded in a mathematical framework? Before this question can be answered definitively another short detour is required.

§ 6.4 The Composition of Sensation

Shortly after the death of Descartes, one of the most astute young admirers of his work formulated a profound metaphysical system of his own; a work which – in the opinion of its author (and many succeeding students of philosophy) – vanquished Descartes' mind-body dualism while preserving and extending the inspired vision of *Mathematics* as standard-bearer of truth that had ignited Descartes' philosophical ambition (as it had that of Pythagoras, Plato and many others before). Spinoza expounded his philosophy of God/Nature and Human Nature in an axiomatic format similar to that of Euclid's *Elements* and Newton's *Principia*. However, it preceded the latter by more than a decade – was so progressive that it had to be written and published in secrecy – and stands today not only as a continuing font of inspiration and ideas but as a meta-example of the spirit, of the curiosity, courage and rigorous rationality that cradled modern science.

Spinoza's primary focus was psychology (though there would be no such discipline *per se* for another two centuries[7]), and he is not known for any contributions to physics – although he was in communication with almost every notable scientific contemporary save Newton. Leibnitz and Huygens considered him a peer, but were keen to hide both their admiration and intellectual debt (except perhaps for the optical lenses he ground for a living, many of which were apparently used for research).

Spinoza can be deemed the first true cognitive scientist,[8] as he documented the phenomenology of the mind with the formal, deductive method of geometry. One of the most perspicacious aspects of his work is the manner in which the properties of emotional experience are analyzed. It treats emotions as divisible into components, some general and fundamental – *pleasure, pain* and *desire* – and others in turn composite; conjunctions of *pleasure* or *pain* with some corresponding *desire* and *particular state of mind*, the latter being associated with the circumstances that occasion the emotion.

For example, following the *Ethics*,[9] one methodically arrives at an explanation of the emotion of *Love* by the following line of reasoning. First, it is established that the mind can undergo two basic types of transformative experience: quite simply, (a) to a lesser, and (b) to a greater *"power of activity."* What Spinoza thus labels the mind's fundamental capacity is either *diminished* or *increased*, respectively. And the experience associated with such a transition is thus either *pleasurable* or *painful* – positive or negative, respectively.

Spinoza defines *Desire* as the *essence of the mind*; the instinctive drive to avoid negative feelings and pursue the positive – thereby, again, enhancing the *power of activity*. He further establishes that transitions generally occur in two different ways: *actively*, by the power of the mind as it pur-

---

7 The date of 1879, when Wilhelm Wundt established his experimental laboratory, is often referred to as the birth date of the science of psychology; *cf.* Boring, E.G. 1942, 1950. A History of Experimental Psychology. 2nd ed. New York: Appleton-Century-Crofts

8 This high regard is not exclusive to the author. Eminent researchers such as *Antonio Damasio* have expressed similar sentiments; *cf.* Damasio, Antonio R. Looking for Spinoza : joy, sorrow, and the feeling brain. Orlando, Fla.: Harcourt, 2003.

9 Spinoza, Benedictus de and R. H. M. Elwes. Benedict de Spinoza : On the improvement of understanding, the Ethics, Correspondence. New York: Dover, 1955.

sues its rationally sanctioned best interest, and *passively*, by the force of circumstance (such as the state of the body: e.g., disease; or the state of the external environment: for example, the presence of an individual whom one finds sexually attractive, to which the mind reacts reflexively, without thought). Experience of pain or pleasure – *the feeling of an emotion* – is thus characterized as either an *action* or a *passion*, accordingly (the latter term is in this fashion *defined*).

From these definitions and propositions, all of which are proven as mathematical theorems, Spinoza describes the nature of love in the following manner. "By joy" is understood "that passion by which the mind passes to a greater perfection" and "The [emotion] of joy" is called "pleasure or cheerfulness." Specifically, *Love* is a positive feeling, a joy or pleasure, that is "accompanied by the idea of an external cause."[10] Therefore the mind is passive in the experience – the transition – and accordingly: Love (of something external to the mind) is a passion (a transition to a greater state of activity, but passively, by virtue of an external cause outside the mind's control). QED[11]

§ 6.5 The Superposition of Sensation

Based on the considerations of the foregoing discussion, it is possible to propose some general properties that a framework must have in order to provide for a mathematical description of experience. The essential idea is that of a mathematical "skeletal structure," so to speak – again, a many-dimensional mathematical space in which the description of experience is embedded, the backbone of which is the first four spatial-temporal dimensions comprising "inner" space and time. While these are not the only dimensions of the manifold there is – in contradistinction to the *phase space* of physics with its generalized coordinates and symmetries[12] – an essential asymmetry, in that the first four coordinates serve again as a sort of skeleton on which is hung the "flesh" of first person experience – the remaining *degrees of freedom* representing all the other dimensions of that experience, from color and sound to taste and touch (much in the way a

10 *Op. Cit.*
11 Only the words enclosed in quotation marks are Spinoza's, the rest is paraphrased.
12 That is, of generalized position and momenta variables

String theorist might imagine a bundle of curled-up, invisible dimensions of space centered on every point of a three-dimensional spatial continuum). The "extra" dimensions here represent all of the various categories of *qualia* – and thus information – that comprise the phenomenology of first person, sensate experience; beyond the universal *spatial* and *temporal* ones.

For example, the dimensionality associated with the olfactory sense can be addressed in the manner of color, which can be treated as a combination of three *primitives*; red, green and blue. Similarly, aromas can be treated as a combination of *"baseline scents."* The possibility of identifying such *scent primitives* by a combination of human and machine sensors, and describing their various compounds in a multidimensional mathematical space, has begun to be explored (largely in connection with commercial technological efforts, e.g., to communicate scents online). Thus *Harel, Carmel* and *Lancet*[13] have proposed the use of a panel of human judges (such as the scent experts employed in the perfume industry) in conjunction with technologies that identify the molecular composition of chemical substances (as per the gas chromatograph and mass spectrometer) to develop suitable primitives, and the development of an appropriate algorithm for their categorization and manipulation.

Because the variety of scents that humans are able to identify and the concomitant variety of olfactory receptors involved is quite large (in comparison, say, to color discrimination and the corresponding variety of visual receptors), the dimensionality of the scent spectrum must be relatively large. But this is fundamentally no different from the dimensionality distinction that exists between space and time, i.e., three versus one (or again, that which exists between the three "large" dimensions of ponderable space and the six or seven "small" dimensions of certain hypothetical physico-mathematical spaces). As suggested here, these early efforts at describing scents have pursued the artifice of a many dimensional *"odor space"* (rather naturally, it would thus seem).

Because of the *Gestalt* aspect of perception and its epistemological relevance – stressed so emphatically in chapters one through three – it might seem problematic to attempt to break experience down in such a fashion.

13 Harel, D, L Carmel, and D Lancet. "Towards an odor communication system." Computational Biology and Chemistry 27, no. 2 (May 2003): 121-133.

But the fact remains that the analysis of perception is a well established, pragmatic possibility – if merely by the power and success of the natural languages and the everyday, physical modelization of experience *vis-à-vis* "naïve realism" (not to mention its extraordinary extensions in the physical sciences and everyday evidence thereto in the technological by-products of those sciences).

§ 6.6 Computing the Environment: Measuring Subjective Space

One of the fascinating developments in the field of human-computer interface design is the ever-growing capacity of machines to sense, and react sensibly to, subtle physical gestures of an operator – from detecting and tracking the focus of the eyes and positions of the fingers to the detailed articulations and motions of the entire body – even anticipating user action by literally "reading the mind": learning to identify patterns of brain activity as antecedents of specific behaviors, analogous to the manner in which voice and handwriting recognition systems learn to identify idiosyncrasies of speech and writing. (For example, by tracking the firing patterns of an animal's *grid cells* during sleep, and comparing them with those tracked during pre-sleep maze expeditions, it is possible to interpret them as the contents of dreams; the mind/brain experience as it presumably lays down long-term memories of the maze running.) When a computer can sense such delicate details of a user's state – and even anticipate changes in that state – then the problem of engineering a system to provide appropriate feedback (and so create a convincing virtual reality) becomes largely a matter of computational speed and memory capacity – i.e., raw processing power.

But there is another salient possibility for the use of such technologies; a sort of reverse application of that contemplated with respect to the compelling virtual environments and enhanced human-computer interfaces that are rapidly evolving. Beyond the enormously important benefits that accrue to disabled computer users, and the incredible flights of fantasy that become available to adventure seekers, such technologies can also open the "doors of perception" wide – from the outside in – making the details of first-person experience accessible to those seeking to chronicle the events of consciousness. For instead of using the computer to furnish a fantastic artificial environment that responds as desired to user action, it can be employed to simulate any "mundane" experiences that are of interest to the researcher, and thereby record every aspect of the user's state of awareness (and *unawareness*, including the details of unconscious input and feedback to every detail of the simulated environment) in an extremely delicate and well-correlated fashion. In other words, the technology can be utilized to turn the user's experience "inside out" by creating – and thus knowing – the internal state (i.e., the simulated environment), and also by measuring, and thereby knowing, the user's various responses in and to that state. Such a system would constitute a comprehensive diagnostic probe of first-person experience.

Although an individual's perception of a given physical environment must be unique (just as the perception of music or a given spoken language must be – e.g., in the case of the latter, depending on the user's knowledge of the given language), studies of large numbers of people will tend to reveal what the underlying commonalities are. To better understand what this means, consider the level of detail that the system is assumed to possess regarding the elements of the simulated environment.

With respect to visual stimuli, the system can control, and thus know, the optical composition of every projected image down to the level of the anatomical details of the retinae, where the elements of the scene are projected (for instance, by eye-wear or other headgear mounted projectors, which directly target specific segments of the retinae). In turn, the system can track eye movement and focus with similar delicacy. This means it can know exactly which portions of the visual field attention is focused on; when and for how long. By a similar manipulation of the projection of sound, not only can all the aural elements of what the individual hears be known, but even the spatial sense of where the sounds originate (by tracking of head position and correlation of that position with the projected visual and sound images, inasmuch as both visual and aural stimulation can be similarly furnished by head-mounted, high-precision apparatus). And with a very fine array of pressure transducers spanning the entire body (e.g., via close-fitting clothing with an extremely fine mesh of such devices), an enormous range of tactual perceptions can be induced, just as, in turn, detailed kinesthetic and other bodily information can be scanned.

The computer knows all of the details of the environment in which the subject is immersed because that entire environment has been created. It is the unfolding of a program in the computer's memory, every moment and nanometer of which can thus be digitally tracked to extreme precision. As part of the virtual reality experience, every sight, sound, scent and physical sensation that the subject experiences is furnished by the computer. And again, internal physical and physiological parameters can be tracked conjointly with all of this. Thus, continuous scanning of the entire body, but particularly the brain, will record the real-time variances of every trackable bodily process, and correlate those processes with every other event that is recorded.

In addition to tracking and capturing all of the user's physical responses, the system will also take and record explicit verbal descriptions provided by the user in real time. So, for example, a highly detailed, human perceivable 3-D coordinate grid can be superimposed over the visual field – projected by the computer, something like the heads-up display in aircraft, although with a much finer degree of three-dimensional detail – so that the subject can furnish feedback such as, "I hear a sound that seems to emanate from 'such and such' coordinates." This can be enhanced by the use of additional representational imagery. For example, the display can show a detailed, user-manipulable three-dimensional anatomical map of the user's body, by which the user can thereby also provide descriptions of where purely subjective or "internal" feelings/imagery are perceived to occur (and of course, the system can superimpose on such maps both the things that it scans and tracks automatically and those things that it receives verbal input on – which, in turn, the user can then assimilate and provide further feedback on).

The environment thus constructed – the perceived world of the subject – is only limited by the degree of detail that can be engineered into the virtual reality gear. To the extent that bodily sensations of all sorts can be projected by the system, an equally vast array of internal responses to such sensations can be measured. Even internal body space can be probed, and induced to generate sensations – by the use, for example, of convergent sonar or other harmless radiant energy; techniques that might allow for the noninvasive stimulation of nerve endings at virtually any location within the body.

Another important aspect of experience that can be captured and studied by such means is the nature of what might be called "intrinsically-difficult-to-describe-feelings," which are as such because the relevant descriptive concepts do not seem available to linguistic expression. For example, when learning a physical skill, such as playing a musical instrument or riding a bicycle, the feelings of muscular coordination that must be acquired are very hard to verbalize in any significant detail, although conscious attention is evidently a necessary part of the learning experience. Moreover, certain physiological processes usually deemed to be outside conscious control – such as heart rate – are now known to be accessible to volition if proper feedback is available. That is, while one can readily become aware of one's heartbeat, mere awareness of the rhythm is not ordinarily sufficient to put it under conscious control. However, when appropriate visual or visual/aural representations of the relevant physiological processes are available – by means of a biofeedback machine – most people can indeed learn, quite readily in fact, to control the rate at which their heart beats. (While it is possible to acquire this skill absent the guiding perceptual feedback furnished by such a machine, this typically requires inordinate immersion in the practice of introspective/meditative techniques over an extended period of time – perhaps years.)

Because this sort of feedback is necessarily furnished in perceptual forms – visual, aural – the mind acquires a perceptual but non-linguistic representational tool, i.e., a *non-verbal concept*,[14] for the representation of otherwise difficult to conceptualize "inner" feelings. Such non-verbal concepts, in turn, can be given names – and thus made available to linguistic (and of course mathematical) description – while the physiological processes that accompany the experiences (muscular, neurological) can be correspondingly tracked by the system, thereby providing additional sets of correlations (in a sense, "computer concepts").

---

14 In this sense, new concepts of altogether new types can be created. This is an important topic that demands attention, but is outside the scope of the present treatise

In this connection, consider the capacity of well crafted poetry and prose to express extraordinarily suggestive representations of subtle life experiences, and to evoke highly nuanced feelings and thoughts. That which seems to be "indescribable" is often really only difficult to describe. And so, just as the poet gives expression to subtle concepts that the listener or reader might never have otherwise given thought, the user of a biofeedback system, working together with a system programmer (who of course may also be the user), can form perceivable representations, just like words or symbols generally, that are suitable to the experience of subtle, internal states – including, perhaps, such states as immediately precede the initiation of voluntary actions (it is known that relevant physiological processes generally commence approximately one-half second before people become aware of willing an action).

The upshot is that an elaborate *Subjective Space* can be fabricated – a mathematical construct roughly analogous to the multi-dimensional phase space of physics except that, again, instead of the position and momenta variables of many particles this construct will comprise the three dimensions of inner space, a single dimension of personal time, and all the other dimensions of *qualia* that are needed to describe the qualitative spectra of sensate experience. That is, for every point of inner space and time there will be corresponding values of every other dimensionality; [at least] three coordinate values of color, each perhaps expressing a number between 0 and 255; x scent coordinate values, perhaps approximately 1600, or maybe only several values, the permutations of which total about 1600; y sound coordinate values, and so on. Of course, under conditions of "normal consciousness" only certain spatial locations will correlate with certain sensate qualities – for example, non-zero values of the coordinate component[s] that represent "bitterness" will likely only exist at space-time locations where parts of the tongue are also found.[15] But in principle, any value of space-time may be matched with any value of the other dimensions comprising the full spectra of the *qualia* of *Subjective Space*. As with "bitterness," certain correlations will occur and certain will not, and experiment will thus reveal the patterns and principals that characterize them.

---

15 It is known that under certain circumstances, such as the influence of psychedelic drugs, people report "hearing colors" and otherwise "confusing" the perception of qualia; phenomena that can likewise be studied.

§ 6.7 Discussion: Correlation of Data

When a sufficient body of data regarding the correlations of Subjective Space parameters is established it will also be possible to determine detailed external correlations – i.e., with neurophysiological and/or other relevant biological processes. And when a sufficient set of such physiological correlations is developed it may be possible to further develop mappings; between neurophysiological processes that do not directly correspond to any known experiences and neurophysiological processes that do. In this fashion, theoretical constructs representing unconscious mental activity may be developed. Such constructs might then make another theoretical leap possible: namely, the correlation of [theoretical representations of] unconscious, biological processes generally: and then, perhaps, the development of theoretical constructs of such "unconscious elements" corresponding to non-biological physical processes. Such a leap would constitute a new way of viewing processes and relations that can now only be conceived as "physical."

All of this is to say that models of something like "subjective experience," but corresponding to generally "non-*experience-able* processes" may be developed as representations of what are usually considered, exclusively, "*physical processes*," but "viewed from the inside" so to speak. This would mean model building of a new kind; that is, a way to represent mental qualities in detail but without invoking physico-spatial concepts (or such concepts exclusively).[16]

For example, as things stand today it is not readily apparent how to describe relations between distinct qualities, such as *color* and *scent*, analytically, without recourse to extra-personal/physical concepts such as frequency and molecular structure. For there seem to be no points of contact between *distinct qualia*, e.g., "red" and "musky." Whereas one can find *intra-qualia* similarities and differences – such as between the colors blue, yellow and green – so that the spectrum of the rainbow is, to a certain extant, susceptible to qualitative analysis, and whereas the same sort of qualitative spectral analysis makes sense for sounds, it is not evident how

---

16 This demands separate treatment, per the reference on page 369 regarding non-verbal concepts

Philosophical Solutions

such an analysis can be effected *inter-qualia*; i.e., for-and-across *categories* of *qualia* generally.[17]

But again, it must be remembered that to expect to find, literally, subjective correlates of "physical things" *per se* is a misconception, inasmuch as such things are nothing more than subjective constructs to begin with. The idea, rather, is that the more extensive the purely subjective model can be made, the more extensive will be the correlations with physical maps, which represent, more or less schematically, the relations and connections among "things at large." Or perhaps it will ultimately be possible to combine Subjective Space and mathematical descriptions of physical space in a single manifold, so that processes in physical space are represented not only by the four coordinates of space-time but also, as with perceptions of physical things in Subjective Space, by coordinates of sensate qualia, or certain elements thereof, which by collective combination and interaction form the multivariate qualia of the totality of all experience in/of the universe, per the concept described above as a *Sensorium*.

However that may be, it is inappropriate to consider abandoning the concepts of space and time as primitives of theory, for all of the reasons hereinabove discussed at length. Moreover, it must be recognized that the possibility does not arise in a fundamental sense with respect to the study of sentience. While physicists may consider the notions of space and time illusory and hence misleading– walking shadows and poor players, signifying nothing – and thus contemplate their dispatch from the foundations of theory, those who would know the mind must perforce give audience to its struts and frets for hours upon the stage; and so space and time must remain principal players indefinitely.

Accordingly, it would seem that the near term purpose and prospect of a framework for the study of consciousness must be a detailed, quantitative elaboration of *Subjective Space,* as described especially in section 6.6. It was there suggested that computer simulations of the phenomenology of sensate experience be constructed in order to create, by the projection of immersive virtual environments, detailed subjective states that can be correlated with the data obtained from similarly delicate physiological scans,

---

17 Again, how to find an *inter-qualia* bridge, such as a common qualitative primitive among such diverse sensations as taste and vision. Of course, once well-developed spectra exist for each sensate category, relevant unifying principles may emerge, perhaps corresponding to the frequency ranges of the underlying neural processes and thus analogous to the physical spectra of light and sound.

further coupled with verbal and other subject feedback, the latter in turn enhanced, as appropriate, by projected virtual coordinate systems and other perceptual aids, available to the subject for such purpose on demand. The data so obtained will furnish the values of the vector variables of *Subjective Space*; superpositions of sensate qualia in a mathematical space analogous, as discussed, to the hypothetical higher-dimensional spaces of, for example, String Theory – except whereas in such spaces as the latter an unobservable higher-dimensional manifold might reside at each point of the observable spacetime, in *Subjective Space* one will find, at every point, a *multitude of manifolds* (*vectors*) – each representing a particular *qualia*, the dimensions of which vary in accordance with the number of variables necessary to its description (e.g., maybe *three* for color – red, green and blue – each in turn with appropriate values [e.g., perhaps 256 gradations]). This superposition of *qualia vectors* mirrors, and is intended to include, compositions of feelings as well, similarly based on "feeling primitives" – elemental sensations such as pain and pleasure together with primal states such as aversion and desire – which are combinatorially taken to yield the entire range of emotional states, as per the discussion of section 6-4 and the example there drawn from Spinoza.[18] And Subjective Space vectors are to be sought for those aspects of cognition that are usually deemed to exist "below the radar" of consciousness; as, again, the feelings of muscular coordination that accompany learning a physical skill, the beating of the heart, the innate rules – perhaps "principles and parameters" – believed by many linguists to govern the formation of verbal expressions, etc. In other words, with the help of biofeedback-related techniques, new conceptual representations can be sought for processes conventionally treated as *unconscious* (and which might even include dream and other sleep states)[19].

In such a manner, the phenomenology of mind may be modeled quantitatively. Of course this alone is salient. As accomplished fact, it could be understood to signify the final overcoming of Psychology's own shallow counterpart of *"Copenhagen-esque" Positivism*, i.e.: *"Skinner-esque" Behaviorism.* Moreover, it would furnish a context for an heuristically appro-

---

18 Also nicely reflected in the writings of Antonio Damasio referenced therein
19 It may turn out that, to a significant extent, what is presently regarded as *consciousness* is identical with those aspects of experience that make their way to memory; i.e., perhaps consciousness can be understood as the focal point or threshold of the memory-formation process, inasmuch as what is not remembered is generally considered "unobserved" and thus part of the *unconscious* mind.

priate invocation of behavioristic concepts and methods, via the establishing of correlations between "subjective" models of first person experience and "objective" (i.e., *physical*) models of information processes and flow.

But more can be expected. If there are common themes to be found between the *Subspace* framework presented in the previous chapter for physics and that of *Subjective Space,* here proposed for psychology, primary among them are these two:

1. What is deemed "internal" to the processes of nature, or for any reason "unobservable" – "gears and wheels"; "subjective states" – should not on that account be excluded from theoretical consideration. Models are *maps*, nothing more. And yet a good map must do more than merely represent the greatest possible quantity of "true" information; it must also be *intelligible – meaningful to a human being.*

2. The nonlinear aspects of *Subspace dynamics* are closely paralleled by those of *Subjective Space.* Remember Wolfgang Köhler's discussion of the dynamics of perception quoted in section 3.6 and, even more to the point, the last paragraph of the Gerhard Werner paper quoted in the same place:

   "This phenomenon has been identified in a variety of biological systems as a mode of self-organization. Phase transitions in nonlinear dynamic systems occur as a matter of principle with very short latency. They would then also account for the short onset latency of the assembly oscillations. These considerations identify the oscillations in the 40 Hz range, and their apparent role in perceptual and motor functions, as manifestation of self-organizing nonlinear system properties of neuronal ensembles."[20]

The analogy with the phase correlations of so-called *entangled* quantum systems could hardly be closer. Who knows, perhaps the designation *Subjective Space* can ultimately be contracted, to share with the physical framework the thus shortened appellation *Subspace…*

---

20 Werner, G. "Perspectives on the Neuroscience of Cognition and Consciousness." Biosystems 87, no. 1 (1, 2007): 82-95.

Appendix A

Mathematica Simulation of Mermin/Bell Scenario

runs = 10000;   (*set number of experimental runs [i.e., iterations of "Do" loop]*)

ΔAHit = 0;   (*set A hit register*)

ΔBHit = 0;   (*set B hit register*)

HitHit = 0;   (*set A-hit & B-hit counter*)

HitMis = 0;   (*set A-hit & B-miss counter*)

MisHit = 0;   (*set A-miss & B-hit counter*)

MisMis = 0;   (*set A-miss & B-miss counter*)

aBat = 0;   (*set A 'bat' position [switch-setting]*)

bBat = 0;   (*set B 'bat' position [switch-setting]*)

aHit = 0;   (*set A hit register*)

bHit = 0;   (*set B hit register*)

bats = 0;   (*set matching-switch counter*)

aMeasure = 0;   (*set A measurement vector*)

bMeasure = 0;   (*set B measurement vector*)

(*Run simulation [runs] times, then print results*)
Do[testMermin[a], {runs}]; Print["|| TOTAL GG&RR: ", N[((HitHit+MisMis)/runs)],"
|| Matching Switch Settings: ", N[bats/runs]," || GG&RR with same switch settings: ",
N[(aHit+bHit)/(bats)]];

testMermin[a_] := (ΔAHit = 0; ΔBHit = 0; *(*re-initialize counters*)*

*(*set A and B bats [switch positions: 120 degrees apart]*)*
aBat = RandomReal[]; If[aBat < (1/3), aBat = 120];
If[(1/3) < aBat < (2/3), aBat = 0];If[(2/3) < aBat < 1, aBat = 240];
bBat = RandomReal[]; If[bBat < (1/3), bBat = 120];
If[(1/3) < bBat < (2/3), bBat = 0]; If[(2/3) < bBat < 1, bBat = 240];

*(*set signal orientation [pitch] in degrees*)*
pitch = RandomReal[] * 360;

*(*Measure signal relative to A/B measurement vectors [bats]*)*
([i.e., determine the square of the cosine of the angle between pitch & bat]*)*
aMeasure = Cos[(aBat - pitch) Degree]^(2);
bMeasure = Cos[(bBat - (pitch-180)) Degree]^(2);

(DETERMINE HIT OR MISS *)*
(i.e., determine whether aMeasure/bMeasure >= .5... signal noise & errors are
accounted for by introduction of a random real number between 0 & 1, average value
of which is .5 *)*
If[aMeasure >= RandomReal[], ΔAHit = 1];
If[bMeasure >= RandomReal[], ΔBHit = 1];

*(*update registers and tabulate results*)*
If[ΔAHit == 1 && ΔBHit == 1, HitHit = HitHit + 1];
If[ΔAHit == 0 && ΔBHit == 0, MisMis = MisMis+1];
If[ΔAHit == 1 && ΔBHit == 1 && aBat == bBat, aHit = aHit+1];
If[aBat == bBat, bats = bats+1];
If[ΔAHit == 0 && ΔBHit == 0 && aBat == bBat, bHit = bHit+1])

Results of a typical run:
|| TOTAL GG & RR: 0.5 || Matching Switch Settings: 0.333239
|| GG & RR with same switch settings: 0.749912

Appendix B

Monk-*e*-business
Puzzle Solution

At sunrise one morning, a monk leaves his mountaintop abode to deliver a special Bible, containing a handwritten codicil, to a customer who has ordered it from his Website. He arrives before sunset the same day. The next morning at sunrise, having concluded his business, the monk returns home along the same trail, and arrives at the mountaintop the same day before sunset. Prove that there exists a point on the trail that he crosses at exactly the same time of day on both trips.

The solution to this problem involves the discovery of a simplifying symmetry. Because the monk travels the same path on both days, the situation is equivalent to one in which *two* monks travel the same path on the *same day*, both leaving at sunrise – one from the mountaintop and one from the bottom.

Once the problem is visualized in this manner, it is obvious that the point in question must exist, simply because the two monks must pass each other *somewhere*.[1]

1The author is not certain where he first heard this puzzle, some twenty five years ago, but recalls that none of the mathematically trained, very bright people whom he passed it along to were able to solve it – whereas the author's then quite young sister, not known for any special mathematical aptitude, saw the solution instantly, evidently in the manner portrayed above. Moreover, inasmuch as the only relevant intellectual commonality, at least at that time, between sister and author (who had the same experience) would seem to be an inclination for visual thinking, this has informed the view, emphasized throughout this work, that visualizable models can be enormously beneficial to scientific research.

Index